Maximum Principles in Differential Equations

Springer
*New York
Berlin
Heidelberg
Barcelona
Hong Kong
London
Milan
Paris
Singapore
Tokyo*

Murray H. Protter
Hans F. Weinberger

Maximum Principles in Differential Equations

With 56 Illustrations

 Springer

Murray H. Protter
Department of Mathematics
University of California
Berkeley, CA 94720-0001
USA

Hans F. Weinberger
Institute for Mathematics and its Applications
University of Minnesota
514 Vincent Hall
206 Church Street
Minneapolis, MN 55455
USA

Mathematics Subject Classification: 35B50

This is a corrected reprint of the second printing of *Maximum Principles in Differential Equations* which was originally published in the Prentice-Hall Partial Differential Equations Series, © 1967 by Prentice-Hall, Englewood Cliffs, New Jersey.

Library of Congress Cataloging-in-Publication Data
Protter, Murray H.
  Maximum principles in differential equations.
  Reprint. Originally published: Englewood Cliffs, N.J., Prentice-Hall, 1967.
(Prentice-Hall partial differential equations series)
  Bibliography: p.
  Includes index.
  1. Differential equations, Partial. 2. Maximum principles (Mathematics)   I. Weinberger, Hans F.
II. Title.
QA377.P76   1984   515.3'53   84-14028

© 1984 Springer-Verlag New York, Inc.
All rights reserved. This work may not be translated or copied in whole or in part without the written permission of the publisher (Springer-Verlag New York, Inc., 175 Fifth Avenue, New York, NY 10010, USA), except for brief excerpts in connection with reviews or scholarly analysis. Use in connection with any form of information storage and retrieval, electronic adaptation, computer software, or by similar or dissimilar methodology now known or hereafter developed is forbidden.
The use of general descriptive names, trade names, trademarks, etc., in this publication, even if the former are not especially identified, is not to be taken as a sign that such names, as understood by the Trade Marks and Merchandise Marks Act, may accordingly be used freely by anyone.

Production managed by A. Orrantia; manufacturing supervised by Jacqui Ashri.
Printed and bound by Braun-Brumfield, Inc., Ann Arbor, MI.
Printed in the United States of America.

9 8 7 6 5 4 3 2 (Corrected second printing, 1999)

ISBN 0-387-96068-6 Springer-Verlag New York Berlin Heidelberg
ISBN 3-540-96068-6 Springer-Verlag Berlin Heidelberg New York   SPIN 10714130

PREFACE

One of the most useful and best known tools employed in the study of partial differential equations is the maximum principle. This principle is a generalization of the elementary fact of calculus that any function $f(x)$ which satisfies the inequality $f'' > 0$ on an interval $[a, b]$ achieves its maximum value at one of the endpoints of the interval. We say that solutions of the inequality $f'' > 0$ satisfy a *maximum principle*. More generally, functions which satisfy a differential inequality in a domain D and, because of it, achieve their maxima on the boundary of D are said to possess a maximum principle.

The study of partial differential equations frequently begins with a classification of equations into various types. The equations most frequently studied are those of elliptic, parabolic, and hyperbolic types. Because equations of these three types arise naturally in many physical problems, mathematicians interested in partial differential equations have tended to concentrate their efforts on those developments which are of both mathematical and physical interest. A reader who learns differential equations by studying physically oriented problems not only parallels the historical development of the subject, but also acquires a clear understanding of the reasons some equations are studied in great detail while others are virtually ignored. Since many problems associated with equations of elliptic, parabolic, and hyperbolic types exhibit maximum principles, we feel that a study of the methods and techniques connected with these principles forms an excellent introduction or supplement to the study of partial differential equations.

There is usually a natural physical interpretation of the maximum principle in those problems in differential equations that arise in physics. In such situations the maximum principle helps us apply physical intuition to mathematical models. Consequently, anyone learning about the maximum principle becomes acquainted with the classically important partial differential equations and, at the same time, discovers the reasons for their importance.

The proofs required to establish the maximum principle are extremely elementary. By concentrating on those applications which can be derived from the maximum principle by elementary methods, we have been able to write this book at a level suitable for the undergraduate science student. Anyone who has completed a course in advanced calculus is qualified to read the entire book. In fact, any student who, in addition to elementary calculus, knows line integration, Green's theorem, and some simple facts on continuity and differentiation should find almost all of the book within his grasp.

The maximum principle enables us to obtain information about solutions of differential equations without any explicit knowledge of the solutions themselves. In particular, the maximum principle is a useful tool in the approximation of solutions, a subject of great interest to many scientists. This book should prove useful, not only to professional mathematicians and students primarily interested in mathematics, but also to those physicists, chemists, engineers, and economists interested in the numerical approximation of solutions of ordinary and partial differential equations and in the determination of bounds for the errors in such approximations.

The maximum principles for partial differential equations can be specialized to functions of one variable, and we have devoted the first chapter to a treatment of this one-dimensional case. The statement of the results and the proofs of the theorems are so simple that the reader should find this introduction to the subject strikingly easy. Of course, the one-dimensional maximum principle is related to second-order ordinary differential equations rather than to partial differential equations. In Chapter 1 we show that portions of the classical Sturm-Liouville theory are a direct consequence of the maximum principle. This chapter is included primarily because it provides an attractive and simple introduction to the various forms of the maximum principle which occur later. It also provides new ways of looking at some topics in the theory of ordinary differential equations.

In Chapter 2 we establish the maximum principle for elliptic operators, state several generalizations, and give a number of applications. Although the maximum principle for Laplace's and some other equations has been known for about a hundred years, it was relatively recently that Hopf established strong maximum principles for general second-order elliptic operators. Many of the important applications which we present make use of these results.

The maximum principle for parabolic operators takes a form quite different from that for elliptic operators. In Chapter 3 we present Nirenberg's strong maximum principle for parabolic operators. We then show, as in the elliptic case, that the principle may be used to yield results on approximation and uniqueness. We conclude the chapter with a section on the maximum principle for a special class of parabolic systems.

The fourth and last chapter treats maximum principles for hyperbolic operators. The forms that these principles take reflect the structure of properly posed problems for hyperbolic equations. Both the statements of the theorems and the methods of proof for hyperbolic operators are quite different from those for elliptic and parabolic operators. In particular, the role of characteristic curves and surfaces becomes evident in the hyperbolic case.

The maximum principle occurs in so many places and in such varied forms that we have found it impossible to discuss some topics which we had originally hoped to treat. For example, the maximum principle for finite difference operators is omitted entirely. We do not mention the maximum principle for the modulus of an analytic function, a subject replete with important and interesting applications. Certain elliptic equations of order higher than the second are known to exhibit a maximum principle. (See, for example, Miranda [1] and Agmon [1].) We decided not to include this topic because advanced techniques of partial differential equations are needed.

Most of the notations and symbols we employ are fairly standard. A **domain** D in Euclidean space is an open connected set. The boundary of D is usually designated ∂D. The symbols $\cup$ and $\cap$ are used for the union and intersection of sets. Boldface letters denote vectors, and the customary notations u_{x_i} and $\partial u/\partial x_i$ are employed for partial derivatives.

We frequently use the letter L followed by brackets to denote a **linear operator** acting on functions. That is, L assigns to each function u of a certain class, a function $L[u]$ of another class. We say that L is **linear** if, whenever $L[u_1]$ and $L[u_2]$ are defined, the quantities $L[\alpha u_1 + \beta u_2]$ and $\alpha L[u_1] + \beta L[u_2]$ are also defined for all constants α and β, and the equation $L[\alpha u_1 + \beta u_2] = \alpha L[u_1] + \beta L[u_2]$ holds.

For those readers who may wish to explore the subject further, we have included at the end of each chapter a bibliographical discussion which contains historical references and a guide to other presentations and further results and applications relevant to the chapter. Since we have a continuing interest in the subject, we would enjoy hearing about results—new or old—which are related to the subject of this book.

We wish to thank the Air Force Office of Scientific Research and the National Science Foundation for their support of investigations leading to a number of results published here for the first time.

M. H. P.
H. F. W.

ACKNOWLEDGMENT

We are grateful to Springer-Verlag for making this reprint edition of our book available once again.

CONTENTS

CHAPTER 1. THE ONE-DIMENSIONAL MAXIMUM PRINCIPLE 1

    1. The maximum principle, 1.   **2.** The generalized maximum principle, 8.   **3.** The initial value problem, 10.   **4.** Boundary value problems, 12.   **5.** Approximation in boundary value problems, 14.   **6.** Approximation in the initial value problem, 24.   **7.** The eigenvalue problem, 37.   **8.** Oscillation and comparison theorems, 42.   **9.** Nonlinear operators, 47.   Bibliographical notes, 49.

CHAPTER 2. ELLIPTIC EQUATIONS 51

    1. The Laplace operator, 51.   **2.** Second-order elliptic operators. Transformations, 56.   **3.** The maximum principle of E. Hopf, 61.   **4.** Uniqueness theorems for boundary value problems, 68.   **5.** The generalized maximum principle, 72.   **6.** Approximation in boundary value problems, 76.   **7.** Green's identities and Green's function, 81.   **8.** Eigenvalues, 89.   **9.** The Phragmèn-Lindelöf principle, 93.   **10.** The Harnack inequalities, 106.   **11.** Capacity, 122.   **12.** The Hadamard three-circles theorem, 128.   **13.** Derivatives of harmonic functions, 137.   **14.** Boundary estimates for the derivatives, 141.   **15.** Applications of bounds for derivatives, 145.   **16.** Nonlinear operators, 149.   Bibliographical notes, 156.

CHAPTER 3. PARABOLIC EQUATIONS 159

    1. The heat equation, 159.   **2.** The one-dimensional parabolic operator, 163.   **3.** The general parabolic operator, 173.   **4.** Uniqueness theorems for boundary value problems, 175.   **5.** A three-curves theorem, 178.   **6.** The Phragmèn-Lindelöf principle, 182.   **7.** Nonlinear operators, 186.   **8.** Weakly coupled parabolic systems, 188.   Bibliographical notes, 193.

CHAPTER 4. HYPERBOLIC EQUATIONS 195

    1. The wave equation, 195.   **2.** The wave operator with lower order terms, 197.   **3.** The two-dimensional hyperbolic operator, 200.

4. Bounds and uniqueness in the initial value problem, 208. **5.** Riemann's function, 210. **6.** Initial-boundary value problems, 213. **7.** Estimates for series solutions, 215. **8.** The two-characteristic problem, 218. **9.** The Goursat problem, 231. **10.** Comparison theorems, 232. **11.** The wave equation in higher dimensions, 234. Bibliographical notes, 239.

BIBLIOGRAPHY 240

INDEX 257

CHAPTER 1

THE ONE-DIMENSIONAL MAXIMUM PRINCIPLE

SECTION 1. THE MAXIMUM PRINCIPLE

A function $u(x)$ that is continuous on the closed interval* $[a, b]$ takes on its maximum at a point on this interval. If $u(x)$ has a continuous second derivative, and if u has a relative maximum at some point c between a and b, then we know from elementary calculus that

$$u'(c) = 0 \text{ and } u''(c) \leq 0. \tag{1}$$

Suppose that in an open interval (a, b), u is known to satisfy a differential inequality of the form

$$L[u] \equiv u'' + g(x)u' > 0, \tag{2}$$

where $g(x)$ is any bounded function. Then it is clear that relations (1) cannot be satisfied at any point c in (a, b). Consequently, whenever (2) holds, the maximum of u in the interval cannot be attained anywhere except at the endpoints a or b. We have here the simplest case of a *maximum principle*.

An essential feature of the above argument is the requirement that the inequality (2) be strict; that is, we assume that $u'' + g(x)u'$ is never zero. In the study of differential equations and in many applications, such a requirement is overly restrictive, and it is important that we remove it if possible. We note, however, that for the nonstrict inequality

$$u'' + g(x)u' \geq 0,$$

the solution $u = $ constant is admitted. For such a constant solution the maximum is attained at every point. We shall prove that this exception is the only one possible.

---

*The symbol $[a, b]$ denotes the closed interval $a \leq x \leq b$; the symbol (a, b) denotes the open interval $a < x < b$.

THEOREM 1. (One-dimensional maximum principle). Suppose $u = u(x)$ satisfies the differential inequality

$$L[u] \equiv u'' + g(x)u' \geq 0 \text{ for } a < x < b, \tag{3}$$

with $g(x)$ a bounded function. If $u(x) \leq M$ in (a, b) and if the maximum M of u is attained at an interior point c of (a, b), then $u \equiv M$.

Proof. We suppose that $u(c) = M$ and that there is a point d in (a, b) such that $u(d) < M$. We shall show this leads to a contradiction. For convenience let $d > c$. We define the function

$$z(x) = e^{\alpha(x-c)} - 1$$

with α a positive constant to be determined. Note that $z(x) < 0$ for $a < x < c$, that $z(x) > 0$ for $c < x < b$, and that $z(c) = 0$. (See Fig. 1.)

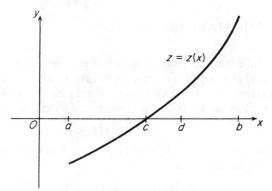

FIGURE 1

A simple computation yields

$$L[z] \equiv z'' + g(x)z' = \alpha[\alpha + g(x)]e^{\alpha(x-c)}.$$

We choose α so large that $L[z] > 0$ for $a < x < d$. That is, we select α so that it satisfies the inequality

$$\alpha > -g(x);$$

we can always do this since $g(x)$ is bounded. We now define

$$w(x) = u(x) + \epsilon z(x),$$

where ϵ is a positive constant chosen so that it satisfies the inequality

$$\epsilon < \frac{M - u(d)}{z(d)}.$$

The assumption $u(d) < M$ and the fact that $z(d) > 0$ make it possible to find such an ϵ. Then, since z is negative for $a < x < c$, we have

$$w(x) < M \text{ for } a < x < c;$$

by the definition of ϵ,
$$w(d) = u(d) + \epsilon z(d),$$
$$< u(d) + M - u(d),$$
so that
$$w(d) < M.$$
At the point c,
$$w(c) = u(c) + \epsilon z(c) = M.$$
Hence w has a maximum greater than or equal to M which is attained at an interior point of the interval (a, d). But
$$L[w] = L[u] + \epsilon L[z] > 0,$$
so that by our previous result concerning the strict inequality (2), w cannot attain its maximum in (a, d). We thereby reach a contradiction.

If $d < c$, we use the auxiliary function
$$z = e^{-\alpha(x-c)} - 1$$
with $\alpha > g(x)$ to reach the same conclusion.

The key to the above proof is the construction of the function $z(x)$ with the properties: (i) $L[z] > 0$; (ii) $z(x) < 0$ for $x < c$; (iii) $z(x) > 0$ for $x > c$; (iv) $z(c) = 0$. [If d is less than c, inequalities (ii) and (iii) are reversed.] The function z is by no means unique. For example, the function
$$z(x) = (x - a)^\alpha - (c - a)^\alpha$$
with α sufficiently large has the same four properties.

By applying Theorem 1 to $(-u)$ we have the *minimum principle* which asserts that a nonconstant function satisfying the differential inequality $L[u] \leq 0$ cannot attain its minimum at an interior point.

The boundedness condition for g in the statement of Theorem 1 may be relaxed. If g is bounded on every interval $[a', b']$ completely interior to (a, b), then the conclusion of Theorem 1 still holds. We simply apply the argument on any subinterval $[a', b']$ containing the points c and d in its interior. Note that it is possible for g to be bounded on every closed subinterval of (a, b) and yet unbounded as x tends to a or b. For example, $g(x) = 1/(1 - x^2)$ is bounded on every closed subinterval of $(-1, 1)$. This may seem to be a minor point, but it turns out that many of the differential equations of mathematical physics have coefficients g which become unbounded at the endpoints of the interval of definition.

The method employed to prove Theorem 1 enables us to obtain additional information about functions which satisfy an inequality such as (3). We might imagine that a solution u of (3) could have the appearance of the function shown in Fig. 2a. That is, the maximum of u on $[a, b]$ is

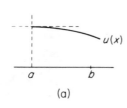 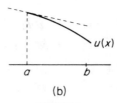 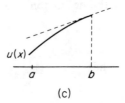

(a)　　　　　　　　(b)　　　　　　　　(c)

FIGURE 2

attained at a and $u'(a) = 0$. In fact, this situation never can occur. If the maximum occurs at the left endpoint, the slope at that point must be negative (Fig. 2b); if the maximum occurs at the right endpoint, the slope at that point must be positive (Fig. 2c). The next theorem establishes the precise result.

THEOREM 2. Suppose u is a nonconstant function which satisfies the inequality $u'' + g(x)u' \geq 0$ in (a, b) and has one-sided derivatives at a and b, and suppose g is bounded on every closed subinterval of (a, b). If the maximum of u occurs at $x = a$ and g is bounded below at $x = a$, then $u'(a) < 0$. If the maximum occurs at $x = b$ and g is bounded above at $x = b$, then $u'(b) > 0$.

Proof. Suppose that $u(a) = M$, that $u(x) \leq M$ for $a \leq x \leq b$, and that for some point d in (a, b) we have $u(d) < M$. Once again we define an auxiliary function

$$z(x) = e^{\alpha(x-a)} - 1 \text{ with } \alpha > 0.$$

We select $\alpha > -g(x)$ for $a \leq x \leq d$ so that $L[z] > 0$. Next, we form the function

$$w(x) = u(x) + \epsilon z(x)$$

with ϵ chosen so that

$$0 < \epsilon < \frac{M - u(d)}{z(d)}.$$

Because $L[w] > 0$, the maximum of w in the interval $[a, d]$ must occur at one of the ends. We have

$$w(a) = M > w(d),$$

so that the maximum occurs at a. Therefore, the one-sided derivative of w at a cannot be positive:

$$w'(a) = u'(a) + \epsilon z'(a) \leq 0.$$

However,

$$z'(a) = \alpha > 0,$$

and therefore
$$u'(a) < 0,$$
which is the desired result.

If the maximum occurs at $x = b$, the argument is similar.

Remarks. (i) If a function u which satisfies (3) has a relative maximum at an interior point c, there is an interval (a_1, b_1) containing c in its interior on which $u(x) \leq u(c)$. Then Theorem 1 shows that $u(x) = u(c)$ on this interval. By applying Theorem 2 to all intervals having c as an endpoint, we see that the value $u(c)$ at the relative maximum is actually the minimum value of u on the interval (a, b).

(ii) If a function u which satisfies (3) has relative minima at two points c_1 and c_2 of the interval (a, b), it must have a relative maximum at some point between c_1 and c_2. It then follows from Remark (i) that $u(c\ ) = u(c_2)$ and that $u(x)$ is constant on the interval (c_1, c_2).

(iii) A function satisfying (3) can have no horizontal point of inflection. (u has a **horizontal point of inflection** at $x = c$ if $u'(c) = 0$ while u is strictly increasing or strictly decreasing in some interval containing c.) If there were such a point, we could select a subinterval with this point as an endpoint (either a right or left endpoint, whichever is appropriate) on which u attains its maximum at c. Then Theorem 2 would be contradicted.

(iv) A result analogous to Theorem 2 holds for solutions of $L[u] \leq 0$, yielding an associated minimum principle. We obtain this principle by applying Theorem 2 to the function $(-u)$.

(v) It is possible to prove Theorem 2 before Theorem 1. Then the following argument yields Theorem 1 immediately. If u has a maximum at an interior point c, then $u'(c) = 0$. Applying Theorem 2 to the intervals (a, c) and (c, b), we conclude that u is constant.

(vi) The boundedness of g is required for the conclusion of Theorems 1 and 2. The equation
$$u'' + g(x)u' = 0$$
with
$$g(x) = \begin{cases} -3/x & \text{for } x \neq 0 \\ 0 & \text{for } x = 0 \end{cases}$$
has the solution
$$u = 1 - x^4.$$
Theorem 1 is clearly violated on the interval $-1 \leq x \leq 1$, as u has a maximum at $x = 0$. Theorem 2 is violated on $[0, 1]$ as $u'(0) = 0$. The results of Theorems 1 and 2 are not applicable because g is not bounded from below in $(0, 1)$.

We now take up the more general differential inequality
$$(L + h)[u] \equiv u'' + g(x)u' + h(x)u \geq 0. \tag{4}$$
The simplest examples show that at best we can only hope for a modified form of the maximum principle; the equation
$$u'' + u = 0$$
has the solution $u = \sin x$ which attains its maximum at $x = \pi/2$. Even the condition $h(x) \leq 0$ is not sufficient to yield an unrestricted maximum principle. We observe that the equation
$$u'' - u = 0$$
has the solution
$$u = -e^x - e^{-x},$$
which attains its maximum value (-2) at $x = 0$. We shall show that a nonconstant solution of (4) with $h \leq 0$ cannot attain a nonnegative maximum at an interior point.

It is easy to see that if the *strict* inequality
$$(L + h)[u] > 0, \text{ with } h \leq 0,$$
holds in an open interval (a, b), then u cannot have a nonnegative maximum in the interior of (a, b). In fact, at any such maximum, we have $u' = 0, u'' \leq 0, hu \leq 0$, contradicting the above strict inequality. This fact enables us to extend Theorems 1 and 2 without altering the argument in any way other than by choosing α so large that $(L + h)[z] > 0$.

The constant α in the function $e^{\alpha(x-c)} - 1$ (or the function $e^{-\alpha(x-c)} - 1$, if d is to the left of c) must only satisfy
$$\alpha^2 + \alpha g(x) + h(x)[1 - e^{-\alpha(x-c)}] > 0$$
$$(\text{or } \alpha^2 - \alpha g(x) + h(x)[1 - e^{\alpha(x-c)}] > 0).$$
Since $h(x) \leq 0$, it is sufficient in either case to select α so that
$$\alpha^2 - \alpha |g(x)| + h(x) > 0.$$
This can certainly be done if $g(x)$ and $h(x)$ are bounded. Again we can show that it suffices for them to be bounded on every closed subinterval of (a, b). In this way we arrive at the next two theorems, which are extensions of Theorems 1 and 2.

THEOREM 3. If $u(x)$ satisfies the differential inequality
$$(L + h)[u] \equiv u'' + g(x)u' + h(x)u \geq 0 \tag{4}$$
in an interval (a, b) with $h(x) \leq 0$, if g and h are bounded on every closed subinterval, and if u assumes a nonnegative maximum value M at an interior point c, then $u(x) \equiv M$.

Note that if h is not identically zero, then the only nonnegative constant M satisfying (4) is $M = 0$.

THEOREM 4. *Suppose that u is a nonconstant solution of the differential inequality (4) having one-sided derivatives at a and b, that $h(x) \leq 0$, and that g and h are bounded on every closed subinterval of (a, b). If u has a nonnegative maximum at a and if the function $g(x) + (x - a)h(x)$ is bounded from below at $x = a$, then $u'(a) < 0$. If u has a nonnegative maximum at b and if $g(x) - (b - x)h(x)$ is bounded from above at $x = b$, then $u'(b) > 0$.*

In extending the proof of Theorem 2 to Theorem 4, we need only observe that

$$(L + h)[e^{\alpha(x-a)} - 1] = e^{\alpha(x-a)}[\alpha^2 + \alpha g + h(1 - e^{-\alpha(x-a)})]$$
$$\geq e^{\alpha(x-a)}[\alpha^2 + \alpha g + \alpha(x-a)h].$$

COROLLARY. *If u satisfies (4) in (a, b) with $h(x) \leq 0$, if u is continuous on $[a, b]$, and if $u(a) \leq 0$, $u(b) \leq 0$, then $u(x) < 0$ in (a, b) unless $u \equiv 0$.*

EXERCISES

1. Prove Theorem 1 by employing the function $z(x) = (x - a)^\alpha - (c - a)^\alpha$ instead of $z(x) = e^{\alpha(x-c)} - 1$.

2. The function $u = \cos x$ satisfies $u'' + g(x)u' = 0$ with $g(x) = -\cot x$, and yet u has a maximum at $x = 0$. Explain. Find a function u which has a horizontal point of inflection at $x = 0$ and which satisfies a differential inequality of the form $u'' + g(x)u' \geq 0$.

3. Show that if $u'' + e^u = -x$ for $0 < x < 1$, then u cannot attain a minimum in $(0, 1)$.

4. Show that a solution of $u'' - 2\cos(u') = 1$ cannot attain a local maximum.

5. Consider the problem
$$u'' + e^x u' = -1 \text{ for } 0 < x < 1,$$
$$u(0) = u(1) = 0.$$
Verify that the solution has no minimum in $(0, 1)$. Also show that $u'(0) > 0$, $u'(1) < 0$.

6. Consider the inequality
$$u'' + (\alpha/x)u' + (\beta/x^2)u \geq 0, \quad 0 < x < 1,$$
with α and β constant. For what values of α and β are Theorems 3 and 4 applicable? Verify by considering solutions of the form $u = x^n$. What is the result if the interval is $-1 < x < 1$?

SECTION 2. THE GENERALIZED MAXIMUM PRINCIPLE

We investigate the differential inequality

$$(L + h)[u] \equiv u'' + g(x)u' + h(x)u \geq 0, \qquad a < x < b, \tag{1}$$

without the requirement that $h(x)$ be nonpositive. Suppose we can find a function w which has a continuous second derivative on $[a, b]$ and which satisfies the inequalities

$$w > 0 \text{ on } [a, b], \tag{2}$$

$$(L + h)[w] \leq 0 \text{ in } (a, b). \tag{3}$$

We define the new dependent variable

$$v = \frac{u}{w}.$$

A simple computation yields

$$(L + h)[u] = (L + h)[vw] = wv'' + (2w' + gw)v' + (L + h)[w]v \geq 0.$$

Dividing by the positive quantity w, we see that v satisfies the differential inequality

$$v'' + \left(2\frac{w'}{w} + g\right)v' + \frac{1}{w}(L + h)[w]v \geq 0. \tag{4}$$

Inequality (4), when taken in conjunction with (2) and (3), shows that $v = u/w$ satisfies Theorems 3 and 4.

The argument above depends on the existence of a function w which satisfies (2) and (3). We shall now show that if $h(x)$ is bounded, if $g(x)$ is bounded from below, and if the interval $[a, b]$ is sufficiently short, then there is a function w which fulfills inequalities (2) and (3). In fact, such a function is given by

$$w = 1 - \beta(x - a)^2, \tag{5}$$

if the constant β is determined suitably. To see this, we compute

$$(L + h)[w] = -2\beta[1 + (x - a)g(x) + \tfrac{1}{2}(x - a)^2 h(x)] + h(x). \tag{6}$$

Since, by assumption, g and h are bounded from below, there are constants G and H such that $g \geq G$ and $h \geq H$. We suppose a and b are so close together that

$$1 + (x - a)G + \tfrac{1}{2}(x - a)^2 H > 0 \text{ for } a \leq x \leq b.$$

Since $h(x)$ is also bounded from above, we can select β so that

$$\beta \geq \frac{1}{2}\left[\frac{h(x)}{1 + (x - a)G + \tfrac{1}{2}(x - a)^2 H}\right].$$

Then, because of (6), we have $(L + h)[w] \leq 0$ in (a, b). If the length $(b - a)$ is also so small that

$$\beta(b-a)^2 < 1,$$

then (5) shows that $w > 0$ on $[a, b]$. In this way, the function w with the desired properties may always be constructed.

The preceding discussion leads to the following **generalized maximum principle**.

THEOREM 5. Suppose the operator $L + h$ is given by (1) with $h(x)$ bounded and with $g(x)$ bounded from below. For any sufficiently short interval $[a, b]$, a function w can be found which satisfies (2) and (3). Then if u is any function satisfying (1) in (a, b), the function u/w satisfies the maximum principles as given in Theorems 3 and 4.

Remark. Theorem 5 shows that a function u which satisfies (1) cannot oscillate too rapidly, for if $u > 0$ between two of its zeros $x = a$ and $x = b$, then u/w must have a positive maximum between them. Hence, Theorem 5 is violated unless the distance $b - a$ between these zeros is so large that this theorem doesn't hold. We thus find that u can have at most two zeros (between which u is negative) in any interval (a, b) where Theorem 5 holds.

If u is a solution of the **equation** $u'' + g(x)u' + h(x)u = 0$, we can apply the same reasoning to both u and $-u$ to find that u can have at most one zero in any interval (a, b) where Theorem 5 holds.

Let $r(x)$ be a solution of the differential equation

$$r'' + g(x)r' + h(x)r = 0, \tag{7}$$

with g and h bounded functions. Suppose that r is not identically zero, and that

$$r(a) = 0.$$

In the light of the remark following Theorem 5, we know that r cannot vanish for some distance to the right of a. If r has any zeros to the right of a, we denote the first one by a^* and call it the **conjugate point** of a. Thus r is of one sign in the interval (a, a^*), and for convenience we assume that

$$r(x) > 0 \text{ for } a < x < a^*.$$

If $w > 0$ on $[a, a^*]$, the function r/w vanishes at a and at a^* and is positive in (a, a^*). Hence it has a maximum in (a, a^*). Therefore by Theorem 5, w cannot satisfy (3). On the other hand, if b is any point in (a, a^*), a function w can be found so that r/w satisfies the maximum principle of Theorem 5. To see this, we observe first that $r(x)$ is bounded from below by a positive number on any subinterval $[c, b]$ contained in (a, a^*). Consequently, for sufficiently small $\epsilon > 0$, the function

$$w(x) = r(x) + \epsilon[2 - e^{\alpha(x-a)}]$$

is positive on $[a, b]$. If α is selected so that $(L + h)[2 - e^{\alpha(x-a)}] \leq 0$ in (a, b), then w is a function for which Theorem 5 holds.

We conclude that **if a^* is the conjugate point of a, there exists a $w > 0$ such that Theorem 5 holds on the interval $[a, b]$ if and only if $b < a^*$.** If $r(x)$ [the solution of (7) which satisfies $r(a) = 0$] has no zero to the right of a, we set $a^* = \infty$, and Theorem 5 holds on every interval $[a, b]$.

If $h(x)$ is unbounded or if g is not bounded from below, there may be no interval $[a, b]$ for which Theorem 5 holds. For example, the function

$$u = x \sin\left(\frac{1}{x}\right)$$

satisfies

$$u'' + x^{-4}u = 0.$$

We see that u vanishes at $x = 1/n\pi$, $n = 1, 2, \ldots$, and so there can be no function $w > 0$ with the property that u/w satisfies a maximum principle in any of the intervals $[0, 1/n\pi]$, $n = 1, 2, \ldots$.

EXERCISES

1. Find the solution of the problem

$$u'' + u = 0, \quad -\frac{\pi}{4} < x < \frac{\pi}{4},$$

$$u\left(-\frac{\pi}{4}\right) = \cos\left(\frac{3\pi}{8}\right), \quad u\left(\frac{\pi}{4}\right) = \cos\left(\frac{\pi}{8}\right).$$

Verify that u attains a maximum in $(-\pi/4, \pi/4)$ but that $u/\cos x$ does not.

2. Suppose that $u'' + (\cos x)u \geq 0$ in $(0, c)$. If $w = 1 - \beta x^2$, find values of β and c such that u/w satisfies Theorem 5 in $(0, c)$.

3. Show that no solution of $u'' + e^u = e$ can attain a minimum value greater than 1 or a maximum value less than 1.

4. Show that, near $x = 0$, $u = x^3$ satisfies an equation of the form $u'' + g(x)u' = 0$ with g unbounded and thus conclude that the boundedness hypothesis in Theorem 5 is essential.

SECTION 3. THE INITIAL VALUE PROBLEM

We shall be interested in a solution of the differential equation

$$u'' + g(x)u' + h(x)u = f(x) \tag{1}$$

which satisfies the **initial conditions**

$$u(a) = \gamma_1, \quad u'(a) = \gamma_2. \tag{2}$$

The functions f, g, and h are given in an interval (a, b), with g and h bounded; γ_1 and γ_2 are prescribed constants. Whenever a solution of (1) in the interval (a, b) is determined which satisfies conditions (2), we say that an **initial value problem** has been solved.

The **existence** of solutions of such initial value problems follows from the general theory of ordinary differential equations. The **uniqueness** of such solutions, although also a consequence of the general theory, follows easily from the generalized maximum principle. The next theorem yields the main result.

THEOREM 6. Suppose that $u_1(x)$ and $u_2(x)$ are solutions of (1) in an interval (a, b) and that u_1 and u_2 satisfy the same initial conditions (2). Then $u_1 \equiv u_2$ in (a, b).

Proof. Define $u(x) = u_1(x) - u_2(x)$. Then u satisfies the equation

$$u'' + g(x)u' + h(x)u = 0$$

with the initial conditions

$$u(a) = u'(a) = 0.$$

Now suppose u is not identically zero in (a, b). We shall reach a contradiction.

According to Theorem 5, there is an $\epsilon > 0$ whose size depends only on g and h and there is a positive function w such that the maximum of u/w in the interval $(a, a + \epsilon)$ must occur at the endpoints. We also observe that $-u$ satisfies the same equation with the same initial conditions. Therefore, according to Theorem 5 again, the maximum of $-u/w$ must occur at one of the endpoints, a or $a + \epsilon$. Hence, either the maximum or the minimum of u/w must occur at a. But at $x = a$

$$\left(\frac{u}{w}\right)' = \frac{u'w - uw'}{w^2} = 0.$$

Since u/w satisfies Theorem 4, we conclude that u/w is constant. Since $u(a) = 0$, this constant is zero. That is, u is identically zero on the interval $[a, a + \epsilon]$. In particular,

$$u(a + \epsilon) = 0, \quad u'(a + \epsilon) = 0.$$

We may now repeat the argument to conclude that $u \equiv 0$ in $(a + \epsilon, a + 2\epsilon)$, ϵ being unchanged, since its size depends only on bounds for g and h in (a, b). We need perform this process only a finite number of times to deduce that $u \equiv 0$ in (a, b).

EXERCISES

1. Show that the problem

$$u'' - \frac{1}{x}u' = 0,$$

$$u(0) = u'(0) = 0,$$

has the two solutions $u \equiv 0$ and $u = x^2$, and conclude that the boundedness of g is needed in Theorem 6.

2. Find a function $h(x)$ such that the equation $u'' + h(x)u = 0$ has two solutions which satisfy $u(0) = u'(0) = 0$.

3. Show that there is at most one solution of the initial value problem

$$u'' + e^u = 1 \text{ for } x > 0,$$

$$u(0) = 1, \quad u'(0) = 0.$$

Hint: Set up a problem satisfied by $u_1 - u_2$ using the mean value theorem

$$e^{u_1} - e^{u_2} = (u_1 - u_2)e^{u_2 + \theta(u_1 - u_2)}$$

for some value θ such that $0 < \theta < 1$.

4. Let $L[u] \equiv u'' + g(x)u' + h(x)u$, $a \leq x \leq b$ with g and h bounded. Suppose u_1 satisfies $L[u_1] \geq 0$ and u_2 satisfies $L[u_2] \leq 0$ in (a, b). If $u_1 \leq u_2$ on $[a, b]$ and if $u_1(a) = u_2(a)$, $u_1'(a) = u_2'(a)$, show that $u_1 \equiv u_2$ in (a, b).

SECTION 4. BOUNDARY VALUE PROBLEMS

The simplest boundary value problem concerns the determination of a solution of the equation

$$u'' + g(x)u' + h(x)u = f(x) \tag{1}$$

in an interval (a, b) subject to the **boundary conditions**

$$u(a) = \gamma_1, \quad u(b) = \gamma_2. \tag{2}$$

Questions about the uniqueness of solutions of (1) which satisfy (2) can be answered by use of the maximum principle. The situation here, however, is not as straightforward as in the case of initial value problems. The simple equation

$$u'' + u = 0$$

has the solutions $u_1 = \sin x$ and $u_2 \equiv 0$ for $0 \leq x \leq \pi$, and both solutions satisfy the boundary conditions $u(0) = u(\pi) = 0$. The next result gives one of the simplest uniqueness theorems for boundary value problems.

THEOREM 7. Suppose that $u_1(x)$ and $u_2(x)$ are solutions of (1) which satisfy the boundary conditions (2). If $h(x) \leq 0$ in (a, b), then $u_1 \equiv u_2$.

Proof. Let $u(x) = u_1(x) - u_2(x)$. Then u satisfies the equation

$$u'' + g(x)u' + hu = 0 \tag{3}$$

and the boundary conditions

$$u(a) = u(b) = 0.$$

According to Theorem 3, we know that $u(x) \leq 0$ in (a, b). Since the function $-u(x)$ satisfies the same equation with the same boundary conditions, we may apply Theorem 3 to $-u(x)$ to conclude that $-u \leq 0$ in (a, b). Therefore $u \equiv 0$ in (a, b).

We now take up a more general boundary value problem, one which contains boundary conditions (2) as a special case. We consider solutions of (1) which satisfy the boundary conditions

$$\left.\begin{aligned} -u'(a)\cos\theta + u(a)\sin\theta &= \gamma_1, \\ u'(b)\cos\phi + u(b)\sin\phi &= \gamma_2, \end{aligned}\right\} \tag{4}$$

where $\gamma_1, \gamma_2, \theta$, and ϕ are prescribed constants with $0 \leq \theta \leq \pi/2$, $0 \leq \phi \leq \pi/2$. Note that conditions (4) reduce to (2) when $\theta = \phi = \pi/2$.

THEOREM 8. Suppose that $u_1(x)$ and $u_2(x)$ are two solutions of (1) which satisfy the boundary conditions (4). If $h(x) \leq 0$ in (a, b), then $u_1 \equiv u_2$ unless $h \equiv 0$, $\theta = \phi = 0$, in which case u_1 and u_2 may differ by a constant.

Proof. As before, define $u = u_1 - u_2$. Then u satisfies (3) and the boundary conditions

$$\left.\begin{aligned} -u'(a)\cos\theta + u(a)\sin\theta &= 0, \\ u'(b)\cos\phi + u(b)\sin\phi &= 0. \end{aligned}\right\} \tag{5}$$

The function $u \equiv M$, a nonzero constant, satisfies these conditions if and only if $h \equiv 0$, $\theta = 0$, and $\phi = 0$. If we suppose u to be a nonconstant solution which is positive at some point, we shall reach a contradiction. By Theorem 3, u attains its positive maximum at a or b. Suppose the maximum occurs at a. We can apply Theorem 4, which asserts that $u'(a) < 0$. Since $0 \leq \theta \leq \pi/2$ and $u(a) > 0$, the first condition in (5) is violated. Similarly, if the maximum occurs at b, the second condition in (5) is violated. We conclude that any nonconstant solution can never be positive. The same reasoning applied to $-u$ shows that u can never be negative. Thus $u \equiv 0$ on $[a, b]$.

Although it is possible to establish uniqueness theorems for boundary

value problems without the restriction that $h(x)$ be nonpositive, care must be exercised in a discussion of the precise conditions involved. For example, the simple equation
$$u'' + u = 0$$
with the boundary conditions $u(a) = u(b) = 0$ has only the solution $u \equiv 0$ so long as $b - a < \pi$. On the other hand, we know the result is false if $b - a = \pi$. The specific result in such cases is obtained by application of the generalized maximum principle as given in Theorem 5.

THEOREM 9. *Suppose that $u_1(x)$ and $u_2(x)$ are solutions of (1) which satisfy the same boundary conditions (2). If $b < a^*$, where a^* is the conjugate point of a, then $u_1 \equiv u_2$.*

The proof of Theorem 9 is a repetition of the proof of Theorem 7, except that the maximum principle as given in Theorem 5 is employed. Note that by definition of conjugate point, uniqueness fails when $b = a^*$.

EXERCISES

1. Find the solution of the problem
$$u'' - u = 0, \quad 0 < x < 1,$$
$$-u'(0) + u(0) = 0,$$
$$u(1) = 1.$$

2. Find the point conjugate to $a = 1$ with respect to the differential equation
$$u'' + 2u' + 2u = 0.$$

3. Find all the intervals on the x-axis in which Theorem 9 may be applied to solutions of the equation
$$(x^2 + 1)u'' - 2xu' + 2u = 0$$
by determining the conjugate point of an arbitrary point a.

SECTION 5. APPROXIMATION IN BOUNDARY VALUE PROBLEMS

Suppose that we seek a solution of the equation
$$(L + h)[u] \equiv u'' + g(x)u' + h(x)u = f(x) \text{ for } a < x < b \quad (1)$$
which satisfies the boundary conditions
$$u(a) = \gamma_1, \quad u(b) = \gamma_2. \quad (2)$$

In most cases it is impossible to find such a solution explicitly. It is frequently desirable to approximate a solution in such a way that an explicit bound for the error is known. Such an approximation is equivalent to the determination of both upper and lower bounds for the values of the solution.

We shall assume that the functions f, g, and h are bounded and, at first, that $h \leq 0$ in (a, b). Under these circumstances it is possible to use the maximum principle in Theorem 3 to obtain a bound for a solution u without any actual knowledge of u itself.

Suppose we can find a function $z_1(x)$ with the properties:

$$(L + h)[z_1] \leq f(x) \text{ for } a < x < b, \tag{3}$$

$$z_1(a) \geq \gamma_1, \quad z_1(b) \geq \gamma_2. \tag{4}$$

Then the function

$$v_1(x) \equiv u(x) - z_1(x)$$

satisfies

$$(L + h)[v_1] \geq 0$$

and

$$v_1(a) \leq 0, \quad v_1(b) \leq 0.$$

The maximum principle as given in Theorem 3 in Section 1 may be applied to v_1, and we conclude that $v_1 \leq 0$ on $[a, b]$. That is,

$$u(x) \leq z_1(x) \text{ for } a \leq x \leq b.$$

The function $z_1(x)$ is an upper bound for $u(x)$.

Similarly, a lower bound for u may be obtained by finding a function $z_2(x)$ with the properties

$$(L + h)[z_2] \geq f(x)$$

and

$$z_2(a) \leq \gamma_1, \quad z_2(b) \leq \gamma_2.$$

The maximum principle applied to $z_2(x) - u(x)$ shows that

$$u(x) \geq z_2(x) \text{ for } a \leq x \leq b.$$

Functions $z_1(x)$, $z_2(x)$ with the desired properties are easily constructed. We may use polynomials, rational functions, exponentials, and so forth. For example, we may set

$$z_1(x) = A\{2 - e^{-\alpha(x-a)}\}$$

and try to select A and α so that (3) and (4) are satisfied. We choose α so large that

$$(L + h)[e^{-\alpha(x-a)}] = (\alpha^2 - \alpha g + h)e^{-\alpha(x-a)} > 0$$

for $a \leq x \leq b$.

We define a constant k by the relation
$$k = \min_{a \leq x \leq b} [(\alpha^2 - \alpha g + h)e^{-\alpha(x-a)}]$$
and select A so that
$$A = \max\left[\gamma_1, \gamma_2, \frac{1}{k}\max_{a \leq x \leq b}\{-f(x)\}, 0\right].$$

That is, A is the largest of the four numbers in the brackets. With the selections of A and α as just described, the function $z_1(x) = A\{2 - e^{-\alpha(x-a)}\}$ satisfies (3) and (4).

To determine a lower bound, we choose
$$z_2(x) = B\{2 - e^{-\alpha(x-a)}\},$$
where α is chosen as in $z_1(x)$ and B is the smallest of four numbers:
$$B = \min\left[\gamma_1, \gamma_2, \frac{1}{k}\min_{a \leq x \leq b}\{-f(x)\}, 0\right].$$

Then
$$B\{2 - e^{-\alpha(x-a)}\} \leq u(x) \leq A\{2 - e^{-\alpha(x-a)}\} \quad \text{for} \quad a \leq x \leq b.$$

In particular, we have
$$|u(x)| \leq 2 \max\left[|\gamma_1|, |\gamma_2|, \frac{1}{k}\max_{a \leq x \leq b}|f(x)|\right] \tag{5}$$
for $a \leq x \leq b$.

The uniqueness theorem of Section 4 for boundary value problems with $h \leq 0$ follows from (5), since $f \equiv 0$, $\gamma_1 = \gamma_2 = 0$ imply that $u \equiv 0$ for $a \leq x \leq b$.

If u is a solution of (1), (2) and $\bar{u}$ is a solution of the related problem
$$\bar{u}'' + g\bar{u}' + h\bar{u} = \bar{f}(x),$$
$$\bar{u}(a) = \bar{\gamma}_1, \quad \bar{u}(b) = \bar{\gamma}_2,$$
then the difference $u - \bar{u}$ satisfies
$$(u - \bar{u})'' + g(u - \bar{u})' + h(u - \bar{u}) = f - \bar{f},$$
$$u(a) - \bar{u}(a) = \gamma_1 - \bar{\gamma}_1, \quad u(b) - \bar{u}(b) = \gamma_2 - \bar{\gamma}_2.$$

Inequality (5) shows that
$$|u(x) - \bar{u}(x)| \leq 2\max\left[|\gamma_1 - \bar{\gamma}_1|, |\gamma_2 - \bar{\gamma}_2|, \frac{1}{k}\max_{a \leq x \leq b}|f(x) - \bar{f}(x)|\right].$$

Therefore, if the quantities
$$|\gamma_1 - \bar{\gamma}_1|, \quad |\gamma_2 - \bar{\gamma}_2|, \quad \max_{a \leq x \leq b}|f(x) - \bar{f}(x)|$$
are all small, then $|u(x) - \bar{u}(x)|$ is small for all x in the interval (a, b). Under these circumstances, we say that the solution u of the problem (1), (2) **depends continuously** on $f(x)$ and the boundary values γ_1, γ_2.

Sec. 5   *Approximation in Boundary Value Problems*   17

We give an example to show how explicit bounds for a solution may actually be obtained.

Example. Estimate the value at $x = \frac{1}{2}$ of the solution of
$$u'' - xu = 0 \text{ for } 0 < x < 1$$
with
$$u(0) = 0, \quad u(1) = 1.$$

In this case it is convenient to select polynomials for the comparison functions z_1 and z_2. We let
$$z_1 = x \text{ and } z_2 = x - \beta x(1 - x).$$
Then clearly $z_1(0) = z_2(0) = 0$, $z_1(1) = z_2(1) = 1$ and
$$(L + h)[z_1] \leq 0, \quad (L + h)[z_2] = -x^2 + \beta[2 + x^2(1 - x)] \geq 0 \text{ for } \beta \geq \tfrac{1}{2}.$$
We choose $\beta = \tfrac{1}{2}$ and find
$$\tfrac{1}{2}(x + x^2) \leq u(x) \leq x.$$
In particular, at $x = \tfrac{1}{2}$,
$$\tfrac{3}{8} \leq u(\tfrac{1}{2}) \leq \tfrac{1}{2}.$$
That is, $u(\tfrac{1}{2}) = 0.4375 \pm 0.0625$ so that we have the solution within 15%. Actually, $u(\tfrac{1}{2}) = 0.4726$.

We now take up the question of approximation of solutions of the more general two-point boundary value problem discussed in Section 4. We suppose that u is a solution of the equation
$$(L + h)[u] \equiv u'' + g(x)u' + h(x)u = f(x) \text{ for } a < x < b \tag{6}$$
which satisfies the boundary conditions
$$\left.\begin{array}{r} -u'(a) \cos \theta + u(a) \sin \theta = \gamma_1, \\ u'(b) \cos \phi + u(b) \sin \phi = \gamma_2. \end{array}\right\} \tag{7}$$
The quantities θ and ϕ are preassigned constants. We assume that
$$0 \leq \theta \leq \pi/2, \quad 0 \leq \phi \leq \pi/2, \quad h(x) \leq 0.$$
In analogy with the development of the simpler two-point boundary problem, we seek a function $z_1(x)$ with the properties:
$$(L + h)[z_1] \leq f(x) \text{ for } a < x < b, \tag{8}$$
$$\left.\begin{array}{r} -z_1'(a) \cos \theta + z_1(a) \sin \theta \geq \gamma_1, \\ z_1'(b) \cos \phi + z_1(b) \sin \phi \geq \gamma_2. \end{array}\right\} \tag{9}$$
Then the function $v_1 \equiv u - z_1$ satisfies
$$(L + h)[v_1] \geq 0,$$
$$-v_1'(a) \cos \theta + v_1(a) \sin \theta \leq 0,$$
$$v_1'(b) \cos \phi + v_1(b) \sin \phi \leq 0.$$

If v_1 is ever positive, Theorem 3 states that its positive maximum occurs at a or at b. If it occurs at a, we have $v_1(a) > 0$, $v_1'(a) \leq 0$. Since

$$-v_1'(a) \cos \theta + v_1(a) \sin \theta \leq 0,$$

this can only occur if $\theta = 0$ and $v_1'(a) = 0$. Theorem 4 then states that $v_1(x)$ is a positive constant, thus implying that $h \equiv 0$.

Similarly, v_1 cannot have a positive maximum at b unless $\phi = 0$ and $h \equiv 0$, since

$$v_1'(b) \cos \phi + v_1(b) \sin \phi \leq 0.$$

We conclude that unless both θ and ϕ are zero and $h \equiv 0$, $v_1(x) \leq 0$. That is,

$$u(x) \leq z_1(x).$$

Similarly, if z_2 satisfies the inequalities

$$(L + h)[z_2] \geq f(x), \tag{10}$$

$$\left.\begin{array}{r}-z_2'(a) \cos \theta + z_2(a) \sin \theta \leq \gamma_1, \\ z_2'(b) \cos \phi + z_2(b) \sin \phi \leq \gamma_2,\end{array}\right\} \tag{11}$$

and if either h is not identically zero or θ and ϕ are not both zero, then

$$u(x) \geq z_2(x).$$

Thus we obtain the following result on approximation.

THEOREM 10. *Suppose that $u(x)$ is a solution of (6) satisfying the boundary conditions (7), and that $h(x) \leq 0$, $0 \leq \theta \leq \frac{1}{2}\pi$, and $0 \leq \phi \leq \frac{1}{2}\pi$. Suppose also that not all the equalities $\theta = 0$, $\phi = 0$, $h \equiv 0$ hold. If $z_1(x)$ satisfies conditions (8), (9) and if $z_2(x)$ satisfies conditions (10), (11), then*

$$z_2(x) \leq u(x) \leq z_1(x).$$

Specific functions z_1, z_2 fulfilling the conditions of Theorem 10 are easily found in the form of polynomials, exponentials, and so forth. We give an example showing how Theorem 10 may be applied.

Example. Find upper and lower bounds for the value at $x = \frac{1}{2}$ of the solution of the boundary value problem

$$u'' - xu = 0 \text{ for } 0 < x < 1.$$

$$-u'(0) + u(0) = 0,$$

$$u(1) = 1.$$

We set

$$z_1 = \tfrac{1}{2}(1 + x).$$

Then
$$(L+h)[z_1] = -\tfrac{1}{2}x(1+x) \le 0,$$
$$-z_1'(0) + z_1(0) = 0, \qquad z_1(1) = 1.$$

For z_2 we select the exponential
$$z_2 = e^{x-1}$$
and find
$$(L+h)[z_2] = (1-x)e^{x-1} \ge 0,$$
$$-z_2'(0) + z_2(0) = 0, \qquad z_2(1) = 1.$$

Hence
$$e^{x-1} \le u(x) \le \tfrac{1}{2}(1+x).$$

In particular at $x = \tfrac{1}{2}$,
$$0.6065 \le u(\tfrac{1}{2}) \le 0.7500,$$
or $u(\tfrac{1}{2}) = 0.6782 \pm 0.0718$. [Actually $u(\tfrac{1}{2}) = 0.6783$.] We can, of course, find better bounds by choosing more complicated functions for z_1 and z_2.

We now drop the hypotheses $h \le 0$, $0 \le \theta \le \pi/2$, $0 \le \phi \le \pi/2$. By multiplying the boundary conditions in (7) by (-1) if necessary, we may always make $\cos\theta \ge 0$ and $\cos\phi \ge 0$. Thus we assume without loss of generality that $-\pi/2 < \theta \le \pi/2$, $-\pi/2 < \phi \le \pi/2$.

In order to use the generalized maximum principle of Section 2, we suppose that we can find a positive function $w(x)$ which satisfies the inequalities
$$(L+h)[w] \le 0 \text{ in } (a,b), \tag{12}$$
$$\left.\begin{array}{l} -w'(a)\cos\theta + w(a)\sin\theta \ge 0, \\ w'(b)\cos\phi + w(b)\sin\phi \ge 0. \end{array}\right\} \tag{13}$$

We set
$$v = \frac{u}{w},$$
and find that v must satisfy
$$v'' + \left(\frac{2w'}{w} + g\right)v' + \frac{1}{w}(L+h)[w]v = \frac{f}{w},$$
$$-v'(a)w(a)\cos\theta + v(a)[-w'(a)\cos\theta + w(a)\sin\theta] = \gamma_1,$$
$$v'(b)w(b)\cos\phi + v(b)[w'(b)\cos\phi + w(b)\sin\phi] = \gamma_2.$$

We can write these equations in the form of (6) and (7),
$$(\bar{L} + H)[v] \equiv v'' + Gv' + Hv = \frac{f}{w}, \tag{14}$$

$$-v'(a)\cos\bar{\theta} + v(a)\sin\bar{\theta} = \bar{\gamma}_1,$$
$$v'(b)\cos\bar{\phi} + v(b)\sin\bar{\phi} = \bar{\gamma}_2,$$
(15)

where $H = (L + h)[w]/w \leq 0$, $G = (2w'/w) + g$,

$$\tan\bar{\theta} = \frac{1}{w(a)\cos\theta}[-w'(a)\cos\theta + w(a)\sin\theta] \geq 0,$$

$$\tan\bar{\phi} = \frac{1}{w(b)\cos\phi}[w'(b)\cos\phi + w(b)\sin\phi] \geq 0,$$

$\bar{\gamma}_1 = \gamma_1\cos\bar{\theta}/w(a)\cos\theta$, and $\bar{\gamma}_2 = \gamma_2\cos\bar{\phi}/w(b)\cos\phi$.

By (13) we may choose $\bar{\theta}$ and $\bar{\phi}$ in the range $0 \leq \bar{\theta} \leq \tfrac{1}{2}\pi$, $0 \leq \bar{\phi} \leq \tfrac{1}{2}\pi$. (If $\theta = \pi/2$ we put $\bar{\theta} = \pi/2$, and if $\phi = \pi/2$ we put $\bar{\phi} = \pi/2$.)

Suppose there is a function $w(x)$, positive on $[a, b]$, which satisfies the conditions (12) and (13). If z_1 and z_2 satisfy conditions (8), (9) and (10), (11), respectively, then the functions

$$\frac{z_1}{w} \quad \text{and} \quad \frac{z_2}{w}$$

satisfy the analogous conditions with respect to equation (14) with boundary conditions (15). Hence, by Theorem 10, we have the inequalities

$$\frac{z_2(x)}{w(x)} \leq \frac{u(x)}{w(x)} \leq \frac{z_1(x)}{w(x)}$$

unless $\bar{\theta} = \bar{\phi} = 0$ and $H \equiv 0$. This exceptional situation occurs when the inequalities (12) and (13) are equations. Therefore, if there is a positive function $w(x)$ satisfying (12) and (13) but such that not all the inequalities are equations, we obtain the bounds

$$z_2(x) \leq u(x) \leq z_1(x),$$
(16)

as before.

If w satisfies (12) and (13) with equality signs rather than inequality signs, we may add any multiple of w to a solution u of (6) and (7) to obtain another solution. That is, the solution is not unique. Of course, there may be no solution at all, but if there is at least one, then there are many. Certainly not all of them will satisfy (16).

If inequality (16) holds for the solution of (6), (7), then the solution w of

$$(L + h)[w] = 0 \text{ in } (a, b)$$
(17)

which satisfies the boundary conditions

$$-w'(a)\cos\theta + w(a)\sin\theta = 1,$$
$$w'(b)\cos\phi + w(b)\sin\phi = 1,$$
(18)

must be nonnegative. This is easily seen by selecting $z_2 \equiv 0$. If $w = 0$ at an interior point, we also have $w' = 0$ there. The uniqueness theorem for the initial value problem implies that $w \equiv 0$, which violates the boundary conditions (18). So w cannot vanish at an interior point. If w vanishes at an endpoint, say at a, then the first condition in (18) becomes $w'(a)\cos\theta = -1$; the nonnegativity of w implies that $w'(a) \geq 0$, a contradiction. Therefore $w(a) > 0$ and, similarly, $w(b) > 0$. Hence $w > 0$ on $[a, b]$.

We have established the following result, under the hypothesis that the problem (17), (18) has a solution.

THEOREM 11. Let u be a solution of (6), (7) with $-\pi/2 < \theta \leq \pi/2$, $-\pi/2 < \phi \leq \pi/2$. Let z_1 and z_2 satisfy the inequalities (8), (9) and (10), (11), respectively. Then the bounds

$$z_2(x) \leq u(x) \leq z_1(x) \tag{19}$$

hold in (a, b) if and only if there exists a positive function w on $[a, b]$ which satisfies inequalities (12) and (13) in such a way that not all the inequalities in (12) and (13) are equalities.

If $h \leq 0$, $0 \leq \theta \leq \pi/2$, and $0 \leq \phi \leq \pi/2$, then the function $w \equiv 1$ satisfies the conditions (12), (13) and Theorem 10 follows immediately.

The function w doesn't appear in the inequalities (19). Therefore it is of interest to obtain a theorem which eliminates w entirely and which provides conditions on z_1 and z_2 guaranteeing that they form upper and lower bounds. The next result gives a necessary and sufficient condition for this to be the case.

THEOREM 12. Suppose $z_1(x)$ and $z_2(x)$ satisfy inequalities (8), (9) and (10), (11), respectively, in such a way that equality does not hold in all the conditions. Let $g(x)$ be bounded below on each interval $[a, c]$ and bounded above on each interval $[c, b]$ with $a < c < b$. Let $u(x)$ be a solution of (6), (7). Then the bounds

$$z_2(x) \leq u(x) \leq z_1(x) \tag{20}$$

hold if and only if $z_2(x) \leq z_1(x)$ for $a \leq x \leq b$.

Proof. If (20) holds, then clearly $z_2(x) \leq z_1(x)$. We now assume that $z_1(x) - z_2(x)$ is nonnegative, and we must show that (20) holds. If

$$q(x) \equiv z_1(x) - z_2(x)$$

is strictly positive on $[a, b]$, then we may select q as the function w of Theorem 11. All the requirements are fulfilled and (20) holds. Therefore, we need only investigate the possibility that q has a zero on $[a, b]$.

According to (8), (9), (10), and (11), q satisfies the inequalities

$$(L + h)[q] \leq 0, \tag{21}$$

$$-q'(a) \cos \theta + q(a) \sin \theta \geq 0,$$
$$q'(b) \cos \phi + q(b) \sin \phi \geq 0,$$
(22)

and, in view of the hypotheses, equality cannot hold in all the conditions (21), (22).

First suppose $q(c) = 0$ at an interior point c. Then q has a local minimum at c and so $q'(c) = 0$. We conclude from Theorem 6 that $q \equiv 0$. Then equality holds in all the conditions (21), (22), contrary to our hypothesis.

The only remaining possibility is that $q > 0$ in (a, b) but that $q = 0$ at an endpoint, say at $x = a$. Then according to Theorem 5, $q'(a) > 0$. But then the first inequality in (22) is violated unless $\theta = \pi/2$. Similarly, if q vanishes at b, ϕ must equal $\pi/2$. If q vanishes at one or both endpoints, then it does not satisfy the conditions required of w in Theorem 11. Under these circumstances we shall show either that all equalities in (21), (22) hold or that we can find a function $w(x)$, positive on $[a, b]$, which can be used as an auxiliary function in Theorem 11.

We first consider the case $q(b) = 0$ and $q(a) > 0$. As we saw above, we must have $\phi = \pi/2$. We construct the function $r(x)$ to satisfy

$$(L + h)[r] = 0,$$
$$r(a) = \cos \theta, \qquad r'(a) = \sin \theta.$$

If $\theta < \pi/2$, then r is positive at a, and if $\theta = \pi/2$ we have $r(a) = 0$ and $r > 0$ near $x = a$.

We now define the function

$$v(x) = \frac{r(x)}{q(x)} \text{ for } a \leq x < b$$

and note that $v(a) \geq 0$ and that

$$v'(a) = \frac{q(a)r'(a) - r(a)q'(a)}{[q(a)]^2} = \frac{q(a) \sin \theta - q'(a) \cos \theta}{[q(a)]^2} \geq 0.$$

Since $(L + h)[r] = 0$ and $(L + h)[q] \leq 0$, v satisfies a second-order differential equation with the coefficient of v nonpositive. Then, according to Theorems 3 and 4 applied to any subinterval of the form $[a, c]$, either $v(c) > v(a)$ and $v'(c) > 0$, or $v(x) \equiv v(a)$ on the whole subinterval $[a, c]$. In either case, $r(x) > 0$ for x in (a, b). We shall show that also $r(b) > 0$.

If $v(x) \equiv v(a)$ on $[a, b]$, q is proportional to r; hence $(L + h)[q] = 0$ and $-q'(a) \cos \theta + q(a) \sin \theta = 0$. Since equality does not hold in all the conditions (21) and (22), it then follows that $q(b) > 0$, contrary to hypothesis. Thus for some number c in (a, b), we must have

$$v(c) > v(a) \geq 0 \text{ and } v'(x) > 0 \quad \text{for} \quad x \geq c.$$

Now

$$v'(x) = \frac{r'q - q'r}{q^2}.$$

Putting

$$\psi(x) \equiv r'q - q'r,$$

we find from the differential equation for r and the differential inequality for q that

$$\psi'(x) + g\psi(x) = -r(L+h)[q] \geq 0.$$

Then if $g \leq M$ on the interval $[c, b]$, we see that

$$\frac{d}{dx} \log \psi \geq -M;$$

consequently,

$$\psi(x) \geq \psi(c) e^{-M(x-c)}.$$

In particular

$$\psi(b) \geq \psi(c) e^{-M(b-c)} > 0.$$

Since $\psi = r'q - q'r$ and since $q(b) = 0$, $q'(b) < 0$, it follows that $r(b) > 0$. We have thus shown that $r > 0$ on the whole interval $(a, b]$. Moreover, $(L+h)[r] = 0$ and

$$-r'(a) \cos \theta + r(a) \sin \theta = 0.$$

Since $\phi = \pi/2$, we have $r'(b) \cos \phi + r(b) \sin \phi = r(b) > 0$. Thus the function $w(x) = q(x) + r(x)$ satisfies the requirements of Theorem 11.

If $q(a) = 0$ and $q(b) > 0$, we show similarly that the solution of

$$(L+h)[s] = 0$$
$$s(b) = \cos \phi$$
$$s'(b) = -\sin \phi$$

is positive in $[a, b)$, so that $w = q + s$ satisfies the conditions of Theorem 11. Finally, if $\theta = \phi = \pi/2$ and $q(a) = q(b) = 0$, we find that $r > 0$ except at $x = a$, and $s > 0$ except at $x = b$, so that $w = r + s$ satisfies the conditions of Theorem 11.

Remarks. (i) If equality holds for all conditions (8), (9) and (10), (11), both z_1 and z_2 satisfy all conditions of (6), (7). If they are distinct, we know the solution is not unique; if $z_1 \equiv z_2$, the solution may or may not be unique.

(ii) It is easily seen from the proof that the boundedness of g at the ends can be replaced by g bounded on each closed subinterval of (a, b), $\int_{c_1}^{x} g(\xi) \, d\xi$ bounded above for $x > c_1$ and $\int_{x}^{c_2} g(\xi) \, d\xi$ bounded below for $x < c_2$, where c_1 and c_2 are some numbers in (a, b).

EXERCISES

1. If
$$u'' - (1 + x^2)u = 0 \text{ for } 0 < x < 1,$$
$$u(0) = 1, \quad u(1) = 0,$$
find upper and lower bounds for $u(\tfrac{1}{2})$.

2. If
$$u'' + 3u' - xu = 0 \text{ for } 0 < x < 1,$$
$$-u'(0) + u(0) = 0, \quad u(1) = 1,$$
find upper and lower bounds for $u(\tfrac{1}{2})$.

3. If
$$u'' - (1 + x^2)u = 0 \text{ for } 0 < x < 1,$$
$$u(0) = 1, \quad u(1) = 0,$$
find upper and lower bounds for $u'(0)$.

Hint: Choose z_1 and z_2 so that $z_1(0) = z_2(0) = 1$; hence, find bounds for the difference quotient $[u(x) - u(0)]/x$.

4. Suppose that u is a solution of $u'' + g(x)u' + h(x)u = 0$ for $a < x < b$ with $h(x) \leq 0$. Show how to construct a constant K depending only on bounds for the coefficients g and h and on $b - a$ such that if $|u| \leq M$ on $[a, b]$, then $|u'(a)| \leq MK$.

Hint: Construct z_1 and z_2 so that $z_1(a) = z_2(a) = u(a)$; hence, obtain a bound for the difference quotient $[u(x) - u(a)]/(x - a)$.

5. Select for $z_1(x)$ a polynomial of degree three and obtain an improved upper bound for $u(\tfrac{1}{2})$ in the example following Theorem 10 (p. 18).

6. If $u'' + xu' - h(x)u = 0$ for $0 < x < 1$, $u(0) = 0$, $u(1) = 1$, and
$$h(x) = \begin{cases} 1 \text{ for } 0 < x \leq \tfrac{1}{2} \\ 2 \text{ for } \tfrac{1}{2} < x \leq 1 \end{cases},$$
find an upper bound for $u(\tfrac{1}{2})$.

SECTION 6. APPROXIMATION IN THE INITIAL VALUE PROBLEM

In Section 3 we discussed the uniqueness of the solution of the initial value problem. That is, we showed that there is at most one solution of the equation

Approximation in the Initial Value Problem

$$(L + h)[u] \equiv u'' + g(x)u' + h(x)u = f(x) \text{ for } x > a \tag{1}$$

which satisfies the initial conditions

$$u(a) = \gamma_1, \quad u'(a) = \gamma_2. \tag{2}$$

As in the case of the boundary value problem, it is usually not possible to solve the initial value problem explicitly. Therefore it is important to be able to find an approximate solution and to bound the error in the approximation. We shall first do this under the assumption that

$$h(x) \leq 0$$

throughout an interval $[a, b]$ in which the solution of (1), (2) is to be approximated.

Suppose we can find a function $z_1(x)$ with the properties

$$(L + h)[z_1] \geq f(x) \quad \text{for} \quad a \leq x \leq b, \tag{3}$$

$$z_1(a) \geq \gamma_1, \quad z_1'(a) \geq \gamma_2. \tag{4}$$

Defining the function

$$v_1(x) \equiv z_1(x) - u(x),$$

we have

$$(L + h)[v_1] \geq 0$$

and

$$v_1(a) \geq 0, \quad v_1'(a) \geq 0.$$

Since $v_1(a) \geq 0$, the function v_1 has a nonnegative maximum on any subinterval $[a, x_0]$ of $[a, b]$. By the maximum principle as given in Theorem 3, this maximum must occur either at a or at x_0. Since $v_1'(a) \geq 0$, we conclude additionally from Theorem 4 that the maximum cannot occur at a unless v_1 is constant throughout (a, x_0). Thus Theorem 3 shows that for every $x_0 > a$, we have

$$v_1(x_0) \geq v_1(a), \tag{5}$$

and

$$v_1'(x_0) \geq 0. \tag{6}$$

Inequality (5) with x in place of x_0 states that

$$u(x) \leq \gamma_1 + z_1(x) - z_1(a) \text{ for } x \geq a, \tag{7}$$

and inequality (6) states that

$$u'(x) \leq z_1'(x) \text{ for } x \geq a. \tag{8}$$

Since $z_1(a) - \gamma_1 \geq 0$, inequality (7) implies that $u(x) \leq z_1(x)$. We observe that the function $z_1(x) - [z_1(a) - \gamma_1]$ again satisfies the conditions (3) and (4), but is equal to u at $x = a$. The inequality $u(x) \leq z_1(x)$ with

$z_1(x)$ replaced by $z_1(x) - [z_1(a) - \gamma_1]$ is (7) again. Thus there is no loss of generality in replacing (7) by the simpler inequality $u(x) \leq z_1(x)$.

Lower bounds may be obtained in a similar way. Suppose we can find a function z_2 such that the inequalities

$$(L + h)[z_2] \leq f(x) \quad \text{for} \quad a \leq x \leq b \tag{9}$$

and

$$z_2(a) \leq \gamma_1, \quad z_2'(a) \leq \gamma_2 \tag{10}$$

hold. Then, by Theorems 3 and 4, we find

$$u(x) \geq \gamma_1 + z_2(x) - z_2(a)$$

and

$$u'(x) \geq z_2'(x).$$

Without loss of generality we may replace the first inequality by $u(x) \geq z_2(x)$.

We have established the following result.

THEOREM 13. If $u(x)$ is a solution of (1) satisfying the initial conditions (2) and if $z_1(x)$ and $z_2(x)$ satisfy (3), (4) and (9), (10), respectively, then

$$z_2(x) \leq u(x) \leq z_1(x) \tag{11}$$

and

$$z_2'(x) \leq u'(x) \leq z_1'(x). \tag{12}$$

Note that inequalities (3) and (9) are the reverse of those required for upper and lower bounds in boundary value problems. If functions $z_1(x)$ and $z_2(x)$ as described above can be found, then inequalities (11) and (12) show that $u(x)$ and $u'(x)$ are approximated by these functions. This approximation may be made without any specific knowledge of the solution u itself. The accuracy of the approximation will depend on how well we can choose the functions z_1 and z_2.

Example. Find bounds for the value $u(1)$ of the solution of the initial value problem

$$(L + h)[u] \equiv u'' + \frac{1}{x} u' - u = 0,$$

$$u(0) = 1, \quad u'(0) = 0.$$

This solution $u(x)$ is called the Bessel function of order zero with imaginary argument and is usually denoted $I_0(x)$.

It is convenient to select polynomials for $z_1(x)$ and $z_2(x)$. We set

$$z_1(x) = c_1 x^2 + 1.$$

Then $z_1(0) = 1$, $z_1'(0) = 0$. We choose the constant c_1 so that

Sec. 6 Approximation in the Initial Value Problem

$$(L + h)[z_1] = c_1(4 - x^2) - 1$$

is nonnegative for $0 \leq x \leq 1$. The value $c_1 = \frac{1}{3}$ is adequate, and we find that

$$u(x) \leq \tfrac{1}{3}x^2 + 1 \text{ for } 0 \leq x \leq 1.$$

In particular,

$$u(1) \leq \tfrac{4}{3}.$$

Similarly, if we set

$$z_2(x) = c_2 x^2 + 1$$

and select $c_2 = \tfrac{1}{4}$, then

$$(L + h)[z_2] \leq 0$$

and $z_2(0) = 1$, $z_2'(0) = 0$. Therefore $u(x) \geq \tfrac{1}{4}x^2 + 1$ for $0 \leq x \leq 1$. The upper and lower bounds for $u(1)$ are

$$1.250 \leq u(1) \leq 1.333. \tag{13}$$

We now exhibit some devices for improving the upper and lower bounds in (13). To improve these bounds we make use of the portion of Theorem 13 giving bounds for $u'(x)$ in terms of $z_1'(x)$ and $z_2'(x)$.

To improve the upper bound, we consider first a subinterval $[0, t]$ where $t < 1$. Choosing the same function

$$z_1(x) = c_1 x^2 + 1 \quad \text{for} \quad 0 \leq x \leq t,$$

we satisfy the inequality

$$(L + h)[z_1] = c_1(4 - x^2) - 1 \geq 0 \quad \text{for} \quad 0 \leq x \leq t$$

by taking

$$c_1 = \frac{1}{4 - t^2}.$$

In this way we find

$$u(x) \leq \frac{x^2}{4 - t^2} + 1$$

and

$$u'(x) = \frac{2x}{4 - t^2}.$$

for $0 \leq x \leq t$. In particular,

$$u'(t) \leq \frac{2t}{4 - t^2}.$$

The trick now is to integrate this last inequality with respect to t between the limits 0 and 1. We find

$$u(1) - u(0) \leq \log\left(\tfrac{4}{3}\right) \leq 0.288.$$

Thus we have the upper bound
$$u(1) \leq u(0) + 0.288 = 1.288. \tag{14}$$

This device does not yield any improvement in the lower bound for $u(1)$ with a function of the form $z_2 = c_2 x^2 + 1$. However, another method, that of subdivision of the interval, may be exploited to improve the lower bound.

We divide the interval $(0, 1)$ into two parts and define $z_2(x)$ separately on each part. For the interval $0 \leq x \leq \frac{1}{2}$, we select z_2 as before:
$$z_2(x) = \tfrac{1}{4} x^2 + 1.$$
In this way we get the bounds
$$u(\tfrac{1}{2}) \geq z_2(\tfrac{1}{2}) = \tfrac{17}{16},$$
$$u'(\tfrac{1}{2}) \geq z_2'(\tfrac{1}{2}) = \tfrac{1}{4}.$$
For the interval $\frac{1}{2} \leq x \leq 1$, we define
$$z_2(x) = c_3(x - \tfrac{1}{2})^2 + \tfrac{17}{16} + \tfrac{1}{4}(x - \tfrac{1}{2}).$$
With this definition z_2 and z_2' are continuous for $[0, 1]$. We shall choose c_3 so that
$$(L + h)[z_2] \equiv c_3 \left[2 + \frac{2x-1}{x} - \left(x - \tfrac{1}{2}\right)^2 \right]$$
$$+ \frac{1}{4x} - \frac{17}{16} - \frac{1}{4}\left(x - \tfrac{1}{2}\right) \leq 0$$
for $\frac{1}{2} \leq x \leq 1$. This inequality will hold if
$$c_3 \leq \frac{4x^2 + 15x - 4}{4(-4x^3 + 4x^2 + 15x - 4)} \quad \text{for} \quad \tfrac{1}{2} \leq x \leq 1.$$
It is not hard to see that the right side is increasing; thus the selection
$$c_3 = \tfrac{9}{32}$$
suffices. In this way we obtain
$$u(1) \geq z_2(1) = \tfrac{161}{128} \geq 1.258.$$
Combining this inequality with (14), we get
$$1.258 \leq u(1) \leq 1.288$$
or
$$u(1) = 1.273 \pm 0.015$$
(actually $u(1) = 1.2661$).

Further improvements in the lower bound can be made by dividing the interval $(0, 1)$ into several subintervals and by defining $z_2(x)$ separately on each subinterval. The lower bounds z_2 and z_2' for u and u' at the end of each interval serve as initial values of z_2 and z_2' on the next interval. This choice for the initial values has the effect of making the function $z_2(x)$ have a continuous first derivative but not necessarily a continuous second derivative on [0, 1]. The upper bound may be improved in the same manner.

The method of subdivision in the above example suggests the following general scheme for obtaining upper and lower bounds. Suppose we divide the interval [a, b] into N subintervals

$$a = x_0 < x_1 < \cdots < x_{N-1} < x_N = b.$$

We shall select $z_1(x)$ to be a quadratic polynomial in each subinterval and choose the coefficients of the polynomial so that $z_1(a) = \gamma_1$, $z_1'(a) = \gamma_2$, and z_1 and z_1' are continuous throughout [a, b]. Also, z_1 will be selected so that inequality (3) holds in each subinterval (x_{i-1}, x_i). We set

$$z_1(x) = c_i(x - x_i)^2 + d_i(x - x_i) + e_i \text{ for } x_i \leq x \leq x_{i+1},$$
$$i = 0, 1, 2, \ldots, N-1.$$

The constants $c_i, d_i, e_i, i = 0, 1, \ldots, N-1$ and the number N of subintervals will be chosen so that all the required conditions are satisfied. We proceed in a step-by-step manner starting with the interval (x_0, x_1). The initial conditions

$$z_1(a) = \gamma_1, \qquad z_1'(a) = \gamma_2$$

require that $e_0 = \gamma_1$ and $d_0 = \gamma_2$. Therefore we have in (x_0, x_1),

$$z_1(x) = c_0(x - x_0)^2 + \gamma_2(x - x_0) + \gamma_1.$$

In this interval, the inequality

$$(L + h)[z_1] \geq f(x)$$

becomes

$$c_0[2 + 2g(x)(x - x_0) + h(x)(x - x_0)^2] + g(x)\gamma_2 + h(x)[\gamma_2(x - x_0) + \gamma_1]$$
$$\geq f(x). \qquad (15)$$

If $g(x)$ and $h(x)$ are bounded, then x_1 can be selected so close to x_0 that the coefficient of c_0 in (15) is positive for $x_0 \leq x \leq x_1$. If, in addition, f is bounded, then c_0 can be taken so large that (15) holds for all x in (x_0, x_1).

We now turn to the interval (x_1, x_2), with $z_1(x)$ defined by

$$z_1(x) = c_1(x - x_1)^2 + d_1(x - x_1) + e_1 \quad \text{for} \quad x_1 \leq x \leq x_2.$$

To insure the continuity of z_1 and z_1' at x_1, we choose

$$\left. \begin{array}{l} e_1 = c_0(x_1 - x_0)^2 + \gamma_2(x_1 - x_0) + \gamma_1, \\ d_1 = 2c_0(x_1 - x_0) + \gamma_2. \end{array} \right\} \qquad (16)$$

In the interval (x_1, x_2), the inequality corresponding to (15) becomes
$$c_1[2 + 2g(x)(x - x_1) + h(x)(x - x_1)^2] + g(x)d_1 + h(x)[d_1(x - x_1) + e_1]$$
$$\geq f(x), \qquad (17)$$
where d_1 and e_1 are constants determined by (16). We select x_2 so close to x_1 that the coefficient of c_1 in (17) is positive. Then we take c_1 so large that inequality (17) holds throughout the interval (x_1, x_2).

Proceeding in this fashion, we determine each d_i, e_i so that z_1 and z_1' are continuous everywhere, and we always take the interval (x_i, x_{i+1}) so small and the constant c_i so large that $(L + h)[z_1] \geq f(x)$ holds everywhere. In fact, the quantities e_i, d_i are determined by the recursion formulas
$$e_i = c_{i-1}(x_i - x_{i-1})^2 + d_{i-1}(x_i - x_{i-1}) + e_{i-1},$$
$$d_i = 2c_{i-1}(x_i - x_{i-1}) + d_{i-1}.$$
In an actual computation to determine the c_i, it is convenient to replace f by its maximum in the ith subinterval and to replace g and h by either their maxima or minima, whichever may be appropriate for making $(L + h)[z_1] \geq f$ throughout.

In a similar manner we may construct lower bounds. The constants d_i, e_i are selected in precisely the same way, and the quantities $-c_i$ are taken so large that $(L + h)[z_2] \leq f(x)$ everywhere.

If f, g, and h are continuous, it can be shown that as the maximum length of the subintervals tends to zero, the upper and lower bounds both tend to the solution u.

Thus far in this section we have assumed that $h(x) \leq 0$. We now take up the problem of approximating the solution of the equation
$$(L + h)[u] \equiv u'' + g(x)u' + h(x)u = f(x)$$
with initial conditions
$$u(a) = \gamma_1, \qquad u'(a) = \gamma_2$$
when the function $h(x)$ may be positive. Under these circumstances we employ the generalized maximum principle as given in Theorem 5. To do so, we suppose there is a function w which is positive on $[a, b]$ and which has the property that
$$(L + h)[w] \leq 0 \text{ for } a < x < b.$$
For example, the function
$$w = 1 - \beta(x - a)^2,$$
with β sufficiently large [assuming that $f(x)$ is bounded], has the desired properties if the interval $[a, b]$ is small enough.

Sec. 6 Approximation in the Initial Value Problem

We saw in Section 2 that $v \equiv u/w$ satisfies an equation of the form

$$(\bar{L} + H)[v] \equiv v'' + G(x)v' + H(x)v = \frac{f}{w},$$

with $G(x) \equiv (2w'/w) + g$ and $H(x) \equiv (L + h)[w]/w \leq 0$. The comparison functions $z_1(x)$ and $z_2(x)$ are now defined so that z_1/w and z_2/w provide bounds for u/w. We make z_1 and z_2 satisfy the inequalities

$$(L + h)[z_1] \geq f(x),$$

$$z_1(a) \geq \gamma_1, \qquad z_1'(a)w(a) - z_1(a)w'(a) \geq \gamma_2 w(a) - \gamma_1 w'(a),$$

and

$$(L + h)[z_2] \leq f(x),$$

$$z_2(a) \leq \gamma_1, \qquad z_2'(a)w(a) - z_2(a)w'(a) \leq \gamma_2 w(a) - \gamma_1 w'(a).$$

Then, at $x = a$,

$$\frac{z_2}{w} \leq \frac{u}{w} \leq \frac{z_1}{w} \text{ and } \left(\frac{z_2}{w}\right)' \leq \left(\frac{u}{w}\right)' \leq \left(\frac{z_1}{w}\right)'.$$

Moreover, it is easily seen by computation that

$$(\bar{L} + H)\left[\frac{z_2}{w}\right] \leq (\bar{L} + H)\left[\frac{u}{w}\right] \leq (\bar{L} + H)\left[\frac{z_1}{w}\right].$$

Hence we find that for $a \leq x \leq b$,

$$\frac{z_2(x)}{w(x)} \leq \frac{u(x)}{w(x)} \leq \frac{z_1(x)}{w(x)}$$

and

$$\left(\frac{z_2}{w}\right)' \leq \left(\frac{u}{w}\right)' \leq \left(\frac{z_1}{w}\right)'.$$

The first of these sets of inequalities gives the bounds

$$z_2(x) \leq u(x) \leq z_1(x). \tag{18}$$

The second set yields

$$w(x)z_2'(x) - z_2(x)w'(x) \leq w(x)u'(x) - u(x)w'(x) \leq w(x)z_1'(x) - z_1(x)w'(x).$$

Since w is positive on $[a, b]$, we find

$$z_2'(x) + \frac{w'(x)}{w(x)}[u(x) - z_2(x)] \leq u'(x) \leq z_1'(x) - \frac{w'(x)}{w(x)}[z_1(x) - u(x)]. \tag{19}$$

If $w'(x) \leq 0$, we may insert the upper bound for $u(x)$ as given in (18) on the left side of (19) and we may insert the lower bound on the right side. If $w'(x) \geq 0$, we use the lower bound on the left and the upper bound on the right. We thus find that

$$z_2'(x) - \frac{[-w'(x)]}{w(x)}[z_1(x) - z_2(x)] \leq u'(x)$$

$$\leq z_1'(x) + \frac{[-w'(x)]}{w(x)}[z_1(x) - z_2(x)] \quad \text{if} \quad w'(x) \leq 0, \quad (20)$$

$$z_2'(x) \leq u'(x) \leq z_1'(x) \quad \text{if} \quad w'(x) \geq 0.$$

Inequalities (18) and (20) give bounds for $u(x)$ and $u'(x)$ which are precise when $z_1(x) - z_2(x)$ and $z_1'(x) - z_2'(x)$ are small.

While it is always possible to find a positive function w which satisfies $(L + h)[w] \leq 0$ on a sufficiently small interval, there is, in general, no such function if the interval is too large. Once more we resort to breaking up the interval and piecing together functions defined on subintervals. Let $w > 0$ and $(L + h)[w] \leq 0$ on an interval $[a, x^*]$, and let w^* be another positive function which satisfies $(L + h)[w^*] \leq 0$ on the interval $[x^*, b]$. We wish to find bounds for the solution u of the initial value problem (1), (2) on the whole interval $[a, b]$.

Let $z_1(x)$ and $z_2(x)$ satisfy the conditions

$$(L + h)[z_2] \leq f(x) \leq (L + h)[z_1]$$

on the interval $[a, x^*]$, and

$$z_2(a) \leq \gamma_1 \leq z_1(a),$$
$$z_2'(a)w(a) - z_2(a)w'(a) \leq \gamma_2 w(a) - \gamma_1 w'(a) \leq z_1'(a)w(a) - z_1(a)w'(a).$$

Then

$$z_2 \leq u \leq z_1,$$
$$\left(\frac{z_2}{w}\right)' \leq \left(\frac{u}{w}\right)' \leq \left(\frac{z_1}{w}\right)' \quad \text{for} \quad a \leq x \leq x^*.$$

We are thus given bounds for $u(x^*)$ and $u'(x^*)$ and, in addition, a function w^* which satisfies $(L + h)[w^*] \leq 0$ on $[x^*, b]$. We can then find bounds for u/w^* and $(u/w^*)'$, as before. Let the functions z_1^* and z_2^* be defined on $[x^*, b]$ and suppose that they satisfy

$$(L + h)[z_2^*] \leq (L + h)[u] \leq (L + h)[z_1^*] \quad \text{in} \quad (x^*, b),$$
$$z_2^*(x^*) \leq u(x^*) \leq z_1^*(x^*),$$
$$\left(\frac{z_2^*}{w^*}\right)' \leq \left(\frac{u}{w^*}\right)' \leq \left(\frac{z_1^*}{w^*}\right)' \quad \text{at} \quad x = x^*.$$

Then we find, as we did previously, that

$$z_2^*(x) \leq u(x) \leq z_1^*(x),$$
$$\left(\frac{z_2^*}{w^*}\right)' \leq \left(\frac{u}{w^*}\right)' \leq \left(\frac{z_1^*}{w^*}\right)' \quad \text{for} \quad x^* \leq x \leq b.$$

While we do not know $u(x^*)$ and $(u/w^*)'$ at x^*, we have bounds for them; therefore, we can give explicit conditions on the values of z_1^*, $(z_1^*)'$,

z_2^*, and $(z_2^*)'$ at x^* which assure that the above inequalities are satisfied. If $w^{*\prime}/w^* \geq w'/w$ (which is usually the case), these conditions are

$$z_1^* \geq z_1, \quad w^*\left(\frac{z_1^*}{w^*}\right)' \geq w\left(\frac{z_1}{w}\right)' - \left(\frac{w^{*\prime}}{w^*} - \frac{w'}{w}\right) z_2 \quad \text{at} \quad x = x^*,$$

$$z_2^* \leq z_2, \quad w^*\left(\frac{z_2^*}{w^*}\right)' \leq w\left(\frac{z_2}{w}\right)' - \left(\frac{w^{*\prime}}{w^*} - \frac{w'}{w}\right) z_1 \quad \text{at} \quad x = x^*.$$

If $w^{*\prime}/w^* \leq w'/w$, we replace z_2 by z_1 in the coefficient of $(w^{*\prime}/w^* - w'/w)$ in the first row of the inequalities and z_1 by z_2 in the second. If these conditions are satisfied, we have the bounds

$$z_2^*(x) \leq u(x) \leq z_1^*(x),$$

$$\left(\frac{z_2^*}{w^*}\right)' \leq \left(\frac{u}{w^*}\right)' \leq \left(\frac{z_1^*}{w^*}\right)'.$$

We now consider w^* as an extension of w to the interval $(x^*, b]$, z_1^* as an extension of z_1, and z_2^* as an extension of z_2. Then these extended functions are, in general, discontinuous at x^*. However, the above inequalities connecting w, w^*, z_1, z_2, z_1^*, and z_2^* at $x = x^*$ relate the right and left limits at points of discontinuity. It may, of course, be necessary or desirable to divide the interval $[a, b]$ into more than two subintervals.

The above considerations then lead to the following theorem:

THEOREM 14. Let $z_1(x)$, $z_2(x)$ and $w(x)$ be piecewise continuous functions with piecewise continuous first and second derivatives on the interval $[a, b]$ with the following properties:

(a) $w > 0$ on $[a, b]$.
(b) $z_2(a) \leq \gamma_1 \leq z_1(a)$.
(c) $z_2'(a)w(a) - z_2(a)w'(a) \leq \gamma_2 w(a) - \gamma_1 w'(a) \leq z_1'(a)w(a) - z_1(a)w'(a)$.
(d) $(L + h)[w] \leq 0$, $(L + h)[z_2] \leq f(x) \leq (L + h)[z_1]$ at all points where the derivatives occurring in these formulas are continuous.
(e) At each point of discontinuity x^* the functions $z_1, -z_2$, and w'/w have nonnegative jumps, the jump in $w(z_1/w)'$ is at least $-z_2(x^* - 0)$ times the jump in w'/w, and the jump in $w(z_2/w)'$ is at most $-z_1(x^* - 0)$ times the jump in w'/w.

Then

$$z_2(x) \leq u(x) \leq z_1(x),$$

$$\left[\frac{z_2(x)}{w(x)}\right]' \leq \left[\frac{u(x)}{w(x)}\right]' \leq \left[\frac{z_1(x)}{w(x)}\right]' \quad \text{on } [a, b].$$

Remarks: (i) By subtracting a suitable multiple of w from z_1 in each interval where z_1 and w are continuous, we can make $z_1(a) = \gamma_1$ and z_1 continuous throughout $[a, b]$. This improves the upper bound. Similarly, we may make $z_2(a) = \gamma_1$ and z_2 continuous.

(ii) Our previous results for $h \leq 0$ can be obtained by choosing $w \equiv 1$.

Example. Find upper and lower bounds for the value at $x = 1$ of the solution of

$$(L + h)[u] \equiv u'' + (4 + x)u = 0$$

which satisfies the initial conditions

$$u(0) = 0, \quad u'(0) = 1.$$

We first note that $h(x) = 4 + x$ is positive on $[0, 1]$ and so we seek an auxiliary function $w(x)$. We observe that

$$w = 1 - 2.1x^2$$

is positive and satisfies $(L + h)[w] \leq 0$ for $0 \leq x \leq \frac{1}{2}$. On this interval, the functions

$$z_1 = x$$
$$z_2 = x - \tfrac{18}{19} x^2$$

fulfill the required conditions for upper and lower comparison functions. From (18) and (20) we get

$$\tfrac{5}{19} \leq u(\tfrac{1}{2}) \leq \tfrac{1}{2},$$
$$-\tfrac{359}{361} \leq u'(\tfrac{1}{2}) \leq \tfrac{739}{361}.$$

On the interval $[\tfrac{1}{2}, 1]$, we can take

$$w = 1 - \tfrac{5}{2}(x - \tfrac{1}{2})^2$$

and

$$z_1 = -\tfrac{9}{8}(x - \tfrac{1}{2})^2 + \tfrac{739}{361}(x - \tfrac{1}{2}) + \tfrac{1}{2},$$
$$z_2 = -\tfrac{45}{76}(x - \tfrac{1}{2})^2 - \tfrac{359}{361}(x - \tfrac{1}{2}) + \tfrac{5}{19}.$$

Then we find that

$$-0.38 \leq u(1) \leq 1.24.$$

Actually $u(1) = -0.09$.

As in the case of equations with h nonpositive, we can obtain more accuracy by subdividing the given interval into a larger number of parts.

Theorem 14 can be used to show that the solution u of the initial value problem (1), (2) depends continuously on the data γ_1, γ_2, and $f(x)$.

We let $w = 1 - \beta^2[x - a - (k/2\beta)]^2$ on the interval $[a + (k/2\beta), a + (k + 1)/2\beta]$, $k = 0, 1, \ldots, k^*$, where $k^* < 2\beta(b - a) \leq k^* + 1$, with β so large that

$$(L + h)[w] = -2\beta^2 \left[1 + \left(x - a - \frac{k}{2\beta}\right) g + \frac{1}{2}\left(x - a - \frac{k}{2\beta}\right)^2 h \right] + h \leq 0$$

Approximation in the Initial Value Problem

on this interval. This can be done by making

$$\beta^2 = \tfrac{1}{4} \max_{a \leq x \leq b} [g^2 + 3h],$$

provided max $h > 0$. If $h \leq 0$, we let $\beta = 0$ so that $w \equiv 1$. On the first interval $[a, a + (1/2\beta)]$, we choose

$$z_1 = C_0 e^{\alpha(x-a)},$$

where α is chosen so that

$$(L + h)[e^{\alpha(x-a)}] = (\alpha^2 + \alpha g + h)e^{\alpha(x-a)} \geq 1.$$

We may take, for example,

$$\alpha = \max_{a \leq x \leq b} |g^2 - 2(h-1)|^{1/2}.$$

We select

$$C_0 = \max \left[|\gamma_1|, \frac{|\gamma_2|}{\alpha}, \max_{a \leq x \leq b} |f(x)| \right].$$

Then $z_1 = C_0 e^{\alpha(x-a)}$ satisfies the conditions on z_1 as stated in Theorem 14, and $z_2 = -C_0 e^{\alpha(x-a)}$ satisfies the conditions on z_2 on the interval $[a, a + (1/2\beta)]$. We find that at $x = a + (1/2\beta)$

$$|u| \leq C_0 e^{\alpha/2\beta}$$

$$\left| \left(\frac{u}{w}\right)' \right| \leq \frac{\tfrac{3}{4}\alpha + \beta}{(\tfrac{3}{4})^2} C_0 e^{\alpha/2\beta},$$

so that

$$\left| u'\left(a + \frac{1}{2\beta}\right) \right| \leq \left(\alpha + \frac{8\beta}{3}\right) C_0 e^{\alpha/2\beta}.$$

We now let $z_1 = -z_2 = C_1 e^{\alpha[x-a-(1/2\beta)]}$ on $[a + (1/2\beta), a + (2/2\beta)]$ with

$$C_1 = \left(1 + \frac{8\beta}{3\alpha}\right) C_0 e^{\alpha/2\beta}$$

to find that

$$\left| u\left(a + \frac{2}{2\beta}\right) \right| \leq \left(1 + \frac{8\beta}{3\alpha}\right) C_0 e^{2\alpha/2\beta},$$

$$\left| u'\left(a + \frac{2}{2\beta}\right) \right| \leq \alpha \left(1 + \frac{8\beta}{3\alpha}\right)^2 C_0 e^{2\alpha/2\beta}.$$

Proceeding in this fashion, we find that for all x in $[a, b]$

$$|u(x)| \leq C_0 e^{\rho(x-a)},$$

$$|u'(x)| \leq \alpha \left[1 + \left(\frac{8\beta}{3\alpha}\right)\right] C_0 e^{\rho(x-a)},$$

where $\rho = \alpha + 2\beta \log [1 + (8\beta)/(3\alpha)]$.

If now u_1 and u_2 are solutions of initial value problems for the same differential operator L and nearby data, we let $u(x) = u_1(x) - u_2(x)$ to find that

$$|u_1(x) - u_2(x)| \leq C_0 e^{\rho(x-a)},$$

$$|u_1'(x) - u_2'(x)| \leq \alpha C_0 \left[1 + \left(\frac{8\beta}{3\alpha}\right)\right] e^{\rho(x-a)},$$

where α, β, and ρ depend only on bounds for g and h, and where

$$C_0 = \max \left\{|u_1(a) - u_2(a)|, \frac{|u_1'(a) - u_2'(a)|}{\alpha}, \max_{a \leq x \leq b} |(L+h)[u_1] - (L+h)[u_2]|\right\}.$$

Thus, if the initial values $u_1(a)$, $u_1'(a)$ are close to $u_2(a)$ and $u_2'(a)$ and if the given functions $(L+h)[u_1]$ and $(L+h)[u_2]$ are close together, then C_0 is small; hence $u_1(x)$ is close to $u_2(x)$ and $u_1'(x)$ is close to $u_2'(x)$ throughout the interval $[a, b]$. In other words, the solution u of the initial value problem (1), (2) depends continuously on the data γ_1, γ_2, and $f(x)$.

EXERCISES

1. If $u'' - e^x u = 0$ for $x > 0$,
$$u(0) = 0, \; u'(0) = 1,$$
find upper and lower bounds for $u(1)$ and $u'(1)$.

2. If $u'' + xu' - x^2 u = 1$ for $x > 0$,
$$u(0) = 0, \; u'(0) = 0,$$
find upper and lower bounds for $u(1)$ and $u'(1)$.

3. If $u'' + (10 + x^2)u = 0$ for $x > 0$,
$$u(0) = 1, \; u'(0) = -1,$$
find upper and lower bounds for $u(1)$ and $u'(1)$.

4. Find improved upper and lower bounds for $u(1)$ in the example following Theorem 13 by subdividing the interval $[0, 1]$ into three parts.

5. Find improved upper and lower bounds for $u(1)$ in the example following Theorem 14 by subdividing the interval $[0, 1]$ into three parts.

6. Suppose that u is a solution of $u'' + g(x)u' + h(x)u = f(x)$, and that g, h, and f are bounded from one side. That is, assume that either $-\infty < g(x) \leq M$ or $m \leq g(x) < \infty$, but not both; similarly, suppose h and f are unbounded in one direction. By considering g, h, and f separately, show what portions of Theorems 13 and 14 remain valid in the unbounded case. Obtain inequalities from one side when appropriate.

SECTION 7. THE EIGENVALUE PROBLEM

Suppose that u is a solution of the equation
$$u'' + g(x)u' + [h(x) + \lambda k(x)]u = 0 \text{ for } a < x < b \tag{1}$$
which satisfies the boundary conditions
$$\left.\begin{array}{r}-u'(a)\cos\theta + u(a)\sin\theta = 0, \\ u'(b)\cos\phi + u(b)\sin\phi = 0.\end{array}\right\} \tag{2}$$
We take θ and ϕ so that $-\pi/2 < \theta \leq \pi/2$, $-\pi/2 < \phi \leq \pi/2$. It is clear that $u \equiv 0$ satisfies (1) and (2) for every value of λ. Any number λ for which there is a solution of (1) and (2) which is not identically zero is called an **eigenvalue of the equation** (1) **with boundary conditions** (2). The corresponding solution u is called an **eigenfunction**. It is determined only to within a multiplicative constant, since for any number A the function Au satisfies (1) and (2) whenever u does.

We shall suppose that the functions g, h, and k are bounded and that there is a positive number η such that
$$k(x) \geq \eta > 0 \text{ for } a \leq x \leq b.$$
If λ is sufficiently small (possibly negative), then
$$h + \lambda k \leq 0 \text{ for } a \leq x \leq b;$$
if also, $\theta \geq 0$, $\phi \geq 0$, it follows from Theorem 10 in Section 5 that such a λ cannot be an eigenvalue. That is, all eigenvalues of (1), (2) are larger than the number*
$$\inf_{a \leq x \leq b} \left[\frac{-h(x)}{k(x)} \right]. \tag{3}$$

We shall now seek a more general lower bound for the eigenvalues of (1), (2). Let $w(x)$ be a twice continuously differentiable function, positive on $[a, b]$, which satisfies the inequalities
$$\left.\begin{array}{r}-w'(a)\cos\theta + w(a)\sin\theta \geq 0, \\ w'(b)\cos\phi + w(b)\sin\phi \geq 0.\end{array}\right\} \tag{4}$$
If w also satisfies
$$w'' + g(x)w' + [h(x) + \lambda k(x)]w < 0, \tag{5}$$
then it follows from Theorem 11 in Section 5 with $z_1 \equiv z_2 \equiv 0$ that λ is not an eigenvalue. Thus we are led to the following result.

---

*The infimum of a function $f(x)$, inf $f(x)$, is its minimum if it is attained, or otherwise its greatest lower bound. The supremum sup $f(x)$ is defined similarly, so that sup $f(x) = -\inf[-f(x)]$.

THEOREM 15. If $w(x)$ is positive on $[a, b]$ and satisfies inequalities (4), then no eigenvalue of (1), (2) can lie below the quantity

$$\inf_{a \leq x \leq b} \left[-\frac{w'' + g(x)w' + h(x)w}{k(x)w} \right]. \tag{6}$$

Note that if $\theta \geq 0$, $\phi \geq 0$, then (6) is an improvement over (3), since the function $w \equiv 1$ satisfies all the hypotheses; in this latter case (6) reduces to (3).

We shall now show that the maximum principle may be used to establish the existence of a smallest or first eigenvalue. That is, we shall show that there is a number λ_1 which is an eigenvalue and which has the property that no number λ smaller than λ_1 is an eigenvalue. For this purpose we first establish the following lemma.

LEMMA 1. For each λ let $r(x; \lambda)$ be the solution of the initial value problem

$$\left. \begin{array}{l} (L + h + \lambda k)[r] \equiv r'' + g(x)r' + [h(x) + \lambda k(x)]\, r = 0, \\ r(a) = \cos \theta, \quad r'(a) = \sin \theta, \end{array} \right\} \tag{7}$$

with $-\pi/2 < \theta \leq \pi/2$. Suppose that $g(x)$, $h(x)$, and $k(x)$ are bounded and that $k(x)$ is positive and bounded away from zero on $[a, b]$. Then there is a number $\bar{\lambda}$ such that for all $\lambda > \bar{\lambda}$ the solution $r(x; \lambda)$ changes sign in $a < x < b$.

Proof. Suppose that for arbitrarily large values of λ there is an r satisfying (7) and such that $r(x; \lambda) > 0$ in (a, b). Then the function $w(x) = r(x; \lambda)$ can be used to establish Theorem 11 for the problem

$$(L + h + \lambda k)[u] = 0 \quad \text{for} \quad c - \epsilon < x < c + \epsilon,$$

$$u(c - \epsilon) = u(c + \epsilon) = 0,$$

provided $c - \epsilon > a$ and $c + \epsilon < b$. This problem clearly has the solution $u(x) \equiv 0$. We shall reach a contradiction by constructing a function z_2 which can be used as a lower bound if Theorem 11 in Section 5 holds, but which is clearly greater than u. Let

$$z_2(x) \equiv e^{-\alpha(x-c)^2} - e^{-\alpha \epsilon^2},$$

where α is a positive constant to be determined. We see that

$$z_2(c - \epsilon) = z_2(c + \epsilon) = 0,$$

and we compute

$$(L + h + \lambda k)[z_2] = 2\alpha[2\alpha(x - c)^2 - 1 - (x - c)g(x)]e^{-\alpha(x-c)^2}$$
$$+ (h + \lambda k)[e^{-\alpha(x-c)^2} - e^{-\alpha \epsilon^2}].$$

We now choose α so large that

$$\tfrac{1}{2}\alpha\epsilon^2 \geq 1 + (x-c)g(x)$$

on the interval $[c-\epsilon, c+\epsilon]$. Then if λ is so large that $h + \lambda k \geq 0$, we see that $(L + h + \lambda k)[z_2] \geq 0$ for $|x-c| \geq \tfrac{1}{2}\epsilon$. To reach the same conclusion for $|x-c| < \tfrac{1}{2}\epsilon$, we must simply take λ so large that

$$(h + \lambda k) \geq 2\alpha |1 + (x-c)g(x)|/[1 - e^{-3\alpha\epsilon^2/4}]. \qquad (8)$$

For such sufficiently large values of λ, then, $(L + h + \lambda k)[z_2] \geq 0$; therefore if Theorem 11 holds, $z_2 \leq 0$. But by definition, $z_2 > 0$ in $(c-\epsilon, c+\epsilon)$, so that Theorem 11 is violated. Hence for λ so large that $h + \lambda k \geq 0$ and that (8) is satisfied, $r(x;\lambda)$ must change sign on $[c-\epsilon, c+\epsilon]$. Thus, the lemma is proved with

$$\bar{\lambda} = \sup_{c-\epsilon \leq x \leq c+\epsilon} \left[-\frac{h(x)}{k(x)} + \frac{2\alpha |1 + (x-c)g(x)|}{(1 - e^{-3\alpha\epsilon^2/4})k(x)} \right],$$

provided $a < c - \epsilon$, and $c + \epsilon < b$.

Remark: For this lemma it is sufficient to assume that $g(x)$ is bounded, that $h(x)$ has a lower bound, and that $k(x)$ has a positive lower bound on some subinterval $[c-\epsilon, c+\epsilon]$.

We now establish a lemma in the opposite direction which describes the situation for small values of λ.

LEMMA 2. Let $r(x;\lambda)$ be the solution of the initial value problem (7). Then there exists a number λ' such that when $\lambda < \lambda'$ we have $r(x;\lambda) > 0$ in $a < x \leq b$ and $r'(b;\lambda)\cos\phi + r(b;\lambda)\sin\phi > 0$.

Proof. Let $w(x)$ be any twice continuously differentiable function which is positive on $[a,b]$, and which satisfies

$$-w'(a)\cos\theta + w(a)\sin\theta \geq 0,$$
$$w'(b)\cos\phi + w(b)\sin\phi \geq 0.$$

Let

$$\lambda' = \inf\left\{ -\frac{(L+h)[w]}{kw} \right\}.$$

Then for $\lambda < \lambda'$ the function w makes Theorem 11 in Section 5 valid for the problem

$$(L + h + \lambda k)[u] = 0 \text{ in } (a, c)$$
$$-u'(a)\cos\theta + u(a)\sin\theta = 0$$
$$u(c) = 0$$

for any c in (a, b). This problem has the solution $u \equiv 0$. Hence if $r(c;\lambda) = 0$ for some c in $(a, b]$, we find from Theorem 11 that $r(x;\lambda) \leq 0$ for x on $[a, c]$. But this contradicts the initial condition for r, since either

$r(a, \lambda) = \cos \theta > 0$ or $r(a, \lambda) = 0$, $r'(a, \lambda) = 1$. Hence for $\lambda < \lambda'$, the function $r(x; \lambda)$ cannot vanish in $(a, b]$.

Moreover, Theorem 11 applied to the problem

$$(L + h + \lambda k)[u] = 0$$
$$-u'(a) \cos \theta + u(a) \sin \theta = 0$$
$$u'(b) \cos \phi + u(b) \sin \phi = 0$$

would show that $r(x; \lambda) \leq 0$ if $r'(b; \lambda) \cos \phi + r(b; \lambda) \sin \phi \leq 0$. Hence for $\lambda < \lambda'$, we have $r(x; \lambda) > 0$ in $(a, b]$ and $r'(b; \lambda) \cos \phi + r(b; \lambda) \sin \phi > 0$.

With the aid of the above lemmas we can establish the following result pertaining to the eigenvalue problem (1), (2).

THEOREM 16. There exists an eigenvalue λ_1 and a corresponding eigenfunction $r(x; \lambda_1)$ such that: (i) no number $\lambda < \lambda_1$ is an eigenvalue, and (ii) $r(x; \lambda_1)$ does not change sign in (a, b).

Proof. Let λ^* be the least upper bound of values of λ for which $r(x; \lambda) > 0$ in $(a, b]$. We see from the two lemmas that there is such a number λ^*, and that $\lambda' \leq \lambda^* \leq \bar{\lambda}$.

We observe that for any numbers λ and μ the difference $q(x) \equiv r(x; \lambda) - r(x, \mu)$ satisfies the initial value problem

$$(L + h + \lambda k)[q] = (\mu - \lambda)kr(x; \mu),$$
$$q(a) = 0,$$
$$q'(a) = 0.$$

It follows from the results of Section 6 that when $|\lambda - \mu|$ is small, $q(x)$ and $q'(x)$ are uniformly small. That is, $r(x; \lambda)$ and $r'(x; \lambda)$ are continuous in λ, uniformly in x. Then since $r(x; \lambda) > 0$ for all $\lambda < \lambda^*$, we conclude that $r(x; \lambda^*) \geq 0$. Moreover, if $r(x; \lambda^*)$ were strictly positive in $(a, b]$, the same would be true by continuity for some $\lambda > \lambda^*$. Hence, $r(x; \lambda^*)$ must vanish somewhere in $(a, b]$. If it vanished at an interior point c, we would have $r(c; \lambda^*) = r'(c; \lambda^*) = 0$. But then by Theorem 6 $r(x; \lambda^*) \equiv 0$, which would contradict the initial conditions. We conclude that $r(x; \lambda^*) > 0$ in (a, b) and that $r(b; \lambda^*) = 0$. Then $r'(b; \lambda^*) < 0$, and hence

$$r'(b; \lambda^*) \cos \phi + r(b; \lambda^*) \sin \phi \leq 0,$$

We now recall from the conclusion of Lemma 2 that

$$r'(b; \lambda) \cos \phi + r(b; \lambda) \sin \phi > 0.$$

for $\lambda < \lambda'$.

Since $r(b; \lambda)$ and $r'(b; \lambda)$ are continuous functions of λ, we see that there must be a value λ_1, $\lambda' \leq \lambda_1 \leq \lambda^*$, such that $r'(b;\lambda) \cos \phi + r(b;\lambda) \sin \phi > 0$ for $\lambda < \lambda_1$, and

$$r'(b;\lambda_1) \cos \phi + r(b;\lambda_1) \sin \phi = 0.$$

Then $r(b;\lambda_1)$ satisfies the eigenvalue problem

$$(L + h + \lambda_1 k)[r] = 0,$$
$$-r'(a;\lambda_1) \cos \theta + r(a;\lambda_1) \sin \theta = 0,$$
$$r'(b;\lambda_1) \cos \phi + r(b;\lambda_1) \sin \phi = 0.$$

Hence λ_1 is an eigenvalue of the problem (1), (2), with $r(x;\lambda_1)$ the corresponding eigenfunction. We observe that since $\lambda_1 \leq \lambda^*$, the eigenfunction $r(x;\lambda_1)$ is positive in (a, b). It vanishes at a if and only if $\theta = \pi/2$ and at b if and only if $\phi = \pi/2$.

Remarks: (i) For $-\pi/2 < \theta < \pi/2$ and for any $\lambda < \lambda_1$, the function $w = r(x;\lambda)$ satisfies all the conditions of Theorem 15. Since $(L + h + \lambda k)[w] = 0$, the lower bound (6) is precisely λ. For $\theta = \pi/2$ and $\lambda < \lambda_1$, we choose $w(x) = r(x;\lambda) + s(x;\lambda)$, where $s(x;\lambda)$ is the solution in (a, b) of the analogous initial value problem at b. The lower bound is again λ. Therefore, in all cases the lower bound (6) can be made arbitrarily close to λ_1 by an appropriate selection of the function w. In particular, it follows that λ_1 is the lowest eigenvalue of the problem.

(ii) In the light of the above remark, if λ_1 is positive, the function $w = r(x;0)$ [or $w = r(x;0) + s(x;0)$, if $\theta = \pi/2$] satisfies the conditions of Theorem 11 in Section 5.

(iii) If $\lambda_1 \leq 0$ and w is positive on $[a, b]$, then

$$(L + h + \lambda_1 k)[w] \leq (L + h)[w].$$

Thus, if w satisfies the hypotheses of Theorem 11 for the operator $L + h$, it also does for $L + h + \lambda_1 k$. Choosing $z_1 \equiv z_2 \equiv 0$ in Theorem 11, we see that $u \equiv 0$ when $\lambda = \lambda_1$. Hence the fact that λ_1 is an eigenvalue is contradicted. We conclude that if $\lambda_1 \leq 0$, there does not exist a function w fulfilling the conditions required in Theorem 11.

Remarks (ii) and (iii) form the content of the following result.

THEOREM 17. **The upper and lower bounds of the boundary value problem as stated in Theorem 11 hold if and only if the smallest eigenvalue λ_1 of the problem (1), (2) is positive.**

We observe that no eigenfunction corresponding to an eigenvalue larger than λ_1 can be of one sign in (a, b). For if $r(x;\lambda_k)$ were positive with $\lambda_k > \lambda_1$, we would reach a contradiction by applying Theorems 3 and 4 to the quantity $r(x;\lambda_1)/r(x;\lambda_k)$.

EXERCISES

1. Find a lower bound for the lowest eigenvalue of the problem
$$u'' + xu' + \lambda(1 + x^2)u = 0 \text{ for } 0 < x < 1,$$
$$u(0) = u(1) = 0.$$

2. Find a lower bound for the lowest eigenvalue of the problem
$$u'' - u' + u + \lambda(1 + x^2)u = 0 \text{ for } 0 < x < 1,$$
$$-u'(0) + u(0) = 0,$$
$$u'(1) + u(1) = 0.$$

3. Find an upper bound for the lowest eigenvalue of the problem
$$u'' + xu' + \lambda(1 + x^2)u = 0 \text{ for } 0 < x < 1,$$
$$u(0) = u(1) = 0.$$

4. Consider the two problems:
$$(L + h + \lambda k)[u] \equiv u'' + g(x)u' + [h + \lambda k(x)]u = 0, \quad a < x < b,$$
$$u(a) = u(b) = 0.$$
$$(L + h + \lambda \bar{k})[\bar{u}] \equiv \bar{u}'' + g(x)\bar{u}' + [h + \lambda \bar{k}(x)]\bar{u} = 0, \quad a < x < b,$$
$$\bar{u}(a) = \bar{u}(b) = 0.$$

Suppose that $k(x) \geq \bar{k}(x)$ for $a \leq x \leq b$. What can be said about the relative sizes of λ_1 and $\bar{\lambda}_1$, the lowest eigenvalues of the two problems? Describe how the lowest eigenvalue changes as the functions g and h vary.

SECTION 8. OSCILLATION AND COMPARISON THEOREMS

The maximum principle can be used to derive relations between solutions of the same or related equations. Let u and w be nontrivial solutions (that is, $u \not\equiv 0$, $w \not\equiv 0$) of the same equation
$$(L + h)[u] = 0, \quad (L + h)[w] = 0.$$

We establish the following result:

THEOREM 18. Between any two consecutive zeros of the function w, the function u can have at most one zero.

Proof. Let a and b be two consecutive zeros of w so that $w(a) = w(b) = 0$, $w \neq 0$ for $a < x < b$. By replacing w by $(-w)$, if necessary, we can assume that $w(x) > 0$ in the interval (a, b). Then the function

$$v(x) = \frac{u(x)}{w(x)}$$

satisfies the differential equation

$$v'' + \left(2\frac{w'}{w} + g\right)v' = 0.$$

By Theorem 1, v can have no maxima or minima in (a, b) unless v is a constant, in which case u is a constant multiple of w. It follows that v, and consequently u, can change sign at most once in the open interval (a, b).

It follows from the uniqueness of the initial value problem as given in Theorem 6 (Section 3) that any zero of u must be simple. The same theorem also shows that if $u(a) = 0$, or $u(b) = 0$, u must be a constant multiple of w.

The above result is called a **Sturmian oscillation theorem.** Fig. 3 shows the typical behavior of two solutions of the same equation.

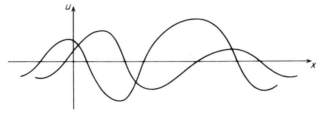

FIGURE 3

We now consider a comparison theorem for solutions of two related equations. Let u satisfy the equation

$$(L + h_1)[u] = 0$$

and w the equation

$$(L + h_2)[w] = 0,$$

with

$$h_2(x) \geq h_1(x) \text{ for } a \leq x \leq b.$$

We again suppose that

$$w > 0 \text{ for } a < x < b.$$

The function v defined by the relation

$$v = \frac{u}{w}$$

satisfies the equation

$$v'' + \left(2\frac{w'}{w} + g\right)v' + (h_1 - h_2)v = 0.$$

Consequently, v satisfies a maximum principle. If u had two zeros in the interval (a, b), then v would also. It would follow from the uniqueness result of Theorem 7 that $v \equiv 0$ between the two zeros. Then, by Theorem 6, $v \equiv 0$; hence $u \equiv 0$ in (a, b). Thus, if u is nontrivial, it can have only one zero in (a, b). In fact, the same is true on the closed interval $[a, b]$ unless $w=0$ at a or b.

Suppose now that $w(a) = u(a) = 0$. By Theorem 6, we have $w'(a) \neq 0$, $u'(a) \neq 0$. From the differential equations for u and w, we find that

$$\frac{u''(a)}{u'(a)} = \frac{w''(a)}{w'(a)} = -g(a). \tag{1}$$

We apply l'Hôpital's rule to see that

$$v(a) = \frac{u'(a)}{w'(a)} \neq 0,$$

and without loss of generality we may assume that $v(a) > 0$. To find $v'(a)$, we write

$$v'(x) = \frac{wu' - uw'}{w^2}$$

and apply l'Hôpital's rule, taking (1) into account. A computation shows that

$$v'(a) = 0.$$

Suppose there were a point x_0 in (a, b) with $v(x_0) < v(a)$. It would follow from Theorem 4 applied to the interval (a, x_0) that $v'(a) < 0$, a contradiction. Thus we have

$$0 < v(a) \leq v(x) \text{ for } a \leq x \leq b.$$

Hence $u(x) \neq 0$ for $a < x < b$. If $u(b) = w(b) = 0$, we find in the same way that $u \neq 0$ in (a, b). Finally, if $w(a) = w(b) = u(a) = u(b) = 0$, we deduce by the above reasoning that $v(a) \leq v(x)$ and $v(x) \geq v(b)$ for $a \leq x \leq b$. Hence $v(a) = v(b)$ and v must be constant. However, this result is possible only if $h_1 - h_2 \equiv 0$. We have proved the following comparison theorem.

THEOREM 19. Let u and w be solutions of the equations

$$(L + h_1)[u] = 0, \quad (L + h_2)[w] = 0,$$

with $h_2 \geq h_1$. **If $w(x) \neq 0$ in the open interval (a, b), then u can have at most one zero on the closed interval $[a, b]$ unless $w(a) = w(b) = 0$, $h_1 \equiv h_2$, and u is a constant multiple of w.**

Roughly speaking, Theorem 19 states that decreasing h increases the distance between zeros of solutions. The theorem can also be given another form by saying that between any two consecutive zeros of u, the function w must have a zero, unless $h_1 \equiv h_2$ and w is a constant multiple of u.

Theorem 19 may be extended to yield a comparison theorem for two different differential operators. Let u be a solution of

$$(L_1 + h_1)[u] \equiv u'' + g_1 u' + h_1 u = 0,$$

and let w be a solution of

$$(L_2 + h_2)[w] \equiv w'' + g_2 w' + h_2 w = 0.$$

We suppose that g_1 and g_2 are continuously differentiable. Introducing the function $\bar{u}(x)$ by the relation

$$\bar{u}(x) = u(x) e^{(1/2) \int_a^x [g_1(\xi) - g_2(\xi)] d\xi}, \tag{2}$$

we find that $\bar{u}$ satisfies the equation

$$(L_2 + H)[\bar{u}] \equiv \bar{u}'' + g_2 \bar{u}' + H(x)\bar{u} = 0$$

with

$$H(x) = \tfrac{1}{2}[g_2'(x) - g_1'(x)] + \tfrac{1}{4}\{[g_2(x)]^2 - [g_1(x)]^2\} + h_1(x).$$

If

$$h_2(x) \geq H(x),$$

then the conclusion of Theorem 19 relating the zeros of w and $\bar{u}$ is applicable. Since (2) shows that u and $\bar{u}$ have the same zeros, we obtain a theorem comparing the zeros of solutions of $(L_1 + h_1)[u] = 0$ with solutions of $(L_2 + h_2)[w] = 0$. Specifically, we find

THEOREM 20. Let $u(x)$ and $w(x)$ be solutions of

$$u'' + g_1(x)u' + h_1(x)u = 0$$

and

$$w'' + g_2(x)w' + h_2(x)w = 0,$$

respectively, and let

$$h_2(x) - \tfrac{1}{2}g_2'(x) - \tfrac{1}{4}[g_2(x)]^2 \geq h_1(x) - \tfrac{1}{2}g_1'(x) - \tfrac{1}{4}[g_1(x)]^2.$$

Then if $w(x) \neq 0$ in the open interval (a, b), $u(x)$ can have at most one zero on the closed interval $[a, b]$ unless $w(a) = w(b) = 0$,

$$h_2(x) - \tfrac{1}{2}g_2'(x) - \tfrac{1}{4}[g_2(x)]^2 \equiv h_1(x) - \tfrac{1}{2}g_1'(x) - \tfrac{1}{4}[g_1(x)]^2,$$

and $u(x)$ is a constant multiple of

$$w(x) e^{-(1/2) \int_a^x [g_1(\xi) - g_2(\xi)] d\xi}.$$

It is interesting to consider the following special case of Theorem 20. We set
$$g_2(x) \equiv 0 \text{ and } h_2(x) \equiv M,$$
where M is a positive constant. Then
$$w = \sin \sqrt{M}\,(x - a)$$
is a solution of
$$w'' + Mw = 0.$$
The function w has zeros at a and $a + (\pi/\sqrt{M})$. We thus obtain the following result:

COROLLARY. If u is a solution of
$$(L + h)[u] \equiv u'' + g(x)u' + h(x)u = 0$$
with
$$h(x) - \tfrac{1}{2}g'(x) - \tfrac{1}{4}[g(x)]^2 \leq M,$$
then the distance between zeros of $u(x)$ is at least $\pi/\sqrt{M}$. If $h(x) - \tfrac{1}{2}g'(x) - \tfrac{1}{4}[g(x)]^2 \geq M' > 0$, then the distance between consecutive zeros of $u(x)$ is at most $\pi/\sqrt{M'}$.

EXERCISES

1. Show that any solution of the equation
$$u'' + (1 + e^{-x})u = 0 \text{ for } x > 0,$$
has infinitely many zeros.

2. Show that any solution of the equation
$$u'' + e^{-x}u = 0$$
has at most one zero in the interval $(0, \infty)$.

3. Compare the zeros of solutions of the equations
$$u'' + (x^2 + 2)u' + (1 + x)u = 0$$
and
$$w'' + \tfrac{1}{2}x^2 w' + (10 + 2x)w = 0.$$

4. Show that any solution of
$$x^2 u'' + (2n + 1)xu' + (n^2 + a)u = 0, \qquad a > 0$$
has infinitely many zeros in the interval (a, ∞).

5. Consider solutions of

$$u'' + g(x)u' + [h(x) + \lambda k(x)]u = 0$$

for various values of λ. Describe the behavior of the zeros of such solutions as $\lambda \to \infty$ and as $\lambda \to -\infty$. What boundedness conditions are required on the functions g, h, k?

SECTION 9. NONLINEAR OPERATORS

A number of results established for linear differential operators can be extended to nonlinear operators. Let $u(x)$ be a solution of the nonlinear equation

$$u'' + H(x, u, u') = 0 \qquad (1)$$

on an interval $a \leq x \leq b$. The functions

$$H(x, y, z), \quad \frac{\partial H(x, y, z)}{\partial y}, \quad \frac{\partial H(x, y, z)}{\partial z}$$

are assumed to be continuous functions of x, y, and z throughout their domains of definition. In addition, we suppose that for each x and z,

$$H(x, y_1, z) \leq H(x, y_2, z) \text{ for } y_1 \geq y_2, \qquad (2)$$

or, equivalently,

$$\frac{\partial H}{\partial y} \leq 0.$$

We observe that if

$$H(x, u, u') = g(x)u' + h(x)u - f(x),$$

then (1) is a linear equation. Condition (2) then becomes $h(x) \leq 0$.

Suppose $w(x)$ satisfies the differential inequality

$$w'' + H(x, w, w') \geq 0 \qquad (3)$$

in (a, b). We consider the function

$$v = w - u$$

and subtract (1) from (3), getting

$$v'' + H(x, w, w') - H(x, u, u') \geq 0.$$

Applying the mean value theorem to H above, we find

$$v'' + \frac{\partial H}{\partial z}v' + \frac{\partial H}{\partial y}v \geq 0.$$

The quantities $\partial H/\partial y$ and $\partial H/\partial z$ are evaluated at $(x, u + \alpha(w - u), u' + \alpha(w' - u'))$ with α a quantity between 0 and 1. The function v satisfies a linear equation, and the maximum principle as given in Theorem 3 applies. We conclude

THEOREM 21. Suppose that

$$w'' + H(x, w, w') \geq u'' + H(x, u, u') \qquad (4)$$

for $a < x < b$ where H, $\partial H/\partial y$, $\partial H/\partial z$ are continuous and $\partial H/\partial y \leq 0$. If $w(x) - u(x)$ attains a nonnegative maximum M in (a, b), then

$$w(x) - u(x) \equiv M.$$

Note that if H depends on x and z but not on y, we may add any constant to w without changing (4). In this case, the assumption in Theorem 21 that M is nonnegative may be dropped.

Theorem 4 may be generalized in a similar way.

We state the following generalization of Theorem 10 in Section 5. The proof, which follows the lines of the proof of Theorem 10, is omitted.

THEOREM 22. Let $u(x)$ be a solution of the boundary value problem

$$u'' + H(x, u, u') = 0 \text{ for } a < x < b, \qquad (5)$$

$$\left. \begin{array}{l} -u'(a) \cos \theta + u(a) \sin \theta = \gamma_1, \\ u'(b) \cos \phi + u(b) \sin \phi = \gamma_2, \end{array} \right\} \qquad (6)$$

where $0 \leq \theta \leq \pi/2$, $0 \leq \phi \leq \pi/2$ and θ and ϕ are not both zero. Suppose that H, $\partial H/\partial y$, $\partial H/\partial z$ are continuous and that $\partial H/\partial y \leq 0$. If $z_1(x)$ satisfies

$$z_1'' + H(x, z_1, z_1') \leq 0,$$
$$-z_1'(a) \cos \theta + z_1(a) \sin \theta \geq \gamma_1,$$
$$z_1'(b) \cos \phi + z_1(b) \sin \phi \geq \gamma_2,$$

and if $z_2(x)$ satisfies

$$z_2'' + H(x, z_2, z_2') \geq 0,$$
$$-z_2'(a) \cos \theta + z_2(a) \sin \theta \leq \gamma_1,$$
$$z_2'(b) \cos \phi + z_2(b) \sin \phi \leq \gamma_2,$$

then the upper and lower bounds

$$z_2(x) \leq u(x) \leq z_1(x)$$

are valid.

This theorem implies that a solution of (5) which satisfies boundary conditions (6) must be unique. For if u and $\tilde{u}$ are solutions, we can let $z_1 = z_2 = \tilde{u}$ to find $u \equiv \tilde{u}$. We give an illustration showing how Theorem 22 may be used to obtain a numerical bound for a solution of (5).

Example. Find bounds for the solution $u(x)$ of

$$u'' - u^3 = 0 \text{ for } 0 < x < 1$$

which satisfies the boundary conditions
$$u(0) = 0, \quad u(1) = 1.$$
We choose
$$z_1 = x, \quad z_2 = x^{(1+\sqrt{5})/2}.$$
Then
$$z_1'' - z_1^3 = -x^3 \leq 0,$$
$$z_2'' - z_2^3 = x^{-(3-\sqrt{5})/2} - x^{3(1+\sqrt{5})/2} \geq 0.$$
Therefore
$$x^{(1+\sqrt{5})/2} \leq u(x) \leq x.$$

We can also obtain an approximation theorem for the initial value problem as described in Section 6. The following result is a generalization of Theorem 13.

THEOREM 23. Suppose that $u(x)$ satisfies
$$u'' + H(x, u, u') = 0 \tag{7}$$
and the initial conditions
$$u(a) = \gamma_1, \quad u'(a) = \gamma_2, \tag{8}$$
that H, $\partial H/\partial y$, $\partial H/\partial z$ are continuous, and that $\partial H/\partial y \leq 0$. If $z_1(x)$ satisfies
$$z_1'' + H(x, z_1, z_1') \geq 0,$$
$$z_1(a) \geq \gamma_1, \quad z_1'(a) \geq \gamma_2,$$
and if $z_2(x)$ satisfies
$$z_2'' + H(x, z_2, z_2') \leq 0,$$
$$z_2(a) \leq \gamma_1, \quad z_2'(a) \leq \gamma_2,$$
then we have the upper and lower bounds
$$z_2(x) + \gamma_2 - z_2(a) \leq u(x) \leq z_1(x) + \gamma_1 - z_1(a),$$
$$z_2'(x) \leq u'(x) \leq z_1'(x).$$

The uniqueness of a solution u of (7), (8) again follows by choosing $z_1 = z_2 = \tilde{u}$ where $\tilde{u}$ is any other supposed solution of (7), (8). Then $\tilde{u} \equiv u$.

EXERCISES

1. If $u'' - u^3 = 0$ for $0 < x < 1$,
$$u(0) = 1, \quad u'(1) = 0,$$
find upper and lower bounds for $u(1)$.

2. If $u'' - e^u = 0$ for $0 < x < 1$,
$$u(0) = 1, \quad u(1) = 0,$$
find upper and lower bounds for $u(\tfrac{1}{2})$.

3. If $u'' - u^3 = 1$ for $x > 0$,
$$u(0) = u'(0) = 0,$$
find upper and lower bounds for $u(1)$.

4. If $u'' - xu' - u^3 = 0$ for $0 < x < 1$,
$$u(0) = 0, \quad u(1) = 1,$$
find upper and lower bounds for $u(\tfrac{1}{2})$.

5. Prove Theorem 22.

BIBLIOGRAPHICAL NOTES

For existence and uniqueness theorems for initial and boundary value problems as discussed in Sections 3 and 4 see, for example, the texts of Coddington and Levinson [1] or Hartman [1]. Hartman also discusses certain monotony properties of solutions of first order systems of ordinary differential equations.

Comparison theorems for second order equations have been obtained by many authors. We refer to the texts mentioned above and to a result of Leighton [1].

Problems on approximation in boundary and initial value problems as well as error estimates in such approximations are treated thoroughly in the book of Collatz [4] where extensive references to the literature may be found. The books of Szarski [5] and Walter [3] give extensive surveys of the treatment of initial value problems for nonlinear systems and higher order equations.

Other results on approximation have been obtained by Gloistehn [1], Lees [1], Müller [1], Pucci [1], Schröder [2], Szarski [1, 2, 5], Uhlmann [1], Walter [1, 2], and Ważewski [1].

CHAPTER 2

ELLIPTIC EQUATIONS

SECTION 1. THE LAPLACE OPERATOR

Let $u(x_1, x_2, \ldots, x_n)$ be a twice continuously differentiable function defined in a domain D in n-dimensional Euclidean space. The **Laplace operator** or **Laplacian** Δ is defined as

$$\Delta \equiv \frac{\partial^2}{\partial x_1^2} + \frac{\partial^2}{\partial x_2^2} + \cdots + \frac{\partial^2}{\partial x_n^2}.$$

If the equation $\Delta u = 0$ is satisfied at each point of a domain D, we say that u is **harmonic** in D or, simply, that u is a **harmonic function**.

Suppose that u has a local maximum at an interior point of D. Then we know from elementary calculus that at this point

$$\frac{\partial u}{\partial x_1} = 0, \quad \frac{\partial u}{\partial x_2} = 0, \quad \ldots, \quad \frac{\partial u}{\partial x_n} = 0$$

and

$$\frac{\partial^2 u}{\partial x_1^2} \leq 0, \quad \frac{\partial^2 u}{\partial x_2^2} \leq 0, \quad \ldots, \quad \frac{\partial^2 u}{\partial x_n^2} \leq 0.$$

Therefore, at a local maximum, the inequality

$$\Delta u \leq 0$$

must hold. The simple reasoning above leads to the assertion: **If a function u satisfies the strict inequality**

$$\Delta u > 0, \tag{1}$$

at each point of a domain D, then u cannot attain its maximum at any interior point of D. Suppose $b_1(x_1, x_2, \ldots, x_n), b_2(x_1, x_2, \ldots, x_n), \ldots, b_n(x_1, x_2, \ldots, x_n)$ are any bounded functions defined in D. Without any change in the argument above, we conclude that if u satisfies the strict inequality

$$\Delta u + b_1 \frac{\partial u}{\partial x_1} + b_2 \frac{\partial u}{\partial x_2} + \cdots + b_n \frac{\partial u}{\partial x_n} > 0 \tag{2}$$

in D, then u cannot attain its maximum at an interior point.

In Chapter 1, where we considered the one-dimensional case, we saw that it is important to establish a maximum principle for differential inequalities which are not strict. We shall show that the above maximum principle is valid even when equality is allowed in the relations (1) and (2). In particular, harmonic functions satisfy a maximum principle.

We shall first prove a maximum principle for functions of two variables which satisfy the inequality

$$\Delta u \equiv \frac{\partial^2 u}{\partial x^2} + \frac{\partial^2 u}{\partial y^2} \geq 0. \tag{3}$$

In doing so, we shall use a number of special properties of the Laplace operator. Later, in considering more general differential operators, we shall see how the methods of Chapter 1 may be modified to yield a maximum principle.

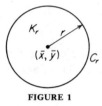

FIGURE 1

Let $(\bar{x}, \bar{y})$ be a point of D and let K_r be a disk situated in D with center at $(\bar{x}, \bar{y})$ and with radius r (Fig. 1). Denote the boundary of K_r by C_r. We recall that the **gradient** of a scalar function v is the vector function defined by

$$\operatorname{grad} v \equiv \frac{\partial v}{\partial x}\mathbf{i} + \frac{\partial v}{\partial y}\mathbf{j},$$

where $\mathbf{i}, \mathbf{j}$ are the customary orthogonal unit vectors in the plane. The **divergence** of a vector function $\mathbf{w} = a(x, y)\mathbf{i} + b(x, y)\mathbf{j}$ is the scalar function defined by the formula

$$\operatorname{div} \mathbf{w} = \frac{\partial a}{\partial x} + \frac{\partial b}{\partial y}.$$

Thus we have

$$\Delta u = \operatorname{div}(\operatorname{grad} u). \tag{4}$$

We apply the divergence theorem* to the Laplacian of u in the disk K_r and obtain

$$\iint_{K_r} \Delta u \, dx \, dy \equiv \iint_{K_r} \operatorname{div}(\operatorname{grad} u) \, dx \, dy = \oint_{C_r} \frac{\partial u}{\partial r} \, ds,$$

where s is arc length along C_r and $\partial u/\partial r$ is the normal derivative taken on the boundary C_r. In polar coordinates, $ds = r \, d\theta$, and so

$$\iint_{K_r} \Delta u \, dx \, dy = r \int_0^{2\pi} \frac{\partial u}{\partial r} \, d\theta.$$

---

*The **divergence theorem** states that if D is a bounded domain with smooth boundary ∂D, then $\iint_D \operatorname{div} \mathbf{w} \, dx \, dy = \oint_{\partial D} \mathbf{w} \cdot \mathbf{n} \, ds$, where $\mathbf{n}$ is the outward normal unit vector. The analogous result is valid in higher dimensions [see Section 7, equation (1)].

It follows that if $\Delta u \geq 0$ in D, then

$$\int_0^{2\pi} \frac{\partial u}{\partial r} d\theta \geq 0. \tag{5}$$

Now we allow r to vary between 0 and a fixed number R, each K_r being a disk with center at $(\bar{x}, \bar{y})$. The number R is taken sufficiently small so that K_R is entirely contained in D. Integrating inequality (5) from 0 to R and interchanging the order of integration, we obtain

$$\int_0^R \int_0^{2\pi} \frac{\partial u}{\partial r} d\theta \, dr = \int_0^{2\pi} u(R, \theta) \, d\theta - 2\pi u(\bar{x}, \bar{y}) \geq 0,$$

or

$$u(\bar{x}, \bar{y}) \leq \frac{1}{2\pi R} \int_0^{2\pi} u(R, \theta) R \, d\theta = \frac{1}{2\pi R} \oint_{C_R} u \, ds. \tag{6}$$

The right side of (6) is the **mean value** of u taken over C_R, the circle of radius R with center at $(\bar{x}, \bar{y})$. Therefore (6) asserts that the value of u at any point of D is bounded by its mean value over any circle in D having that point as center. If $\Delta u = 0$, this inequality is true for both u and $-u$, and consequently we obtain the following result.

THEOREM 1. (Mean value theorem). **If u is harmonic in D, then $u(\bar{x}, \bar{y})$ is equal to its mean value taken over any circle in D with center at $(\bar{x}, \bar{y})$. That is,**

$$u(\bar{x}, \bar{y}) = \frac{1}{2\pi R} \oint_{C_R} u \, ds.$$

To obtain a maximum principle we suppose that u satisfies the inequality (3) in D and that it attains its maximum M at some point (x_0, y_0) in D. Since $u \leq M$ and $u(x_0, y_0) = M$, we conclude from (6) that u must be identically equal to M on every circle centered at (x_0, y_0) and situated in D. Now suppose that there is a point (x_1, y_1) in D at which $u < M$; then the same is true in a neighborhood of (x_1, y_1). We connect (x_1, y_1) with (x_0, y_0) by a curve in D and let (x_2, y_2) be the first point on this curve where $u = M$ (Fig. 2). Then u is not identically equal to M on any sufficiently small circle centered at (x_2, y_2). Thus we contradict inequality (6) and thereby establish the following result.

THEOREM 2. (Maximum principle). **Let**

$$\Delta u \geq 0 \text{ in } D.$$

If u attains its maximum M at any point of D, then $u \equiv M$ in D.

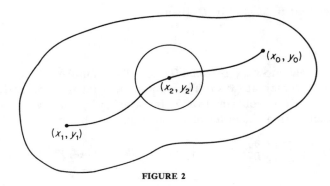

FIGURE 2

We have not assumed that the domain D is bounded. Denoting the boundary of D by ∂D, we observe that if u is continuous on $D \cup \partial D$ and is not constant, then the values of u in D are below the maximum of u on ∂D if D is bounded. If D is unbounded, these values are either below the maximum of u on ∂D or below the limit superior* of u as $(x^2 + y^2)$ tends to infinity.

Since a corresponding minimum principle holds for functions satisfying the inequality

$$\Delta u \leq 0$$

(obtained by applying Theorem 2 to the function $-u$), we can conclude that a nonconstant harmonic function can attain neither its maximum nor its minimum at any interior point of D.

DEFINITION. A function u satisfying $\Delta u \geq 0$ in a domain D is said to be subharmonic in D, or simply a subharmonic function. If $\Delta u \leq 0$ so that $-u$ is subharmonic, we say that u is superharmonic.

Suppose u is subharmonic and v is harmonic in a domain D with boundary ∂D. The function

$$w = u - v$$

will then be subharmonic in D. If u and v coincide on ∂D, then w will vanish on ∂D and, by the maximum principle, will be nonpositive throughout D. We thus obtain the inequality

$$u \leq v \text{ in } D.$$

The term subharmonic comes from the property just described. The values of a subharmonic function u in a domain D are always *below* the values of the harmonic function which coincides with u on the boundary of D.

---

*For the definition of limit superior see the footnote on page 94.

Theorems 1 and 2 have appropriate extensions in any number of dimensions. Let u be a subharmonic function of n variables; that is, suppose $\Delta u \geq 0$. We denote by K_R an n-dimensional ball of radius R with center at $(\bar{x}_1, \bar{x}_2, \ldots, \bar{x}_n)$, and we denote its boundary by C_R. The surface area of the sphere C_R, whose equation is

$$\sum_{k=1}^{n} (x_k - \bar{x}_k)^2 = R^2,$$

is designated by S_R. It is convenient to write

$$S_R = \omega_n R^{n-1},$$

where ω_n is an absolute constant depending only on n. For example, $\omega_2 = 2\pi$ and $\omega_3 = 4\pi$. The mean value inequality asserts that for subharmonic functions

$$u(\bar{x}_1, \bar{x}_2, \ldots, \bar{x}_n) \leq \frac{1}{\omega_n R^{n-1}} \oint_{C_R} u \, dS, \tag{7}$$

where dS is the $(n-1)$-dimensional element of surface area and the integral on the right is an $(n-1)$-fold integral. The proof of (7) follows exactly the same method as the proof of (6). The mean value theorem for harmonic functions and the maximum principle for n-dimensional subharmonic functions are direct consequences of (7).

EXERCISES

1. Let $u(x_1, x_2, \ldots, x_n) = (x_1^2 + x_2^2 + \cdots, x_n^2)^\alpha$. For what values of α is u subharmonic? For what values is u superharmonic?

2. If r, θ are polar coordinates in the plane and n is an integer, show that $r^n \cos n\theta$ and $r^n \sin n\theta$ are harmonic functions.

3. Show that the problem

$$\frac{\partial^2 u}{\partial x^2} + \frac{\partial^2 u}{\partial y^2} = u^3 \text{ for } x^2 + y^2 < 1,$$

$$u = 0 \text{ for } x^2 + y^2 = 1$$

has no solution other than $u \equiv 0$.

4. Let u be a solution to the problem

$$\frac{\partial^2 u}{\partial x^2} + \frac{\partial^2 u}{\partial y^2} = -1 \text{ for } |x| < 1, \quad |y| < 1,$$

$$u = 0 \text{ for } |x| = 1, \quad |y| = 1.$$

Find upper and lower bounds for $u(0, 0)$. *Hint:* Consider the function $v = u + \frac{1}{4}(x^2 + y^2)$.

5. Prove the mean value inequality [formula (7)] for subharmonic functions in n variables.

SECTION 2. SECOND ORDER ELLIPTIC OPERATORS. TRANSFORMATIONS

We shall be concerned with second order differential operators of the form

$$\sum_{i,j=1}^{n} \alpha_{ij}(x_1, x_2, \ldots, x_n) \frac{\partial^2}{\partial x_i \partial x_j}.$$

Since $\partial^2/\partial x_i \partial x_j \equiv \partial^2/\partial x_j \partial x_i$, we may define

$$a_{ij} = \tfrac{1}{2}(\alpha_{ij} + \alpha_{ji})$$

and write the above differential expression as

$$\mathscr{L} \equiv \sum_{i,j=1}^{n} a_{ij}(x_1, x_2, \ldots, x_n) \frac{\partial^2}{\partial x_i \partial x_j}, \qquad a_{ij} = a_{ji}, \qquad i,j = 1, 2, \ldots, n. \tag{1}$$

In other words, there is no loss of generality in supposing that the coefficients of the second-order operator $\mathscr{L}$ are symmetric, an assumption we shall always make.

DEFINITIONS. The operator (1) is called elliptic at a point $\mathbf{x} = (x_1, x_2, \ldots, x_n)$ if and only if there is a positive quantity $\mu(\mathbf{x})$ such that*

$$\sum_{i,j=1}^{n} a_{ij}(\mathbf{x}) \xi_i \xi_j \geq \mu(\mathbf{x}) \sum_{i=1}^{n} \xi_i^2 \tag{2}$$

for all n-tuples of real numbers $(\xi_1, \xi_2, \ldots, \xi_n)$. The operator $\mathscr{L}$ is said to be elliptic in a domain D if it is elliptic at each point of D. It is uniformly elliptic in D if (2) holds for each point of D and if there is a positive constant μ_0 such that $\mu(\mathbf{x}) \geq \mu_0$ for all $\mathbf{x}$ in D.

In matrix language the ellipticity condition asserts that the symmetric matrix

---

*It is customary to define $\mathscr{L}$ to be elliptic if either the coefficients $a_{ij}(\mathbf{x})$ or $-a_{ij}(\mathbf{x})$ satisfy inequality (2). It is more convenient for our purposes to assume that $\mathscr{L}$ has been multiplied by (-1) if necessary so that (2) always holds.

Sec. 2 Second Order Elliptic Operators. Transformations

$$A(\mathbf{x}) = \begin{pmatrix} a_{11} & a_{12} & \cdots & a_{1n} \\ a_{21} & a_{22} & \cdots & a_{2n} \\ \vdots & \vdots & & \vdots \\ a_{n1} & a_{n2} & \cdots & a_{nn} \end{pmatrix}$$

is positive definite at each point x.

We are interested in studying the effect of various transformations of coordinates on the form of $\mathscr{L}$. A linear change of coordinates

$$y_i = \sum_{j=1}^{n} c_{ij} x_j, \qquad i = 1, 2, \ldots, n \tag{3}$$

may be written in matrix notation

$$\mathbf{y} = C\mathbf{x}.$$

In this equation, $\mathbf{x}$ and $\mathbf{y}$ are $n \times 1$ matrices, and C is the $n \times n$ matrix (c_{ij}). The $n \times n$ matrix C is said to be **orthogonal** if and only if its elements satisfy the relations

$$\sum_{i=1}^{n} c_{ij} c_{ik} = \begin{Bmatrix} 1 \text{ if } j = k \\ 0 \text{ if } j \neq k \end{Bmatrix}, \qquad j, k = 1, 2, \ldots, n. \tag{4}$$

It can easily be shown that the condition

$$\sum_{k=1}^{n} c_{ik} c_{jk} = \begin{Bmatrix} 1 \text{ if } i = j \\ 0 \text{ if } i \neq j \end{Bmatrix}, \qquad i, j = 1, 2, \ldots, n \tag{5}$$

is equivalent to (4). In fact, denoting the transpose of C by C^T and the inverse by C^{-1}, we observe that the orthogonality criteria (4) and (5) are both equivalent to the relation

$$C^T = C^{-1}.$$

DEFINITION. The transformation (3) is called a **rotation** or an **orthogonal transformation** if and only if the matrix $C = (c_{ij})$ is an orthogonal matrix.

The next result shows that the ellipticity of an operator is unaffected by orthogonal transformations of coordinates.

THEOREM 3. Suppose the operator

$$\mathscr{L} \equiv \sum_{i,j=1}^{n} a_{ij} \frac{\partial^2}{\partial x_i \partial x_j}$$

is elliptic. Then under the orthogonal transformation (3) the operator $\mathscr{L}$ takes the form

$$\sum_{k,l=1}^{n} b_{kl} \frac{\partial^2}{\partial y_k \partial y_l}, \qquad (6)$$

where

$$b_{kl} = \sum_{i,j=1}^{n} a_{ij} c_{ki} c_{lj}, \qquad k, l = 1, 2, \ldots, n. \qquad (7)$$

Furthermore, the operator (6) is elliptic.

Proof. Since $\partial y_i / \partial x_j = c_{ij}$, we apply the chain rule to $\mathscr{L}$ and obtain

$$\sum_{i,j=1}^{n} a_{ij} \frac{\partial^2}{\partial x_i \partial x_j} = \sum_{i,j,k,l=1}^{n} a_{ij} c_{ki} c_{lj} \frac{\partial^2}{\partial y_k \partial y_l}.$$

Defining the matrix $B = (b_{kl})$ by the relations (7), we get $\mathscr{L}$ in the form (6).

To establish the ellipticity of (6), we consider any n-tuple $(\xi_1, \xi_2, \ldots, \xi_n)$ and write

$$\sum_{k,l=1}^{n} b_{kl} \xi_k \xi_l = \sum_{i,j,k,l=1}^{n} a_{ij} c_{ki} \xi_k c_{lj} \xi_l.$$

Defining

$$\eta_i = \sum_{k=1}^{n} c_{ki} \xi_k,$$

we have

$$\sum_{k,l=1}^{n} b_{kl} \xi_k \xi_l = \sum_{i,j=1}^{n} a_{ij} \eta_i \eta_j.$$

Hence, by the ellipticity condition (2) and by (5),

$$\sum_{k,l=1}^{n} b_{kl} \xi_k \xi_l \geq \mu(\mathbf{x}) \sum_{i=1}^{n} \eta_i \eta_i$$

$$= \mu(\mathbf{x}) \sum_{i,k,l=1}^{n} c_{ki} \xi_k c_{li} \xi_l = \mu(\mathbf{x}) \sum_{k=1}^{n} \xi_k^2.$$

Remarks. (i) We note not only that ellipticity is preserved under an orthogonal transformation of coordinates, but also that the quantity $\mu(\mathbf{x})$ is unchanged.

(ii) If an operator is uniformly elliptic, then the operator will remain so after any orthogonal transformation.

(iii) In order to apply the chain rule in the manner described, it is essential that the elements of C be constant.

(iv) If an operator with constant coefficients is elliptic at a point, then it is uniformly elliptic in all of n-space. In particular, the Laplace operator is elliptic everywhere.

It is known from linear algebra that, for any particular symmetric matrix $A = (a_{ij})$, there exists an orthogonal matrix $C = (c_{ij})$ such that the matrix $B = (b_{ij})$, given by

$$b_{ij} = \sum_{k,l=1}^{n} c_{ik} a_{kl} c_{jl}, \quad i, j = 1, 2, \ldots, n$$

or equivalently by

$$B = CAC^{-1},$$

is a *diagonal* matrix. That is, the matrix B has the property

$$b_{ij} = 0 \text{ if } i \neq j.$$

The elements b_{kk} along the diagonal of B are called the **eigenvalues** of the original matrix (a_{ij}), and the rows of (c_{ij}) are the **eigenvectors** of the corresponding linear transformation.

If we perform such a diagonalization at a particular point $\bar{x}$, we get

$$\sum_{k,l=1}^{n} b_{kl}(\bar{x}) \eta_k \eta_l = \sum_{k=1}^{n} b_{kk}(\bar{x}) \eta_k^2 \geq \mu(\bar{x}) \sum_{k=1}^{n} \eta_k^2$$

for *all* real n-tuples $(\eta_1, \eta_2, \ldots, \eta_n)$. It follows that all the diagonal elements $b_{kk}(\bar{x})$ are positive. In fact, we have

$$b_{kk}(\bar{x}) \geq \mu(\bar{x}).$$

Moreover, if $\mathscr{L}$ is not elliptic, such an inequality cannot hold for any positive number $\mu(\bar{x})$. For, if it did, the inverse transformation would lead to an inequality for the original operator which is the same as ellipticity. Hence if $\mathscr{L}$ is not elliptic, at least one of the eigenvalues $b_{kk}(\bar{x})$ must be nonpositive.

We suppose that the original operator $\mathscr{L}$ has been put in diagonal form at a particular point $\bar{x}$. At this point we may write

$$\mathscr{L} \equiv \sum_{i=1}^{n} b_{ii}(\bar{x}) \frac{\partial^2}{\partial y_i^2}.$$

Now we introduce a second transformation of coordinates, one which consists of a "stretching." We set

$$z_k = \frac{1}{\sqrt{b_{kk}(\bar{x})}} y_k, \quad k = 1, 2, \ldots, n. \tag{8}$$

In terms of the original $\{x_k\}$-coordinates, we have

$$z_k = \frac{1}{\sqrt{b_{kk}}} \sum_{j=1}^{n} c_{kj} x_j \equiv \sum_{j=1}^{n} d_{kj} x_j, \quad k = 1, 2, \ldots, n.$$

This has the effect of putting the operator $\mathscr{L}$ in the form

$$\mathscr{L} \equiv \sum_{k=1}^{n} \frac{\partial^2}{\partial z_k^2}.$$

The transformation (8) can be performed only when all the $b_{kk}(\bar{x})$ are positive. We have established the following result.

THEOREM 4. A second order operator $\mathscr{L}$ is elliptic at a point $\bar{x}$ if and only if there is a linear transformation

$$z_k = \sum_{j=1}^{n} d_{kj} x_j, \qquad k = 1, 2, \ldots, n.$$

such that at $\bar{x}$, $\mathscr{L}$ becomes the Laplacian in the $\{z_k\}$-coordinates.

It should be emphasized that we have assumed throughout that (d_{kj}) is a constant matrix, so that this reduction occurs only at a particular point $\bar{x}$, rather than in a whole domain.*

The operator

$$(L + h) \equiv \sum_{i,j=1}^{n} a_{ij} \frac{\partial^2}{\partial x_i \partial x_j} + \sum_{i=1}^{n} b_i \frac{\partial}{\partial x_i} + h$$

is said to be elliptic at $\bar{x}$ if and only if

$$\mathscr{L} \equiv \sum_{i,j=1}^{n} a_{ij} \frac{\partial^2}{\partial x_i \partial x_j}$$

is elliptic there. It is **uniformly elliptic** in D if $\mathscr{L}$ is uniformly elliptic in D. The operator $\mathscr{L}$ is called the **principal part** of $L + h$.

EXERCISES

1. Verify that the following matrices are orthogonal:

(a) $\begin{pmatrix} \cos \theta & \sin \theta \\ -\sin \theta & \cos \theta \end{pmatrix}$. (b) $\begin{pmatrix} \frac{2}{3} & -\frac{2}{3} & \frac{1}{3} \\ \frac{1}{3} & \frac{2}{3} & \frac{2}{3} \\ -\frac{2}{3} & -\frac{1}{3} & \frac{2}{3} \end{pmatrix}$. (c) $\frac{1}{2}\begin{pmatrix} 1 & 1 & -1 & -1 \\ -1 & 1 & 1 & -1 \\ 1 & 1 & 1 & 1 \\ -1 & 1 & -1 & 1 \end{pmatrix}$.

2. Show that the operator

$$L[u] \equiv (1 - x^2) \frac{\partial^2 u}{\partial x^2} + 2xy \frac{\partial^2 u}{\partial x \partial y} + (1 - y^2) \frac{\partial^2 u}{\partial y^2}$$

is elliptic but not uniformly elliptic for $x^2 + y^2 < 1$.

3. Find the coordinate transformation $(x, y) \to (\xi, \eta)$ which changes the operator L of problem 2 into the Laplacian at a particular point $(\bar{x}, \bar{y})$.

4. Find the transformation which takes the operator

$$L[u] \equiv 5 \frac{\partial^2 u}{\partial x^2} + 4 \frac{\partial^2 u}{\partial y^2} + 3 \frac{\partial^2 u}{\partial z^2} + 4 \frac{\partial^2 u}{\partial x \partial y} + 4 \frac{\partial^2 u}{\partial y \partial z}$$

into the Laplacian.

---

*It can be shown that it is not possible, in general, to reduce a second-order elliptic operator to the Laplace operator in an entire domain by a transformation of coordinates.

SECTION 3. THE MAXIMUM PRINCIPLE OF E. HOPF

Consider the strict differential inequality

$$L[u] \equiv \sum_{i,j=1}^{n} a_{ij} \frac{\partial^2 u}{\partial x_i \partial x_j} + \sum_{i=1}^{n} b_i \frac{\partial u}{\partial x_i} > 0 \tag{1}$$

in a domain D, and assume that L is elliptic in D. If u has a relative maximum at a point $\bar{x} = (\bar{x}_1, \bar{x}_2, \ldots, \bar{x}_n)$, we know from the calculus of several variables that at $\bar{x}$

$$\frac{\partial u}{\partial z_k} = 0 \text{ and } \frac{\partial^2 u}{\partial z_k^2} \leq 0, \quad k = 1, 2, \ldots, n$$

for any coordinates $z_1, z_2, \ldots, z_n$ obtained from the coordinates $x_1, x_2, \ldots, x_n$ by a linear transformation. In particular, if $\mathscr{L}$, the principal part of L, is the Laplace operator in the z-coordinates, then (1) cannot hold at $\bar{x}$. Whenever L is elliptic, we can find a linear transformation of coordinates so that at $\bar{x}$ the operator $\mathscr{L}$ becomes the Laplace operator. We conclude that **if L is elliptic, a function u which satisfies (1) in a domain D cannot have a maximum in D.**

As in the one-dimensional case, we shall extend the maximum principle to include the possibility that $L[u]$ satisfies an inequality which may not be strict.

THEOREM 5. Let $u(x_1, x_2, \ldots, x_n)$ satisfy the differential inequality

$$L[u] \equiv \sum_{i,j=1}^{n} a_{ij}(\mathbf{x}) \frac{\partial^2 u}{\partial x_i \partial x_j} + \sum_{i=1}^{n} b_i(\mathbf{x}) \frac{\partial u}{\partial x_i} \geq 0$$

in a domain D where L is uniformly elliptic. Suppose the coefficients a_{ij} and b_i are uniformly bounded. If u attains a maximum M at a point of D, then $u \equiv M$ in D.

Proof. Assume that $u = M$ at some point P of D and that $u < M$ at some other point Q of D. We shall establish a contradiction. Construct an arc γ in D extending from Q to P. (See Fig. 3.) Denote by R the first point on γ where $u = M$. That is, $u < M$ on the portion of γ between Q and R and $u(R) = M$. Let $d > 0$ be the greatest lower bound of the distances between any point of γ and any point on the boundary of D. We consider a point P_1 of the portion QR of γ at distance less than $\frac{1}{2}d$ from R and construct the largest ball having center at P_1 in which $u < M$. This ball (called K) will have radius less than $\frac{1}{2}d$ and will therefore lie entirely in D. Let S be a point on the boundary ∂K of K such that $u = M$

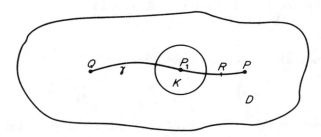

FIGURE 3

at S. By continuity there must be at least one such point. There may be many. Construct the ball K_1 tangent to ∂K at S and otherwise lying completely in K. (See Fig. 4.) Denoting the boundary of K_1 by ∂K_1, we observe that $u < M$ in K_1 and on ∂K_1 except at the single point S where $u = M$. The radius of K_1 will be denoted r_1.

We make yet another construction. A ball K_2, with boundary ∂K_2 and with radius $r_2 = \frac{1}{2} r_1$, is drawn with center at S. (See Fig. 5.) We denote by C_2' the portion of ∂K_2 inside K_1 and on ∂K_1. That is, the surface C_2' includes its boundary, which is the intersection of ∂K_2 and ∂K_1. The part of ∂K_2 outside ∂K_1 we call C_2''.

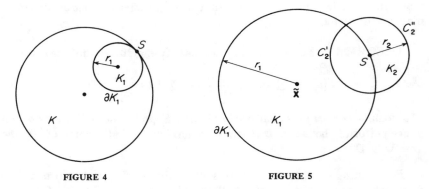

FIGURE 4   **FIGURE 5**

Since $u < M$ on the closed set C_2', there is a constant $\zeta > 0$ such that

$$u \leq M - \zeta \text{ on } C_2'.$$

This follows from the fact that a continuous function on a bounded, closed set assumes its maximum. On the other hand, because $u \leq M$ everywhere we know that

$$u \leq M \text{ on } C_2''.$$

Let the center of K_1 be denoted by $\tilde{x} = (\tilde{x}_1, \tilde{x}_2, \ldots, \tilde{x}_n)$. We define the auxiliary function

Sec. 3   The Maximum Principle of E. Hopf

$$z(\mathbf{x}) = e^{-\alpha \sum_{i=1}^{n} (x_i - \tilde{x}_i)^2} - e^{-\alpha r_1^2}, \tag{2}$$

where α is a positive constant to be determined. Then it is clear that

$$z > 0 \text{ in } K_1,$$
$$z = 0 \text{ on } \partial K_1,$$
$$z < 0 \text{ outside } K_1.$$

We compute

$$L[z] = e^{-\alpha \sum_{i=1}^{n} (x_i - \tilde{x}_i)^2} \left\{ 4\alpha^2 \sum_{i,j=1}^{n} a_{ij}(x_i - \tilde{x}_i)(x_j - \tilde{x}_j) \right.$$
$$\left. - 2\alpha \sum_{i=1}^{n} [a_{ii} + b_i(x_i - \tilde{x}_i)] \right\}.$$

Because of the uniform ellipticity of the operator L, we have

$$\sum_{i,j=1}^{n} a_{ij}(x_i - \tilde{x}_i)(x_j - \tilde{x}_j) \geq \mu_0 \sum_{i=1}^{n} (x_i - \tilde{x}_i)^2.$$

Since $\sum_{i=1}^{n} (x_i - \tilde{x}_i)^2 > \frac{1}{4} r_1^2$ in K_2, we conclude that for $\mathbf{x}$ in K_2,

$$L[z] \geq \alpha e^{-\alpha \sum_{i=1}^{n} (x_i - \tilde{x}_i)^2} \left\{ \alpha \mu_0 r_1^2 - 2 \sum_{i=1}^{n} [a_{ii} + b_i(x_i - \tilde{x}_i)] \right\}.$$

By choosing α sufficiently large, we make the quantity in braces above positive and so obtain

$$L[z] > 0 \text{ in } K_2.$$

As in the proof of Theorem 1 in Chapter 1, we form the function

$$w = u + \epsilon z,$$

with ϵ such that

$$0 < \epsilon < \frac{\zeta}{1 - e^{-\alpha r_1^2}}.$$

The function w has the following properties:

(i) $w < M$ on C_2'. To see this we first note that

$$0 \leq z < 1 - e^{-\alpha r_1^2}.$$

Therefore $\epsilon z < \zeta$ and $u \leq M - \zeta$ on C_2'. Adding these inequalities we get $w < M$ on C_2'.

(ii) $w < M$ on C_2''. This follows because $z < 0$ on C_2'' and $u \leq M$ everywhere.

(iii) $w = M$ at S, the center of K_2. This statement is true because by hypothesis $u = M$ at S, and by construction $z = 0$ at S.

On the basis of properties (i), (ii), and (iii), w has a maximum somewhere in the interior of the ball K_2. On the other hand,

$$L[w] = L[u] + \epsilon L[z] > 0 \text{ in } K_2.$$

Therefore, no maximum can occur in K_2 if L is elliptic. A contradiction has been reached, and the theorem is proved.

Remarks. (i) The uniform ellipticity of the operator L and the boundedness of the coefficients are not essential. It is sufficient to assume that the quantities

$$\frac{\sum_{i=1}^{n} a_{ii}(\mathbf{x})}{\mu(\mathbf{x})} \quad \text{and} \quad \frac{\sum_{i=1}^{n} |b_i(\mathbf{x})|}{\mu(\mathbf{x})}$$

are bounded on every closed ball contained in the interior of D.

(ii) The domain D need not be bounded.

(iii) A minimum principle for functions satisfying $L[u] \leq 0$ is obtained by applying the maximum principle to the function $-u$. Therefore **a nonconstant solution of the elliptic differential equation $L[u] = 0$ can attain neither a maximum nor a minimum at an interior point of D.**

For operators of the form $(L + h)$ we have a result analogous to the one-dimensional case as given in Section 1 of Chapter 1.

THEOREM 6. Let u satisfy the differential inequality

$$(L + h)[u] \geq 0$$

with $h \leq 0$, with L uniformly elliptic in D, and with the coefficients of L and h bounded. If u attains a nonnegative maximum M at an interior point of D, then $u \equiv M$.

The proof of Theorem 6 follows exactly the same procedure as the proof of Theorem 5. The auxiliary function z and the positive quantity ϵ are defined in the same manner. As before, the contradiction is obtained from the observation that a function w satisfying the inequality $(L + h)[w] > 0$ in a domain cannot have a nonnegative maximum there if $h \leq 0$.

Remarks. (i) This proof still holds under the weaker hypotheses that the ratios $\sum_{i=1}^{n} a_{ii}(\mathbf{x})/\mu(\mathbf{x})$, $\sum_{i=1}^{n} |b_i(\mathbf{x})|/\mu(\mathbf{x})$, and $-h(\mathbf{x})/\mu(\mathbf{x})$ are bounded in every closed ball in D.

(ii) The restriction $h \leq 0$ is essential, as counterexamples are easily obtained if $h > 0$. For example, the function $u = e^{-r^2}$ has an absolute maximum at $r = 0$ and is a solution of the equation $\Delta u + (2n - 4r^2)u = 0$ in n dimensions.

Let u satisfy $L[u] \geq 0$ in a domain D with a smooth boundary ∂D. We know that if u takes on a maximum at all it must do so at a boundary point. We shall now suppose that u is continuous and bounded on $D \cup \partial D$ and that there is a point P on ∂D at which u takes on its maximum value. If D is bounded, such a point P will always exist. First of all, we observe that the directional derivative of u at P taken in any direction on the boundary that points outward from D cannot be negative. If it were, the function u would start increasing as we enter the domain D at P, and so the maximum could not occur at P.

Let $\mathbf{n} = (\eta_1, \eta_2, \ldots, \eta_n)$ be the unit normal vector in an outward direction at a point P on the boundary of D. We say that the vector $\nu = (\nu_1, \ldots, \nu_n)$ points outward from D at the boundary point P if

$$\nu \cdot \mathbf{n} > 0.$$

We define the **directional derivative** of u at the boundary point P in the direction ν as*

$$\frac{\partial u}{\partial \nu} \equiv \lim_{\mathbf{x} \to P} [\nu \cdot \text{grad } u(\mathbf{x})] = \lim_{\mathbf{x} \to P} \left(\nu_1 \frac{\partial u}{\partial x_1} + \cdots + \nu_n \frac{\partial u}{\partial x_n} \right),$$

if it exists. The directional derivative is said to be **outward** if ν points outward from D. Then, if u has a maximum at P, we have $\partial u / \partial \nu \geq 0$ at P.

We shall now show that unless u is constant, the strict inequality $\partial u / \partial \nu > 0$ holds at P. This result is clearly an extension of Theorem 2 of Chapter 1.

THEOREM 7. Let u satisfy the inequality

$$L[u] \equiv \sum_{i,j=1}^{n} a_{ij} \frac{\partial^2 u}{\partial x_i \partial x_j} + \sum_{i=1}^{n} b_i \frac{\partial u}{\partial x_i} \geq 0$$

in a domain D in which L is uniformly elliptic. Suppose that $u \leq M$ in D and that $u = M$ at a boundary point P. Assume that P lies on the boundary of a ball K_1 in D. If u is continuous in $D \cup P$ and an outward directional derivative $\partial u / \partial \nu$ exists at P, then

$$\frac{\partial u}{\partial \nu} > 0 \text{ at } P$$

unless $u \equiv M$.

Proof. We proceed as in the proof of Theorem 5. By shrinking K_1 slightly if necessary, we may assume that ∂K_1 lies entirely in $D \cup P$. Construct a ball K_2 with center at P and radius $\frac{1}{2} r_1$ where r_1 is the radius of K_1. (See Fig. 6.) Define the function z according to (2), selecting α so large that $L[z] > 0$ in K_2. The function

$$w = u + \epsilon z$$

---

*This is a slight extension of the usual definition, in which ν is required to be a unit vector.

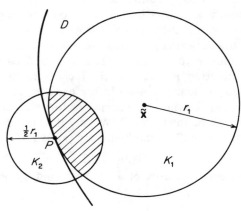

FIGURE 6

is now formed. According to Theorem 5, if $u \not\equiv M$, then $u < M$ in K_1 and on its boundary except at the point P. We recall that $z = 0$ on the boundary of K_1. We select $\epsilon > 0$ so small that $w \leq M$ on the portion of the boundary of K_2 lying in K_1. Then $w \leq M$ on the entire boundary of the shaded region shown in Fig. 6. Because $L[w] > 0$ in this region, the maximum of w occurs at P and $w(P) = M$. Therefore, at P

$$\frac{\partial w}{\partial \nu} = \frac{\partial u}{\partial \nu} + \epsilon \frac{\partial z}{\partial \nu} \geq 0.$$

We shall now show that $\partial z/\partial \nu < 0$ at P, so that $\partial u/\partial \nu > 0$ there. Selecting $\tilde{x}$ as the origin of our coordinate system and letting r represent Euclidean distance from $\tilde{x}$, we have

$$z = e^{-\alpha r^2} - e^{-\alpha r_1^2}.$$

Then

$$\frac{\partial z}{\partial x_i} = -2\alpha x_i e^{-\alpha r^2}$$

and

$$\eta_i = \frac{x_i}{r}.$$

Hence

$$\frac{\partial z}{\partial \nu} = -2\alpha r e^{-\alpha r^2} \sum_{i=1}^{n} \nu_i \eta_i < 0.$$

Therefore $\partial u/\partial \nu > 0$, establishing the conclusion of the theorem.

The proof of Theorem 7 also works for the operator $L + h$ with $h \leq 0$, provided $M \geq 0$ and α is chosen so large that $(L + h)[z] > 0$.

THEOREM 8. Let u satisfy the inequality

$$(L + h)[u] \geq 0,$$

where L is the operator of Theorem 7, and $h(\mathbf{x}) \leq 0$ in D. Suppose that $u \leq M$ in D, that $u = M$ at a boundary point P, and that $M \geq 0$. Assume that P lies on the boundary of a ball in D. If u is continuous in $D \cup P$, any outward directional derivative of u at P is positive unless $u \equiv M$ in D.

Remarks. (i) Two important outward directional derivatives are the **normal derivative** in which $\nu = \mathbf{n}$ and the **conormal derivative** in which $\nu_i = \sum_{j=1}^{n} a_{ij} \eta_j$.

(ii) It is possible to prove Theorem 7 before Theorem 5 if the assumption that $u < M$ in the interior of D is added to the hypotheses of Theorem 7. Then Theorem 5 may be derived as a consequence of Theorem 7. To see this, we need only apply Theorem 7 with the additional hypothesis to the ball K_1 in Figure 4 and its boundary point S where $u = M$. The observation that at an interior maximum the derivative in every direction must vanish contradicts the conclusion of Theorem 7. Thus u has no interior maximum, and Theorem 5 follows.

(iii) A relative maximum of u is an absolute maximum in some subdomain of D. Hence Theorem 5 shows that if $L[u] \geq 0$ in D and u has a relative maximum at an interior point P, then u is constant in a neighborhood of P. By Theorem 6, if $(L + h)[u] \geq 0$, if $h \leq 0$, and if u has a nonnegative relative maximum at P, then u is constant. If this constant value of u is positive, then $h = 0$ in some neighborhood of P.

(iv) The radius of the ball K_2 in the proof of Theorems 7 and 8 may be chosen arbitrarily small. Since the vector $\mathbf{x} - \tilde{\mathbf{x}}$ is then almost parallel to the normal vector at P, we can weaken the hypotheses by requiring only that the ratios $a_{ij}/\sum a_{kl}\eta_k\eta_l$, $b_i/\sum a_{kl}\eta_k\eta_l$, and $h/\sum a_{kl}\eta_k\eta_l$ are bounded in a neighborhood in D of P. The result is still valid if the operator L ceases to be elliptic on the boundary, provided the boundary is not a characteristic surface* of L.

(v) If the derivative $\partial u/\partial \nu$ does not exist at P, the proof of Theorem 7 still shows that**

$$\liminf_{\alpha \to 0+} \frac{[u(\mathbf{x}) - u(\mathbf{x} - \alpha\nu)]}{\alpha} > 0.$$

---

*This sentence may be skipped by readers unfamiliar with the notion of characteristic surface. A brief discussion of characteristic curves is given in Chapter 4.

**For the definition of lim inf see the footnote on page 94.

EXERCISES

1. Suppose that u satisfies $L[u] \geq 0$ in a bounded domain D with L given by (1), and suppose that v is a solution of $L[v] = 0$ in D. If $u \leq v$ on ∂D, prove that $u \leq v$ throughout D. Show that the statement is false if D is unbounded.

2. The function $u = [1 - (x^2 + y^2)]/[(1 - x)^2 + y^2]$ is harmonic and positive for $x^2 + y^2 < 1$. Since $u \equiv 0$ for $x^2 + y^2 = 1$ except at $(1, 0)$, is the maximum principle valid for u? Explain.

3. Show that a solution of

$$\frac{\partial^2 u}{\partial x^2} + \frac{\partial^2 u}{\partial y^2} - u^2 = 0$$

in a domain D cannot attain its maximum in D unless $u \equiv 0$.

4. Find the conormal derivative for the operator

$$L[u] \equiv (1 + x^2)\frac{\partial^2 u}{\partial x^2} - \frac{\partial^2 u}{\partial x \partial y} + (1 + y^2)\frac{\partial^2 u}{\partial y^2}$$

for each point of the boundary of D: $x^2 + y^2 < 1$.

SECTION 4. UNIQUENESS THEOREMS FOR BOUNDARY VALUE PROBLEMS

We begin with a study of one of the simplest boundary value problems for second-order partial differential equations. On a bounded, two-dimensional domain D with boundary ∂D, we pose the problem of determining a function $v(x, y)$ which is twice differentiable in D, is continuous on $D \cup \partial D$, and satisfies the equation

$$\Delta v \equiv \frac{\partial^2 v}{\partial x^2} + \frac{\partial^2 v}{\partial y^2} = f(x, y) \quad \text{in} \quad D \tag{1}$$

and the boundary condition

$$v = g(s) \quad \text{on} \quad \partial D. \tag{2}$$

Equation (1) is known as the **Poisson equation**. The function f is prescribed throughout D, and the function g, given in terms of the arc length s, is prescribed along ∂D.

The problem of determining such a solution v is known as the **Dirichlet problem**. It is also called the **first boundary value problem**.

We shall not investigate the conditions on D and those on f and g which are sufficient to guarantee the *existence* of a solution of the Dirichlet problem. However, by means of the maximum principle alone, it is possible to show that if a solution of the first boundary value does exist, then it must be *unique*. That is, we prove that **there can be at most one solution of the problem.**

To establish this result, we suppose that v_1 and v_2 are two functions which satisfy (1) and (2) with the same f and g. Defining

$$u = v_1 - v_2,$$

we see that u satisfies

$$\Delta u = 0 \text{ in } D,$$
$$u = 0 \text{ on } \partial D.$$

According to the maximum principle, u cannot have a maximum in the interior of D. However, the maximum of a continuous function on a closed bounded set must be attained. Since u is continuous on $D \cup \partial D$ and since $u = 0$ on ∂D, we conclude that $u \leq 0$ in D. Applying the same reasoning to $-u$, we obtain $u \geq 0$ in D. Hence

$$u = v_1 - v_2 \equiv 0 \text{ in } D.$$

It is essential to assume that the domain D is *bounded*. Otherwise, as the following example shows, the result is false. Let D be the infinite strip (Fig. 7) given by

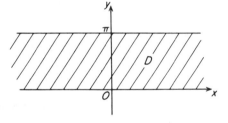

FIGURE 7

$$D: \begin{cases} -\infty < x < \infty \\ 0 \leq y \leq \pi. \end{cases}$$

The function

$$v = e^x \sin y$$

satisfies the Laplace equation in D and, as the relations

$$v(x, 0) = v(x, \pi) = 0$$

show, v vanishes on the boundary of D. Although the function v satisfies the maximum principle, it does not assume its maximum on the boundary. We note that

$$v\left(x, \frac{\pi}{2}\right) = e^x \to +\infty \text{ as } x \to +\infty.$$

To obtain a uniqueness theorem for unbounded domains, an additional condition on the behavior of v must be specified. For example, if the limit of v as $x^2 + y^2 \to \infty$ is prescribed, then the limit of $u = v_1 - v_2$ as $x^2 + y^2 \to \infty$ is zero, and we may deduce the uniqueness result. We will show in Section 9 that much weaker conditions at infinity insure the uniqueness result.

The proof of the uniqueness of the solution of the Dirichlet problem for the Laplace equation in any number of variables is identical with the above proof given in the two-dimensional case.

For an n-dimensional domain D with boundary ∂D, we now consider the problem of determining a function $v(\mathbf{x}) = v(x_1, x_2, \ldots, x_n)$ satisfying in D the equation

$$(L + h)[v] \equiv \sum_{i,j=1}^{n} a_{ij}(\mathbf{x}) \frac{\partial^2 v}{\partial x_i \partial x_j} + \sum_{i=1}^{n} b_i(\mathbf{x}) \frac{\partial v}{\partial x_i} + h(\mathbf{x})v = f \qquad (3)$$

and subject to the boundary conditions

$$\left.\begin{aligned}\frac{\partial v}{\partial \nu} + \alpha(\mathbf{x})v &= g_1 \text{ on } \Gamma_1, \\ v &= g_2 \text{ on } \Gamma_2,\end{aligned}\right\} \qquad (4)$$

where $\partial/\partial \nu$ is a given outward directional derivative at each point on Γ_1. We assume that the operator L is elliptic in D and that Γ_1 and Γ_2 are disjoint sets whose union comprises ∂D, the boundary of D. The sets Γ_1 and Γ_2 may consist of a number of separate pieces, and we do not exclude the possibility that either Γ_1 or Γ_2 is vacuous.

THEOREM 9. Suppose v_1 and v_2 satisfy (3) and (4) in a bounded domain D. Assume that each point of Γ_1 lies on the boundary of a ball in D. If L is uniformly elliptic, $h(\mathbf{x}) \leq 0$ is bounded, and $\alpha(\mathbf{x}) \geq 0$, then $v_1 \equiv v_2$, except when $h \equiv \alpha \equiv 0$ and Γ_2 is vacuous, in which case $v_1 - v_2$ must be constant.

Proof. Define $u = v_1 - v_2$. Then u satisfies

$$(L + h)[u] = 0 \text{ in } D, \qquad (5)$$

$$\left.\begin{aligned}\frac{\partial u}{\partial \nu} + \alpha(\mathbf{x})u &= 0 \text{ on } \Gamma_1, \\ u &= 0 \text{ on } \Gamma_2.\end{aligned}\right\} \qquad (6)$$

If u were ever positive, it would have a positive maximum. By Theorem 6 this maximum must occur at a point P on Γ_1. If u is not constant, $\partial u/\partial \nu > 0$ at P by Theorem 8, which contradicts the first condition in (6). Thus either u is constant or $u \leq 0$ in D. Applying the same argument to $(-u)$, we see that u must be constant.

But no constant other than 0 satisfies (5) and (6) unless $h \equiv \alpha \equiv 0$ and Γ_2 is vacuous, in which case any constant satisfies them.

When Γ_1 is vacuous, Theorem 9 establishes a uniqueness result for the Dirichlet problem. If Γ_2 is vacuous, $\alpha \equiv 0$, and $\partial/\partial \nu$ is the conormal derivative, we call the problem (3), (4), the **Neumann problem** or **second boundary value problem**. Theorem 9 establishes the uniqueness of the solution (up to an additive constant if $h \equiv 0$) for this problem. If α does not vanish identically, we have the **third boundary value problem**. If neither

Γ_1 nor Γ_2 is vacuous, we frequently say that the problem is one with **mixed boundary conditions.**

As in the case of the Laplace equation, a uniqueness theorem for equation (3) when the domain is unbounded requires an additional hypothesis on the behavior of the solutions as $r = (x_1^2 + x_2^2 + \cdots + x_n^2)^{1/2}$ tends to infinity.

As an application of Theorems 5, 7, and 9, we consider the steady state temperature distribution of a solid. Let $T(x, y, z)$ be the temperature at the point $P(x, y, z)$ of a homogeneous, isotropic medium occupying a bounded domain D with boundary surface ∂D. Suppose a part of the boundary surface, denoted Γ_1, is insulated so that no heat flows across it. The temperature distribution is prescribed on the remainder of the boundary, Γ_2. (Γ_1 and Γ_2 may each consist of several pieces; see Fig. 8.) If the temperature of the medium is in equilibrium (that is, if the temperature does not change with time), then T satisfies the equation

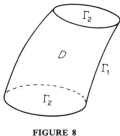

FIGURE 8

$$\Delta T = 0 \text{ in } D$$

and the boundary conditions

$$\frac{\partial T}{\partial \mathbf{n}} = 0 \text{ on } \Gamma_1$$

$$T \text{ prescribed on } \Gamma_2$$

($\partial/\partial \mathbf{n}$ is the outward normal derivative).

According to Theorems 5 and 7, neither the maximum nor the minimum temperature can occur in D or on Γ_1. Thus the maximum principle enables us to establish the result: **The equilibrium temperature always lies between the maximum and minimum values of the prescribed boundary temperature.** Note that the result holds even when the temperature is prescribed over only a portion of the boundary, provided there is no heat flow across the remainder. Since the heat flow into the body is proportional to $\partial T/\partial \mathbf{n}$, Theorem 7 asserts that an uneven temperature distribution can be maintained only by supplying heat at the warmest point on the boundary and removing heat from the coolest point.

The uniqueness of the equilibrium temperature distribution is assured by Theorem 9.

EXERCISES

1. Let D be the square $|x| < 1$, $|y| < 1$. Show that the vector ν with components (x, y) is outward pointing at each point of ∂D. Verify that any constant

satisfies $\Delta u = 0$ in D, $\partial u/\partial v = 0$ on ∂D, but that no constant other than zero satisfies $\Delta u = 0$ in D, $\partial u/\partial v + \alpha(\mathbf{x})u = 0$ on ∂D if $\alpha \not\equiv 0$.

2. Show that the problem $\Delta u = -1$ for $|x| < 1$, $|y| < 1$, $u = 0$ for $|x| = 1$, $(\partial u/\partial x) - (\partial u/\partial y) = 0$ for $|y| = 1$ has at most one solution.

3. Show that if the problem $\Delta u = 0$ for $|x| < 1$, $|y| < 1$, $u = 0$ for $|x| = 1$, $(\partial u/\partial x)^2 + (\partial u/\partial y)^2 = 1$ for $|y| = 1$ has any solution at all, then it has at least two solutions.

SECTION 5. THE GENERALIZED MAXIMUM PRINCIPLE

The condition $h(\mathbf{x}) \leq 0$ in Theorem 8 cannot be removed entirely. For example, the function

$$u = \sin x \sin y$$

is a solution of

$$\frac{\partial^2 u}{\partial x^2} + \frac{\partial^2 u}{\partial y^2} + 2u = 0$$

in the square $S: 0 < x < \pi$, $0 < y < \pi$, and u satisfies the boundary condition

$$u = 0 \text{ on } \partial S.$$

Clearly, u attains its maximum at an interior point.

As in Chapter 1, the methods used to prove a maximum principle with $h \leq 0$ can be extended to establish a generalized maximum principle. Uniqueness theorems for boundary value problems are then a direct consequence of this principle.

Suppose that $u(\mathbf{x})$ satisfies

$$(L + h)[u] \equiv \sum_{i,j=1}^{n} a_{ij}(\mathbf{x}) \frac{\partial^2 u}{\partial x_i \partial x_j} + \sum_{i=1}^{n} b_i(\mathbf{x}) \frac{\partial u}{\partial x_i} + h(\mathbf{x})u \geq 0 \quad (1)$$

in a domain D where L is uniformly elliptic. We do not assume that h is nonpositive. Let $w(\mathbf{x})$ be a given positive function on $D \cup \partial D$ and define

$$v(\mathbf{x}) = \frac{u(\mathbf{x})}{w(\mathbf{x})}.$$

A computation shows that

$$\frac{1}{w}(L+h)[u] = \sum_{i,j=1}^{n} a_{ij}(\mathbf{x}) \frac{\partial^2 v}{\partial x_i \partial x_j} + \sum_{i=1}^{n} \left\{ \frac{2}{w} \sum_{j=1}^{n} a_{ij} \frac{\partial w}{\partial x_j} + b_i \right\} \frac{\partial v}{\partial x_i}$$
$$+ \frac{1}{w}(L+h)[w]v \geq 0$$

in D. Then, if w satisfies the inequality

$$(L + h)[w] \leq 0$$

in D, we may apply the maximum principle as given in Theorems 6 and 8 to the function $v(\mathbf{x})$, obtaining thereby the following theorem.

THEOREM 10. Let $u(\mathbf{x})$ satisfy the differential inequality (1) in a domain D where L is uniformly elliptic. If there exists a function $w(\mathbf{x})$ such that

$$w(\mathbf{x}) > 0 \text{ on } D \cup \partial D, \tag{2}$$

$$(L + h)[w] \leq 0 \text{ in } D, \tag{3}$$

then $u(\mathbf{x})/w(\mathbf{x})$ cannot attain a nonnegative maximum in D unless it is a constant. If $u(\mathbf{x})/w(\mathbf{x})$ attains its nonnegative maximum at a point P on ∂D which lies on the boundary of a ball in D and if u/w is not constant, then

$$\frac{\partial}{\partial \nu}\left(\frac{u}{w}\right) > 0 \text{ at } P,$$

where $\partial/\partial \nu$ is any outward directional derivative.

If a function w with the prescribed properties exists and if $u = 0$ on ∂D, then Theorem 10 implies the uniqueness of solutions of the first boundary problem. We state this conclusion in the next theorem.

THEOREM 11. If there exists a function $w(\mathbf{x}) > 0$ on $D \cup \partial D$ such that $(L + h)[w] \leq 0$ in D and if D is bounded, then the problem

$$(L + h)[u] = f(\mathbf{x}) \text{ in } D$$

$$u = g(\mathbf{x}) \text{ on } \partial D$$

has at most one solution.

Of course, it is not always possible to find a function w satisfying the conditions stated in Theorems 10 and 11. In the example given at the beginning of this section, we see that the function $u = c \sin x \sin y$, where c is any constant, is a solution of $(\partial^2 u/\partial x^2) + (\partial^2 u/\partial y^2) + 2u = 0$ in the square $0 < x < \pi$, $0 < y < \pi$ which satisfies the condition $u = 0$ on the boundary. Hence there is no unique solution, and therefore no function w can exist for this problem.

We now give a specific method for determining a function $w(\mathbf{x})$ having properties (2) and (3) provided the domain D is contained in a sufficiently narrow slab bounded by two parallel hyperplanes. Suppose that the bounded domain D is contained in a slab $a < x_1 < b$, where x_1 is the first coordinate of $\mathbf{x} = (x_1, x_2, \ldots, x_n)$; set

$$w(\mathbf{x}) = 1 - \beta e^{\alpha(x_1 - a)}. \tag{4}$$

The numbers α and β are to be selected so that (2) and (3) hold. We compute
$$(L + h)[w] = -\beta[\alpha^2 a_{11}(\mathbf{x}) + \alpha b_1(\mathbf{x}) + h(\mathbf{x})]e^{\alpha(x_1-a)} + h(\mathbf{x}).$$
By the uniform ellipticity hypothesis, $a_{11}(\mathbf{x}) \geq \mu_0$. We suppose that $h(\mathbf{x})$ is bounded and that $b_1(\mathbf{x})$ is bounded from below; that is,
$$-m \leq h(\mathbf{x}) \leq M,$$
$$-m \leq b_1(\mathbf{x}),$$
where m and M are nonnegative. We choose α so large that
$$\alpha^2 \mu_0 - (\alpha + 1)m > 0.$$
Then we select
$$\beta = \frac{M}{\alpha^2 \mu_0 - (\alpha + 1)m}. \tag{5}$$
Under these circumstances,
$$(L + h)[w] \leq 0 \text{ on } D \cup \partial D.$$
However, to insure that $w > 0$ on $D \cup \partial D$, we must have
$$\beta e^{\alpha(b-a)} < 1.$$
That is, the inequality
$$M < [\alpha^2 \mu_0 - (\alpha + 1)m]e^{-\alpha(b-a)} \tag{6}$$
must be satisfied. We are still free to increase the size of α if we wish. We may choose α so that the right side of (6) is a maximum.* Notice that the right side becomes larger as $b - a$ becomes smaller. Also, condition (6) becomes less restrictive as M, the maximum of $h(\mathbf{x})$, becomes smaller. The maximum principle of Theorem 10 holds for $w(\mathbf{x})$ given by (4) whenever α and β satisfy (5) and (6). Therefore, given the operator $L + h$, there is always a number $b - a$ such that if $D \cup \partial D$ lies in a strip of width $b - a$, the first boundary problem has a unique solution. Alternatively, the condition (6) asserts that, for any bounded domain $D \cup \partial D$, if the positive maximum M of $h(\mathbf{x})$ is sufficiently small, then the solution of the first boundary value problem is unique.

Clearly, the width $b - a$ of the strip can be in any direction, as a rotation of coordinates makes this direction the x_1-direction.

The uniqueness of solutions of other boundary value problems may also be proved by means of the generalized maximum principle. Let u be a solution of

---

*The best possible value of α gives the inequality
$$M < (b-a)^{-2}[2\mu_0 + \{4\mu_0^2 + (m^2 + 4m\mu_0)(b-a)^2\}^{1/2}] \cdot$$
$$\exp\left[-\frac{1}{2\mu_0}(m(b-a) + 2\mu_0 + \{4\mu_0^2 + (m^2 + 4m\mu_0)(b-a)^2\}^{1/2})\right].$$

$$(L+h)[u] \equiv \sum_{i,j=1}^{n} a_{ij}(\mathbf{x})\frac{\partial^2 u}{\partial x_i \partial x_j} + \sum_{i=1}^{n} b_i(\mathbf{x})\frac{\partial u}{\partial x_i} + h(\mathbf{x})u = 0$$

in a domain D. We assume that u satisfies the boundary conditions

$$\frac{\partial u}{\partial \nu} + \alpha(\mathbf{x})u = 0 \text{ on } \Gamma_1,$$

$$u = 0 \text{ on } \Gamma_2.$$

where the boundary ∂D is composed of the parts Γ_1 and Γ_2, and where $\partial u/\partial \nu$ is any outward directional derivative. We again suppose that each point of Γ_1 lies on the boundary of a ball in D. Let $w(\mathbf{x})$ be positive on $D \cup \partial D$, and define

$$v(\mathbf{x}) = \frac{u(\mathbf{x})}{w(\mathbf{x})}.$$

Then v satisfies

$$\sum_{i,j=1}^{n} a_{ij}(\mathbf{x})\frac{\partial^2 v}{\partial x_i \partial x_j} + \sum_{i=1}^{n} \left\{ b_i + \frac{2}{w}\sum_{j=1}^{n} a_{ij}\frac{\partial w}{\partial x_j}\right\}\frac{\partial v}{\partial x_i} + \frac{1}{w}(L+h)[w]v = 0$$

in D; also, v satisfies the boundary conditions

$$\frac{\partial v}{\partial \nu} + \frac{1}{w}\left\{\frac{\partial w}{\partial \nu} + \alpha(\mathbf{x})w\right\}v = 0 \text{ on } \Gamma_1,$$

$$v = 0 \text{ on } \Gamma_2.$$

By Theorem 9 we find that, if $(L+h)[w] \leq 0$ in D and $(\partial w/\partial \nu) + \alpha(\mathbf{x})w \geq 0$ on Γ_1, then $v \equiv 0$ in D unless: (i) $(L+h)[w] \equiv 0$, (ii) $(\partial w/\partial \nu + \alpha(\mathbf{x})w \equiv 0$ on Γ_1, and (iii) Γ_2 is vacuous, in which case v may be any constant.

The above argument establishes the following general uniqueness theorem.

THEOREM 12. Suppose there exists a function $w(\mathbf{x}) > 0$ on $D \cup \partial D$ such that

$$(L+h)[w] \leq 0 \text{ in } D$$

and

$$\frac{\partial w}{\partial \nu} + \alpha(\mathbf{x})w \geq 0 \text{ on } \Gamma_1,$$

where ∂D is composed of two parts Γ_1 and Γ_2. Suppose that each point of Γ_1 lies on the boundary of a ball in D, and that D is bounded. Then there is at most one solution $u(\mathbf{x})$ of the problem

$$(L+h)[u] = f(\mathbf{x}) \text{ in } D,$$

$$\frac{\partial u}{\partial \nu} + \alpha(\mathbf{x})u = g_1(\mathbf{x}) \text{ on } \Gamma_1,$$

$$u = g_2(\mathbf{x}) \text{ on } \Gamma_2,$$

unless (i) $(L + h)[w] \equiv 0$, (ii) $(\partial w/\partial \nu) + \alpha(\mathbf{x})w \equiv 0$ **on** Γ_1, **and** (iii) Γ_2 **is vacuous, in which case u is determined to within a constant multiple of w.**

Theorem 12 follows from the fact that if z_1 and z_2 are two solutions, then $(z_1 - z_2)/w$ satisfies the maximum principle. Note that Theorem 12 contains Theorem 11 as a special case (when Γ_1 is vacuous).

The existence of $w(\mathbf{x})$ satisfying the required conditions means that no special assumptions are needed on $h(\mathbf{x})$ and $\alpha(\mathbf{x})$.

EXERCISES

1. Let $(L + h)[u] \equiv (\partial^2 u/\partial x^2) + (\partial^2 u/\partial y^2) + 2u$, and let D be the square: $|x| < b$, $|y| < b$. If $w(x) = 1 - \beta e^{\alpha(x+b_1)}$, find values for $\alpha, \beta, b,$ and b_1 such that if $(L + h)[u] \geq 0$, then u/w satisfies a maximum principle. What are the values if $w(x, y) = 1 - \beta e^{\alpha[(x+b_1)^2+(y+b_1)^2]}$?

2. Let $(L + h)[u] \equiv (\partial^2 u/\partial x^2) + (\partial^2 u/\partial y^2) + yu$, and let D_γ be the domain: $\{0 < y < \gamma e^{-x^2}, -\infty < x < \infty\}$. Find a function w and a number γ such that if $(L + h)[u] \geq 0$, then u/w satisfies a maximum principle in D_γ.

SECTION 6. APPROXIMATION IN BOUNDARY VALUE PROBLEMS

Let $u(\mathbf{x})$ be a solution of

$$(L + h)[u] = f(\mathbf{x}) \text{ in } D \tag{1}$$

which satisfies the boundary conditions

$$\left.\begin{aligned} \frac{\partial u}{\partial \nu} + \alpha(\mathbf{x})u &= g_1(\mathbf{x}) \text{ on } \Gamma_1, \\ u &= g_2(\mathbf{x}) \text{ on } \Gamma_2. \end{aligned}\right\} \tag{2}$$

As usual, we assume that L is uniformly elliptic, that $\Gamma_1 \cup \Gamma_2$ comprises the boundary ∂D, and that $\partial u/\partial \nu$ is an outward directional derivative at each point of Γ_1. Moreover, each point of Γ_1 lies on the boundary of a ball contained in D, and D is bounded.

The maximum principle can be used to obtain bounds for solutions of (1), (2). The method we employ is a natural extension of that used in Section 5 of Chapter 1 for solutions of ordinary differential equations.

We suppose that L, h, and D are such that there exists a function $w(\mathbf{x})$, positive on $D \cup \partial D$ with the properties:

$$(L + h)[w] \leq 0 \text{ in } D,$$

$$\frac{\partial w}{\partial \nu} + \alpha(\mathbf{x})w \geq 0 \text{ on } \Gamma_1.$$

(If $h(\mathbf{x}) \leq 0$ and $\alpha(\mathbf{x}) \geq 0$, then $w(\mathbf{x}) \equiv 1$ has the desired properties.)

Now we assume that a function $z_1(\mathbf{x})$ can be found which satisfies the inequalities:

$$(L + h)[z_1] \leq f(\mathbf{x}) \text{ in } D, \tag{3}$$

$$\left.\begin{array}{r}\dfrac{\partial z_1}{\partial \nu} + \alpha(\mathbf{x})z_1 \geq g_1(\mathbf{x}) \text{ on } \Gamma_1, \\[4pt] z_1 \geq g_2(\mathbf{x}) \text{ on } \Gamma_2.\end{array}\right\} \tag{4}$$

Under such circumstances, the function

$$v = \frac{u - z_1}{w}$$

satisfies the three inequalities:

$$\frac{1}{w}(L + h)[vw] = \sum_{i,j=1}^{n} a_{ij}(\mathbf{x})\frac{\partial^2 v}{\partial x_i \partial x_j} + \sum_{i=1}^{n}\left\{b_i + \frac{2}{w}\sum_{j=1}^{n} a_{ij}\frac{\partial w}{\partial x_j}\right\}\frac{\partial v}{\partial x_i}$$

$$+ (L + h)[w]v \geq 0 \text{ in } D,$$

$$\frac{\partial v}{\partial \nu} + \frac{1}{w}\left\{\frac{\partial w}{\partial \nu} + \alpha w\right\}v \leq 0 \text{ on } \Gamma_1,$$

$$v \leq 0 \text{ on } \Gamma_2.$$

According to Theorem 6, if v is not identically constant, it can only have a nonnegative maximum at a point on the boundary. By Theorem 8, v cannot have a nonnegative maximum on Γ_1 unless it is a constant. If Γ_2 is vacuous, if $(\partial w/\partial \nu) + \alpha w \equiv 0$ on Γ_1, and if $(L + h)[w] \equiv 0$ in D, then $v \equiv$ constant satisfies all the required conditions. In all other cases we conclude that $v \leq 0$, so that

$$u(\mathbf{x}) \leq z_1(\mathbf{x}) \text{ in } D.$$

In other words, a function $z_1(\mathbf{x})$ satisfying (3) and (4) provides an upper bound for u.

To obtain a lower bound we suppose that a function $z_2(\mathbf{x})$ can be found which satisfies the inequalities:

$$(L + h)[z_2] \geq f(\mathbf{x}) \text{ in } D, \tag{5}$$

$$\left.\begin{array}{r}\dfrac{\partial z_2}{\partial \nu} + \alpha(\mathbf{x})z_2 \leq g_1(\mathbf{x}) \text{ on } \Gamma_1, \\[4pt] z_2 \leq g_2(\mathbf{x}) \text{ on } \Gamma_2.\end{array}\right\} \tag{6}$$

By defining $v = (z_2 - u)/w$ and reasoning in the same way as we did for z_1, we find

$$z_2(\mathbf{x}) \leq u(\mathbf{x}) \text{ in } D.$$

THEOREM 13. Suppose there exists a function $w > 0$ on $D \cup \partial D$ such that

$$(L + h)[w] \leq 0 \text{ in } D,$$

$$\frac{\partial w}{\partial \nu} + \alpha(\mathbf{x})w \geq 0 \text{ on } \Gamma_1,$$

where L is uniformly elliptic and $(\partial u/\partial \nu)$ is an outward directional derivative. We assume that the three conditions (i) $(\partial w/\partial \nu) + \alpha(\mathbf{x})w \equiv 0$ on Γ_1, (ii) $(L + h)[w] \equiv 0$ in D, and (iii) Γ_2 is vacuous, do not all hold. If $z_1(\mathbf{x})$ and $z_2(\mathbf{x})$ satisfy (3), (4) and (5), (6) respectively, then the solution u of the problem (1), (2) satisfies the inequalities

$$z_2(\mathbf{x}) \leq u(\mathbf{x}) \leq z_1(\mathbf{x}) \text{ in } D.$$

Remark. Theorem 12 is an immediate consequence of Theorem 13. If $\bar{u}$ and u are both solutions of (1), (2), we can put $z_1 = z_2 = \bar{u}$ and apply Theorem 13 to find that $\bar{u} \leq u \leq \bar{u}$, so that $u \equiv \bar{u}$.

We now show how Theorem 13 may be used to obtain a numerical bound in a particular case.

Example. Find bounds for the solution of

$$\frac{\partial^2 u}{\partial x^2} + \frac{\partial^2 u}{\partial y^2} = 0$$

in the square

$$S: \begin{cases} 0 < x < 1 \\ 0 < y < 1 \end{cases}$$

which satisfies the boundary conditions (Fig. 9)

$u = 1$ on the segment l_1: $0 < x < 1$, $y = 0$,

$u = 1$ on the segment l_2: $0 < y < 1$, $x = 0$,

$$\frac{\partial u}{\partial x} + u = 0$$

on the segment l_3: $0 < y < 1$, $x = 1$,

$$\frac{\partial u}{\partial y} = 0$$

on the segment l_4: $0 < x < 1$, $y = 1$.

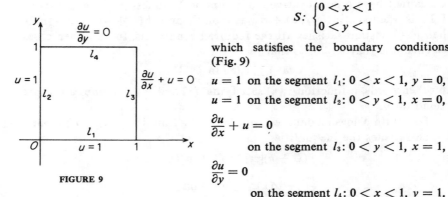

FIGURE 9

It is easily verified that $z_1 \equiv 1$ satisfies (3), (4) and $z_2 = e^{-x}$ satisfies (5), (6). Moreover, the function $w \equiv 1$ satisfies the conditions of Theorem 13. Therefore,

$$e^{-x} \leq u(x, y) \leq 1 \text{ in } S.$$

As in the case of one dimension, we can eliminate the extraneous function w from Theorem 13. If we have a function z_1 satisfying (3), (4) and a function z_2 satisfying (5), (6) such that not all the inequalities given by (3), (4), (5), and (6) are identities, and if $z_1 \geq z_2$, then we can construct a function w such that Theorem 13 holds.

If $z_1 > z_2$, we can take $w = z_1 - z_2$. If $q \equiv z_1 - z_2 \geq 0$, then q cannot vanish at interior points or on Γ_1. If $q = 0$ on a part of Γ_2, we may select $w = q + \epsilon r$, where r is the solution of the problem

$$(L + h)[r] = 0 \text{ in } D,$$

$$\frac{\partial r}{\partial \nu} + \alpha r = 0 \text{ on } \Gamma_1,$$

$$r = 1 \text{ on } \Gamma_2,$$

and ϵ is chosen so small that $w > 0$ on $D \cup \partial D$. The existence of the function r follows if we assume that the problem (1), (2) can be solved for arbitrary continuous boundary values $g_2(\mathbf{x})$.

THEOREM 14. *Let z_1 and z_2 satisfy the conditions (3), (4) and (5), (6), respectively, in such a way that identity does not hold in all of them. If the problem (1), (2) has solutions for arbitrary continuous boundary values $g_2(\mathbf{x})$, and if u is the solution of the particular problem (1), (2), then the bounds*

$$z_2 \leq u \leq z_1$$

are valid if and only if $z_1 \geq z_2$.

Suppose now that there exists a positive function w which satisfies the inequalities

$$\left.\begin{array}{c} (L + h)[w] \leq -1 \text{ in } D \\ \dfrac{\partial w}{\partial \nu} + \alpha(\mathbf{x})w \geq 1 \text{ on } \Gamma_1 \\ w \geq 1 \text{ on } \Gamma_2, \end{array}\right\} \quad (7)$$

which are slightly stronger than those required for Theorem 13. If u is the solution of the problem (1), (2), and if we define

$$A = \max \left\{ \sup_D |f(\mathbf{x})|, \sup_{\Gamma_1} |g_1(\mathbf{x})|, \sup_{\Gamma_2} |g_2(\mathbf{x})| \right\}$$

then Theorem 13 shows that

$$|u(\mathbf{x})| \leq A w(\mathbf{x}).$$

If $\bar{u}$ is the solution of the problem

$$(L + h)[\bar{u}] = \bar{f}(\mathbf{x}) \text{ in } D$$

$$\frac{\partial \bar{u}}{\partial \nu} + \alpha(\mathbf{x})\bar{u} = \bar{g}_1(\mathbf{x}) \text{ on } \Gamma_1$$

$$\bar{u} = \bar{g}_2(\mathbf{x}) \text{ on } \Gamma_2,$$

the above argument shows that

$$|\bar{u}(\mathbf{x}) - u(\mathbf{x})| \leq w(\mathbf{x}) \max\left\{\sup_D|\bar{f} - f|, \sup_{\Gamma_1}|\bar{g}_1 - g_1|, \sup_{\Gamma_2}|\bar{g}_2 - g_2|\right\} \quad (8)$$

throughout D. In particular, if $\bar{f} - f$, $\bar{g}_1 - g_1$, and $\bar{g}_2 - g_2$ are uniformly small, then $\bar{u} - u$ is uniformly small. In other words, the solution of the problem (1), (2) depends continuously on the data.

If Theorem 13 holds and if the problem (1), (2) can be solved for arbitrary continuous data, the solution of the problem

$$(L + h)[w] = -1 \text{ in } D$$

$$\frac{\partial w}{\partial \nu} + \alpha(\mathbf{x})w = 1 \text{ on } \Gamma_1$$

$$w = 1 \text{ on } \Gamma_2$$

satisfies the inequalities (7). Thus we see that **the solution of the problem (1), (2) depends continuously on the data whenever the problem can be solved for arbitrary continuous data and Theorem 13 holds.**

In order to use the inequality (8) to bound the error made in approximating u by $\bar{u}$ it is, of course, necessary to find an explicit function w which satisfies the inequalities (7).

EXERCISES

1. Let D be the square $|x| < 1$, $|y| < 1$, and let u be the solution of

$$\frac{\partial^2 u}{\partial x^2} + \frac{\partial^2 u}{\partial y^2} = 0 \text{ in } D$$

$$u = |x| \text{ for } |y| = 1, \quad u = |y| \text{ for } |x| = 1.$$

Select a harmonic polynomial of the second degree in x and y whose boundary values approximate those of u on ∂D, and thereby obtain bounds for $u(0, 0)$.

2. Let D be the unit disk: $x^2 + y^2 < 1$ and let u be the solution of the problem

$$\frac{\partial^2 u}{\partial x^2} + \frac{\partial^2 u}{\partial y^2} = 0 \text{ in } D$$

$u = 1 \quad \text{for } r = 1,\ 0 \leq \theta \leq \frac{\pi}{2}$ (polar coordinates),

$u = \left(2 - \frac{2}{\pi}\theta\right) \quad \text{for } r = 1,\ \frac{\pi}{2} \leq \theta \leq \pi,$

$u = 0 \quad \text{for } r = 1,\ \pi \leq \theta \leq \frac{3\pi}{2},$

$u = \left(\frac{2}{\pi}\theta - 3\right) \quad \text{for } r = 1,\ \frac{3\pi}{2} \leq \theta \leq 2\pi.$

Let z be the harmonic function $\sum_{k=0}^{2} r^k(a_k \cos k\theta + b_k \sin k\theta)$. Select the quantities $a_k, b_k, k = 0, 1, 2$ so as to approximate the boundary values of u and thereby obtain bounds for u at the point $x = \frac{1}{2}, y = 0$. On the basis of this result, what can be said about the bounds at the point $x = 0, y = \frac{1}{2}$?

SECTION 7. GREEN'S IDENTITIES AND GREEN'S FUNCTION

Let D be a bounded domain in three-space with a piecewise smooth* boundary ∂D. Suppose that $\mathbf{w}$ is a smooth vector field defined in an open set containing $D \cup \partial D$ and that $\mathbf{n}$ is the outward unit normal vector to the boundary surface ∂D. The **divergence theorem** states that

$$\int_D \text{div } \mathbf{w} \, dV = \int_{\partial D} \mathbf{w} \cdot \mathbf{n} \, dS, \tag{1}$$

where dV is the volume element in D and dS is an element of surface on ∂D. Let u and v be scalar functions defined on $D \cup \partial D$, sufficiently smooth so that the operations of differentiation and integration to be applied are always valid. We select $\mathbf{w} = v \text{ grad } u$. Then formula (1) becomes

$$\int_D [v \text{ div grad } u + \text{grad } v \cdot \text{grad } u] dV = \int_{\partial D} v \text{ grad } u \cdot \mathbf{n} \, dS. \tag{2}$$

We use the identity

$$\Delta u = \text{div grad } u$$

and the notation $\partial u / \partial \mathbf{n} = \text{grad } u \cdot \mathbf{n}$ to write (2) in the form

$$\int_D v \Delta u \, dV + \int_D \text{grad } v \cdot \text{grad } u \, dV = \int_{\partial D} v \frac{\partial u}{\partial \mathbf{n}} dS. \tag{3}$$

Equation (3) is known as **Green's first identity**. Interchanging u and v in (3), we get

$$\int_D u \Delta v \, dV + \int_D \text{grad } u \cdot \text{grad } v \, dV = \int_{\partial D} u \frac{\partial v}{\partial \mathbf{n}} dS.$$

Subtracting this equation from (3), we obtain **Green's second identity**

---

*We say that a boundary surface ∂D is *piecewise smooth* if it is composed of a finite number of pieces on each of which one of the coordinates representing the piece of surface can be expressed as a continuously differentiable function of the other two coordinates. Such a surface can have a finite number of edges and cusps which connect the smooth pieces.

$$\int_D (v\,\Delta u - u\,\Delta v)\,dV = \int_{\partial D} \left(v\,\frac{\partial u}{\partial \mathbf{n}} - u\,\frac{\partial v}{\partial \mathbf{n}}\right) dS. \tag{4}$$

With the aid of the two identities of Green, we deduce immediately several interesting facts.

THEOREM 15. If u is harmonic in a bounded domain D and continuously differentiable on the closure $D \cup \partial D$ then

$$\int_{\partial D} \frac{\partial u}{\partial \mathbf{n}}\,dS = 0, \tag{5}$$

where $\partial/\partial \mathbf{n}$ is the derivative in the direction normal to ∂D.

Proof. Setting $v \equiv 1$ in (3), we observe that grad $v \equiv 0$, and hence (5) follows directly.

If u is a harmonic function, then Green's first identity yields uniqueness theorems similar to those obtained in Section 4. For example, if we select $v \equiv u$, then (3) becomes (with u harmonic)

$$\int_D |\text{grad }u|^2\,dV = \int_{\partial D} u\,\frac{\partial u}{\partial \mathbf{n}}\,dS.$$

If, in addition, u vanishes on ∂D, then the right hand side is zero. Since $|\text{grad }u|^2$ is always nonnegative, it must vanish throughout D. Hence $u \equiv 0$ in D. The same reasoning shows that if $\partial u/\partial \mathbf{n} \equiv 0$ on ∂D, then u must be constant throughout D. Such results were already obtained by use of the maximum principle. While Green's identities yield uniqueness theorems simply and quickly, the comparable results obtained by use of the maximum principle impose less restrictive hypotheses on the function and the domain.

By means of Green's second identity (4), we shall now obtain an important representation formula for twice continuously differentiable functions defined in a domain D in three-dimensional Euclidean space. We consider two points $P(x, y, z)$, $Q(\xi, \eta, \zeta)$ and denote the Euclidean distance between them by r_{PQ}; that is,

$$r_{PQ}^2 = (x - \xi)^2 + (y - \eta)^2 + (z - \zeta)^2.$$

A simple computation shows that

$$\Delta\left(\frac{1}{r_{PQ}}\right) = 0 \text{ whenever } P \neq Q.$$

In computing the Laplacian of r_{PQ}^{-1}, we assume that Q is fixed and that differentiation is taken with respect to x, y, and z.

Let Q be a fixed point in a bounded domain D, and suppose $\psi(x, y, z)$ is harmonic throughout D. We define the function

$$W(x, y, z) = \frac{1}{4\pi r_{PQ}} + \psi(x, y, z),$$

which is harmonic in D except at $P = Q$. We would like to apply Green's second identity to the function W. Since W is singular at Q, we exclude this point by forming the domain $D - D_\rho$ where D_ρ is the ball of radius ρ which has its center at Q. (See Fig. 10.) Now employing (4) in the domain $D - D_\rho$ with $v = W$ and u a given twice continuously differentiable function, we find

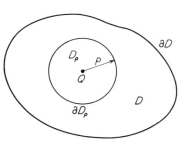

FIGURE 10

$$\int_{D-D_\rho} W \Delta u \, dV = \int_{\partial(D-D_\rho)} \left(W \frac{\partial u}{\partial \mathbf{n}} - u \frac{\partial W}{\partial \mathbf{n}} \right) dS.$$

The boundary of D_ρ is the sphere of radius ρ and center at Q. Therefore the normal derivative on D_ρ pointing outward from $D - D_\rho$ is the *inner* radial derivative. We obtain

$$\int_{D-D_\rho} W \Delta u \, dV = \int_{\partial D} \left(W \frac{\partial u}{\partial \mathbf{n}} - u \frac{\partial W}{\partial \mathbf{n}} \right) dS - \int_{\partial D_\rho} \left(W \frac{\partial u}{\partial r} - u \frac{\partial W}{\partial r} \right) dS. \quad (6)$$

We consider first the integral over ∂D_ρ in (6). Inserting the value for W and using spherical coordinates centered at Q, we get

$$\int_{\partial D_\rho} \left(W \frac{\partial u}{\partial r} - u \frac{\partial W}{\partial r} \right) dS = \frac{1}{4\pi} \int_{r=\rho} \left(\frac{1}{r} \frac{\partial u}{\partial r} - \frac{u}{r^2} \right) r^2 \sin\theta \, d\theta \, d\phi$$

$$+ \int_{\partial D_\rho} \left(\psi \frac{\partial u}{\partial r} - u \frac{\partial \psi}{\partial r} \right) dS. \quad (7)$$

The last integral on the right in (7) may be transformed by Green's second identity applied to the ball D_ρ. Since ψ is harmonic, we obtain

$$\int_{\partial D_\rho} \left(\psi \frac{\partial u}{\partial r} - u \frac{\partial \psi}{\partial r} \right) dS = \int_{D_\rho} \psi \Delta u \, dV. \quad (8)$$

In all the integrals above, we are interested in their behavior as the radius ρ of the ball D_ρ tends to zero. Because ψ and Δu are bounded in D_ρ, we conclude from (8) that

$$\lim_{\rho \to 0} \int_{\partial D_\rho} \left(\psi \frac{\partial u}{\partial r} - u \frac{\partial \psi}{\partial r} \right) dS = 0. \quad (9)$$

We consider separately the two parts of the first integral on the right side of (7). Since $|\text{grad } u|$ is bounded throughout, say by M, we have

$$\left|\frac{1}{4\pi}\int_{r=\rho} r\frac{\partial u}{\partial r}\sin\theta\,d\theta\,d\phi\right|\leq M\rho.$$

Hence

$$\lim_{\rho\to 0}\frac{1}{4\pi}\int_{r=\rho}\frac{1}{r}\frac{\partial u}{\partial r}\cdot r^2\sin\theta\,d\theta\,d\phi=0. \tag{10}$$

Since the point Q is fixed, we may write

$$u(Q)=\frac{1}{4\pi}\int_{r=\rho}u(Q)\sin\theta\,d\theta\,d\phi$$

and, since u is continuous at Q,

$$\lim_{\rho\to 0}\frac{1}{4\pi}\int_{r=\rho}|u(P)-u(Q)|\sin\theta\,d\theta\,d\phi=0.$$

We conclude that

$$u(Q)=\lim_{\rho\to 0}\frac{1}{4\pi}\int_{r=\rho}u(P)\sin\theta\,d\theta\,d\phi. \tag{11}$$

Now we let ρ tend to zero in (6) and make use of (7), (9), (10), and (11). We obtain the representation formula known as **Green's third identity**:

$$u(Q)=\int_{\partial D}\left(W\frac{\partial u}{\partial \mathbf{n}}-u\frac{\partial W}{\partial \mathbf{n}}\right)dS-\int_D W\,\Delta u\,dV. \tag{12}$$

We observe that this formula has a certain arbitrariness in it since the harmonic function ψ is still at our disposal. Usually, Green's third identity is stated with $\psi\equiv 0$, in which case $W=1/4\pi r_{PQ}$.

The representation formula (12) can be used to solve the first boundary value problem for the Laplace operator. Let u be the solution of

$$\Delta u=f(x,y,z)\text{ in }D \tag{13}$$

which satisfies the boundary condition

$$u=g(x,y,z)\text{ on }\partial D. \tag{14}$$

Then formula (12) becomes

$$u(\xi,\eta,\zeta)=\int_{\partial D}\left(W\frac{\partial u}{\partial \mathbf{n}}-g\frac{\partial W}{\partial \mathbf{n}}\right)dS-\int_D Wf\,dV. \tag{15}$$

The right-hand side of (15) is known except for the term containing $\partial u/\partial \mathbf{n}$. In order to eliminate this term we make use of the harmonic function ψ in the expression for W, a function which is still at our disposal. We select ψ so that W vanishes on ∂D; that is, we solve the problem

$$\left.\begin{array}{l}\Delta\psi=0\text{ in }D,\\ \psi=-\dfrac{1}{4\pi r_{PQ}}\text{ on }\partial D.\end{array}\right\} \tag{16}$$

With this selection for ψ (which, we note, depends on Q as a parameter) we are led to the following definition.

DEFINITION. Let D be a bounded domain and Q a fixed point of D. The function $(1/4\pi r_{PQ}) + \psi(P)$ with the properties: (i) ψ is harmonic throughout D, and (ii) $(1/4\pi r_{PQ}) + \psi$ vanishes on ∂D is called the **Green's function** of D with respect to the Laplace equation.

We denote the Green's function by $G(x, y, z; \xi, \eta, \zeta)$, or $G(P; Q)$. When the Green's function is used for W in (15), we find

$$u(\xi, \eta, \zeta) = -\int_{\partial D} g \frac{\partial G}{\partial \mathbf{n}_P} dS_P - \int_D Gf \, dV_P.$$

The subscript P in the above formula indicates that differentiation and integration are carried out with respect to the point P with coordinates (x, y, z).

It is natural to ask whether or not a Green's function exists for all domains and, if so, whether it can be found explicitly. The existence of a Green's function is equivalent to the solvability of problem (16). Although it can be shown that the Green's function actually exists for a wide class of domains, there are only a few special domains for which an explicit formula for it is known. The primary interest in the Green's function results from the information it yields about properties of solutions of the first boundary value problem.

From the definition of the Green's function we know that

$$G(P; Q) \to +\infty \text{ as } P \to Q.$$

Therefore, if D_ρ is a sufficiently small ball with center at Q, the Green's function for a domain D containing D_ρ will be positive on ∂D_ρ. Since G is harmonic in $D - D_\rho$, positive on ∂D_ρ, and equal to 0 on ∂D, it follows from the maximum principle that G is positive throughout $D - D_\rho$. Hence $G > 0$ in D except at Q, where it is undefined. The minimum of G is zero and it occurs at every point of ∂D. By Theorem 7, we find that the outward normal derivative of G is negative at each boundary point P which lies on the boundary of some ball in D. Thus as a direct consequence of the maximum principle, we conclude that for any bounded domain in which the Green's function is defined,

$$G > 0 \text{ in } D,$$

$$\frac{\partial G}{\partial \mathbf{n}} < 0 \text{ on } \partial D.$$

We now take up the more general boundary value problem

$$\Delta u = f(x, y, z) \text{ in } D, \tag{17}$$

$$\left. \begin{array}{l} \dfrac{\partial u}{\partial \mathbf{n}} + \alpha(x, y, z)u = g_1(x, y, z) \text{ on } \Gamma_1, \\ u = g_2(x, y, z) \text{ on } \Gamma_2, \end{array} \right\} \quad (18)$$

where $\alpha \geq 0$ and $\Gamma_1 \cup \Gamma_2 = \partial D$.

As was the case for the boundary value problem (13), (14), we can define the Green's function* for the boundary value problem (17), (18). We do this by solving a problem of the type (17), (18) for a function ψ. We determine $\psi(x, y, z)$ so that

$$\Delta \psi = 0 \text{ in } D,$$

$$\frac{\partial \psi}{\partial \mathbf{n}} + \alpha \psi = -\frac{1}{4\pi}\left[\frac{\partial}{\partial \mathbf{n}}\left(\frac{1}{r_{PQ}}\right) + \alpha\left(\frac{1}{r_{PQ}}\right)\right] \text{ on } \Gamma_1,$$

$$\psi = -\frac{1}{4\pi}\left(\frac{1}{r_{PQ}}\right) \text{ on } \Gamma_2.$$

If we can find a function ψ satisfying the above conditions, we then define

$$G(P; Q) = \frac{1}{4\pi r_{PQ}} + \psi$$

as the **Green's function for the problem** (17), (18). Choosing this Green's function for W in Green's third identity (12), we get the solution formula for (17), (18):

$$u(\xi, \eta, \zeta) = -\int_D G(P; Q)f(P)\,dV_P + \int_{\Gamma_1} G(P, Q)g_1(P)\,dS_P - \int_{\Gamma_2} g_2(P)\frac{\partial G}{\partial \mathbf{n}_P}\,dS_P.$$
(19)

Here again the maximum principle may be used to show that the problem (17), (18) satisfies

$$\left. \begin{array}{l} G > 0 \text{ in } D \cup \Gamma_1, \\ \dfrac{\partial G}{\partial \mathbf{n}} < 0 \text{ on } \Gamma_2. \end{array} \right\} \quad (20)$$

Conversely, the above properties of the Green's function allow us to read off properties of the solution u of (17), (18) from formula (19). For example, if $f \leq 0$ in D, $g_1 \geq 0$ on Γ_1, and $g_2 \geq 0$ on Γ_2, then (19) and (20) show that $u \geq 0$ in D. Moreover, if f, g_1, and g_2 are not all identically zero, then $u > 0$ in D.

---

*The Green's function depends not only on the domain D and on the differential operator but also on the boundary conditions imposed. Until now we considered only the condition $G = 0$ on ∂D, corresponding to the first boundary value problem. However, each type of boundary condition yields a Green's function. For some special types of boundary conditions the corresponding functions are frequently associated with the names of Neumann and Robin.

Sec. 7 Green's Identities and Green's Function

The preceding results for harmonic functions can be extended to solutions of general second order elliptic equations in any number of independent variables. We consider the problem of finding a function u which satisfies

$$(L + h)[u] \equiv \sum_{i,j=1}^{n} a_{ij}(\mathbf{x}) \frac{\partial^2 u}{\partial x_i \partial x_j} + \sum_{i=1}^{n} b_i(\mathbf{x}) \frac{\partial u}{\partial x_i} + h(\mathbf{x})u = f(\mathbf{x}) \quad (21)$$

in a bounded domain D and which fulfills the boundary conditions

$$\left. \begin{array}{c} \dfrac{\partial u}{\partial \nu} + \alpha(\mathbf{x})u = g_1(\mathbf{x}) \text{ on } \Gamma_1, \\[2mm] u = g_2(\mathbf{x}) \text{ on } \Gamma_2. \end{array} \right\} \quad (22)$$

We suppose that $\Gamma_1 \cup \Gamma_2 = D$, and that each point of Γ_1 is on the boundary of a ball in D. The derivative $\partial/\partial \nu$ is the conormal derivative, which is defined by

$$\frac{\partial u}{\partial \nu} = \sum_{i,j=1}^{n} a_{ij} \eta_j \frac{\partial u}{\partial x_i}.$$

If, in particular, $L = \Delta$ and $h(\mathbf{x}) \equiv 0$, then the extension of the results of this section to n independent variables, $n > 2$, is immediate. Because of the invariant character of their derivation, we note that Green's first and second identities are valid in any number of dimensions without change. To obtain the Green's function for the Laplace operator in n dimensions, $n > 3$, we replace $(1/4\pi r_{PQ})$, employed when $n = 3$, by the function $(1/\omega_n r_{PQ}^{n-2})$ where ω_n is the surface area of the unit sphere in n dimensions and r_{PQ} is the Euclidean distance from P to Q in n-space. A computation shows that $\Delta(1/r_{PQ}^{n-2}) = 0$ for $P \neq Q$, and the remainder of the derivation of Green's third identity parallels the case of $n = 3$. In two dimensions a special consideration is required, and we employ the function $(1/2\pi) \log(1/r_{PQ})$ in place of $(1/\omega_n r_{PQ}^{n-2})$.

For a general operator L, we define the Green's function G for the problem (21), (22) by the properties:

(i) $G(P; Q)r_{PQ}^{n-2}$ is a bounded function of Q and has a positive lower bound for Q near P.

(ii) $(L_Q + h(Q))[G(P; Q)] = 0$ in D for $Q \neq P$. The notation L_Q means that we apply the operator L to the coordinates $(\xi_1, \xi_2, \ldots, \xi_n)$ of Q in $G(P; Q)$ and keep $P(x_1, x_2, \ldots, x_n)$ fixed.

(iii) $\partial G/\partial \nu_Q + \alpha(Q)G(P; Q) = 0$ for each Q on Γ_1 and for each fixed point P in D. Differentiation is with respect to ξ.

(iv) $G(P; Q) = 0$ for Q on Γ_2 and for each P in D.

The Green's function with properties (i)–(iv) can be shown to exist for an operator L in a domain D if the coefficients of L and the boundary of D are sufficiently smooth, and if the problem (21), (22) has a unique solution for arbitrary data. Under these circumstances, the Green's identities may be derived and a solution $u(\xi)$ of (21), (22) is given by the formula

$$u(Q) = -\int_D G(P; Q) f(P) \, dV_P + \int_{\Gamma_1} G(P; Q) g_1(P) \, dS_P - \int_{\Gamma_2} g_2(P) \frac{\partial G}{\partial v_P} \, dS_P.$$
(23)

In view of condition (i) for the Green's function, we see that G is positive on a sufficiently small sphere D_ρ of radius ρ about the point P. Then for Q in $D - D_\rho$ we have $(L + h)[G] = 0$, $(\partial G/\partial v) + \alpha G = 0$ on Γ_1, and $G \geq 0$ on $\Gamma_2 \cup \partial D_\rho$. Now suppose there is a function $w > 0$ defined on $D \cup \partial D$ which satisfies the hypotheses of Theorem 13. Then we may apply the maximum principle to G/w in $D - D_\rho$ and conclude that $G > 0$ in $D - D_\rho$. Letting $\rho \to 0$, we obtain $G > 0$ in D. Moreover, we find that $G = 0$ for P on Γ_2 and hence that $\partial G/\partial v_P \leq 0$ on Γ_2. Conversely, if the Green's function has these properties, then we see from the solution formula (23) that the problem

$$(L + h)[w] = -1 \text{ in } D$$

$$\frac{\partial w}{\partial v} + \alpha w = 1 \text{ on } \Gamma_1$$

$$w = 1 \text{ on } \Gamma_2$$

has a solution which satisfies the hypotheses of Theorem 13. We have thereby shown

THEOREM 16. The bounds of Theorem 13 applied to a solution u of (21), (22) are valid if and only if the Green's function for the problem is positive in D.

EXERCISES

1. Verify that

$$G(x, y; \xi, \eta) \equiv \frac{1}{4\pi} \log \frac{(x^2 + y^2)(\xi^2 + \eta^2) - 2(x\xi + y\eta) + 1}{(x - \xi)^2 + (y - \eta)^2}$$

is the Green's function for the problem $\Delta u = f$ in D: $x^2 + y^2 < 1$, $u = g$ for $x^2 + y^2 = 1$.

2. For the Green's function of Exercise 1, verify that the outward normal derivative is negative at each point of ∂D.

3. Verify that

$$G(x, y, z; \xi, \eta, \zeta) = \frac{1}{4\pi} \{[(x - \xi)^2 + (y - \eta)^2 + (z - \zeta)^2]^{-\frac{1}{2}}$$

$$- [(x^2 + y^2 + z^2)(\xi^2 + \eta^2 + \zeta^2) - 2(x\xi + y\eta + z\zeta) + 1]^{-\frac{1}{2}}\}$$

is the Green's function for the problem $\Delta u = f$ in $D: x^2 + y^2 + z^2 < 1$, $u = g$ on ∂D.

4. For the Green's function of Exercise 3, verify that the outward normal derivative is negative at each point of ∂D.

SECTION 8. EIGENVALUES

We consider the problem of determining a solution u of the equation

$$(L + h + \lambda k)[u] \equiv \sum_{i,j=1}^{n} a_{ij}(\mathbf{x}) \frac{\partial^2 u}{\partial x_i \partial x_j} + \sum_{i=1}^{n} b_i(\mathbf{x}) \frac{\partial u}{\partial x_i} + [h(\mathbf{x}) + \lambda k(\mathbf{x})]u = 0 \tag{1}$$

in a domain D with u subject to the boundary conditions

$$\left. \begin{array}{r} \dfrac{\partial u}{\partial \nu} + \alpha(\mathbf{x})u = 0 \text{ on } \Gamma_1, \\ u = 0 \text{ on } \Gamma_2. \end{array} \right\} \tag{2}$$

We suppose that $\Gamma_1 \cup \Gamma_2 = \partial D$, that $k(\mathbf{x}) \geq \eta > 0$ on $D \cup \partial D$, that $\partial/\partial \nu$ is a directional derivative in a direction outward from D, and that λ is a constant. As usual, we assume that each point of Γ_1 lies on the boundary of a ball in D. Since $u \equiv 0$ is always a solution of (1), (2), we restrict our attention to solutions which do not vanish identically. We call such solutions nontrivial.

DEFINITIONS. If for some number λ there is a nontrivial solution $u(\mathbf{x})$ of (1), (2), we call this value of λ an eigenvalue; the corresponding solution u is called an eigenfunction.

It may happen that solutions of (1), (2) exist for which the corresponding eigenvalues are complex numbers. Since the coefficients of L are real, it is clear that the eigenfunction corresponding to a complex eigenvalue will be a complex-valued function of $\mathbf{x}$.

We now show how the maximum principle may be used to obtain a lower bound for the real part of any eigenvalue of (1), (2). We employ the following notation: Re μ and Im μ denote the real and imaginary parts of the complex number μ, respectively; $\bar{\mu} \equiv \text{Re } \mu - i \text{ Im } \mu$ is the complex conjugate of μ; $|\mu|^2 = \mu\bar{\mu}$.

Suppose we can find a real-valued function $w(\mathbf{x})$ which satisfies the conditions

$$(L + h + \beta k)[w] \leq 0 \text{ in } D, \tag{3}$$

$$\left. \begin{array}{r} w(\mathbf{x}) > 0 \text{ on } D \cup \partial D, \\ \dfrac{\partial w}{\partial \nu} + \alpha(\mathbf{x})w \geq 0 \text{ on } \Gamma_1, \end{array} \right\} \tag{4}$$

where β is a real constant. We shall show that if λ is an eigenvalue of (1), (2), then
$$\operatorname{Re} \lambda \geq \beta. \tag{5}$$
That is, β is a lower bound for the real parts of all eigenvalues of (1), (2).

To establish (5), we let $u(\mathbf{x})$ be an eigenfunction of (1), (2) corresponding to λ, and we define
$$v(\mathbf{x}) = \frac{u(\mathbf{x})}{w(\mathbf{x})},$$
where w fulfills the conditions (3), (4). Since u satisfies (1), we have
$$\frac{1}{w}(L + h + \lambda k)[vw] = \left(L + \frac{2}{w}\sum_{i,j=1}^{n} a_{ij}(\mathbf{x})\frac{\partial w}{\partial x_j}\frac{\partial}{\partial x_i}\right)[v] + \frac{(L + h + \lambda k)[w]}{w}v$$
$$= 0.$$
We write
$$L_1 = L + \frac{2}{w}\sum_{i,j=1}^{n} a_{ij}(\mathbf{x})\frac{\partial w}{\partial x_j}\frac{\partial}{\partial x_i}$$
so that v satisfies the equation
$$L_1[v] + \frac{(L + h + \lambda k)[w]}{w}v = 0. \tag{6}$$
We take complex conjugates in (6) to find that $\bar{v}$ satisfies the equation
$$L_1[\bar{v}] + \frac{(L + h + \bar{\lambda}k)[w]}{w}\bar{v} = 0. \tag{7}$$
Now we compute the quantity $L_1(|v|^2)$:
$$L_1(v\bar{v}) = \bar{v}L_1[v] + vL_1[\bar{v}] + 2\sum_{i,j=1}^{n} a_{ij}(\mathbf{x})\frac{\partial v}{\partial x_i}\frac{\partial \bar{v}}{\partial x_j}.$$
Since the a_{ij} are symmetric, we have
$$\sum_{i,j=1}^{n} a_{ij}(\mathbf{x})\frac{\partial v}{\partial x_i}\frac{\partial \bar{v}}{\partial x_j} = \sum_{i,j=1}^{n} a_{ij}\left[\left(\operatorname{Re}\frac{\partial v}{\partial x_i}\right)\left(\operatorname{Re}\frac{\partial v}{\partial x_j}\right) + \left(\operatorname{Im}\frac{\partial v}{\partial x_i}\right)\left(\operatorname{Im}\frac{\partial v}{\partial x_j}\right)\right].$$
Also, because L is elliptic, this last expression is always nonnegative. Therefore
$$L_1[|v|^2] \geq \bar{v}L_1[v] + vL_1[\bar{v}].$$
Substituting in this expression from (6) and (7), we find
$$L_1[|v|^2] \geq -\frac{(L + h + \lambda k)[w]}{w}|v|^2 - \frac{(L + h + \bar{\lambda}k)[w]}{w}|v|^2,$$
or
$$L_1[|v|^2] \geq -2\frac{(L + h + (\operatorname{Re}\lambda)k)[w]}{w}|v|^2.$$

If λ is such that $\text{Re } \lambda \leq \beta$, we see from (3) and (4) that
$$L_1[|v|^2] \geq 0 \text{ in } D.$$
Thus $|v|^2$ satisfies a maximum principle in D. Since $v = u/w$, we know from (2) that
$$v = 0 \text{ on } \Gamma_2.$$
Therefore if $|v|^2$ is not identically zero, its maximum must occur on Γ_1. However, on Γ_1 we have
$$\frac{\partial}{\partial \nu}(v\bar{v}) = v\frac{\partial \bar{v}}{\partial \nu} + \bar{v}\frac{\partial v}{\partial \nu} = v\frac{[w(\partial \bar{u}/\partial \nu) - \bar{u}(\partial w/\partial \nu)]}{w^2} + \bar{v}\frac{[w(\partial u/\partial \nu) - u(\partial w/\partial \nu)]}{w^2}$$
$$= -2\alpha v\bar{v} - 2\frac{v\bar{v}}{w}\frac{\partial w}{\partial \nu}.$$

In other words, $|v|^2$ satisfies on Γ_1 the boundary condition
$$\frac{\partial}{\partial \nu}(|v|^2) + \frac{2|v|^2}{w}\left(\frac{\partial w}{\partial \nu} + \alpha w\right) = 0. \tag{8}$$

If w satisfies (4) on Γ_1, then (8) and the maximum principle as given in Theorem 7 in Section 3 shows that $|v|^2$ cannot have a positive maximum on Γ_1. Hence $v \equiv 0$ whenever $\text{Re } \lambda < \beta$. Therefore, such a value of λ cannot be an eigenvalue of (1), (2). We have shown that β is a lower bound for the real part of any eigenvalue λ.

For a positive function w which is twice continuously differentiable in D, the largest value of β which satisfies (3) is
$$\beta = \inf_{x \in D} \frac{-(L+h)[w]}{k(x)w}. \tag{9}$$

We have established the following result:

THEOREM 17. *Let λ be an eigenvalue of (1), (2). Let $w(x)$ be positive on $D \cup \partial D$ and satisfy*
$$\frac{\partial w}{\partial \nu} + \alpha(x)w \geq 0 \text{ on } \Gamma_1.$$

Then
$$\text{Re } \lambda \geq \inf_{x \in D} \frac{-(L+h)[w]}{kw}. \tag{10}$$

Remarks. (i) If we graph all the eigenvalues of (1), (2), in the complex $z = x + iy$ plane, then Theorem 17 shows that all of them lie in the half-plane $x \geq \beta$ where β is given by (9).

(ii) For elliptic operators of the type we are considering, it can be shown that there is always at least one real eigenvalue which is smallest in the sense that no other eigenvalue has smaller real part (Protter and Weinberger [2]).

(iii) When it is known that all eigenvalues are real, various bounds of the form (10) have been given by Barta [1], Duffin [1], Hersch [1, 2], and Hooker [1]. Pucci [6] has given a related bound in the general case.

Example. Let D be the semi-circle $x^2 + y^2 < 1$, $y > 0$. Find a lower bound for the eigenvalues of

$$\frac{\partial^2 u}{\partial x^2} + \frac{\partial^2 u}{\partial y^2} + \lambda u = 0 \text{ in } D, \tag{11}$$

$$\left. \begin{array}{l} -\dfrac{\partial u}{\partial y} = 0 \text{ on } \Gamma_1: -1 \leq x \leq 1, \quad y = 0, \\ u = 0 \text{ on } \Gamma_2: x^2 + y^2 = 1, \quad y > 0. \end{array} \right\} \tag{12}$$

We set

$$w(x, y) = 1 - \alpha(x^2 + y^2) \text{ with } \alpha < 1.$$

Clearly, w is positive in D and $\partial w/\partial y = 0$ on Γ_1. Also,

$$\Delta w = -4\alpha,$$

and thus we have the lower bound

$$\lambda \geq \inf_D \frac{4\alpha}{1 - \alpha(x^2 + y^2)} = 4\alpha.$$

Since α is any number less than 1, we find

$$\lambda \geq 4.$$

The choice $w = \cos \alpha \sqrt{x^2 + y^2}$ for $\alpha < \pi/2$ leads to the improved bound $\lambda \geq \pi^2/2$. Actually, it can be shown that the smallest eigenvalue is

$$\lambda_1 = 5.76.$$

In the example above, we tacitly assumed that all the eigenvalues of (11), (12) are real. It is a simple matter to prove that this is indeed the case. We recall Green's second identity:

$$\int_D (v \Delta u - u \Delta v) \, dV = \int_{\partial D} \left(v \frac{\partial u}{\partial \mathbf{n}} - u \frac{\partial v}{\partial \mathbf{n}} \right) dS.$$

We assume that u is a solution of (11), (12) and set $v = \bar{u}$. Then we have

$$\int_D (\bar{u} \Delta u - u \Delta \bar{u}) \, dV = \int_{\partial D} \left(\bar{u} \frac{\partial u}{\partial \mathbf{n}} - u \frac{\partial \bar{u}}{\partial \mathbf{n}} \right) dS = 0,$$

or

$$(\bar{\lambda} - \lambda) \int_D |u|^2 \, dV = 0. \tag{13}$$

If λ is not real, then $\bar{\lambda} - \lambda \neq 0$ and we conclude from (13) that $u \equiv 0$, so that λ is not an eigenvalue. Thus all eigenvalues of the special problem (11), (12) must be real.

EXERCISES

1. Let D be the square: $|x| < 1$, $|y| < 1$. Find a lower bound for the first eigenvalue of the problem

$$\frac{\partial^2 u}{\partial x^2} + \frac{\partial^2 u}{\partial x\, \partial y} + \frac{\partial^2 u}{\partial y^2} + \lambda u = 0 \text{ in } D,$$

$$u = 0 \text{ on } \partial D.$$

2. Let D be the square $|x| < \tfrac{1}{2}\pi$, $|y| < \tfrac{1}{2}\pi$. Show that any solution of the problem

$$\frac{\partial^2 u}{\partial x^2} + \frac{\partial^2 u}{\partial y^2} + u = 0 \text{ in } D,$$

$$u = g(x, y) \text{ on } \partial D$$

is unique. Is the same result true for the square $|x| < \pi/\sqrt{2}$, $|y| < \pi/\sqrt{2}$?

SECTION 9. THE PHRAGMÈN-LINDELÖF PRINCIPLE

The uniqueness and approximation theorems which we have proved by means of the maximum principle apply to bounded domains, but not to unbounded domains. For example, the function

$$u(x, y) = e^x \sin y$$

satisfies Laplace's equation in the strip $|y| < \pi$ and vanishes on its boundary $y = \pm \pi$. However, u takes both positive and negative values in the strip, so that neither the maximum nor the minimum value is attained on the boundary. Since the function $u \equiv 0$ also satisfies $\Delta u = 0$ in the strip $|y| < \pi$, $u = 0$ on the boundary, the uniqueness theorem is also violated for this unbounded domain.

In this section we shall establish a class of maximum principles on unbounded domains by imposing some restriction on the growth of the function at infinity. At the same time, we shall also establish maximum principles for solutions in bounded domains when there are gaps in the prescription of boundary data. Such gaps often arise when the specified boundary condition differs from one part of the boundary to another. For example, the boundary function u may be specified over a portion Γ_2 of ∂D, while the combination $(\partial u/\partial v) + \alpha u$ is specified over another portion Γ_1 of ∂D. While we require that the union of the *closures* of Γ_1 and Γ_2 comprises ∂D, it may very well be that no boundary data are prescribed along the common boundary of Γ_1 and Γ_2. As in Sections 4 and 6, these maximum principles lead to uniqueness and approximation theorems.

To illustrate the Phragmèn-Lindelöf principle, we first establish a classical result concerning the growth of subharmonic functions in an unbounded sector in the plane.

Let D be the sector defined by the inequalities $-cx < y < cx$, $x > 0$.

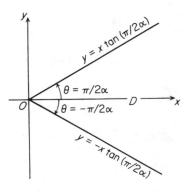

FIGURE 11

It is convenient to introduce polar coordinates (r, θ) and write the equations of the boundary of D as $\theta = \pm \pi/2\alpha$ where $c = \tan(\pi/2\alpha)$. (See Fig. 11.) The function

$$w = r^\alpha \cos \alpha\theta$$

is harmonic in the sector D and vanishes on the boundary of D. Writing the Laplace operator in polar coordinates

$$\Delta \equiv \frac{\partial^2}{\partial r^2} + \frac{1}{r}\frac{\partial}{\partial r} + \frac{1}{r^2}\frac{\partial^2}{\partial \theta^2},$$

we easily verify that w is a harmonic function. Thus we have an example of an unbounded harmonic function which vanishes over the entire boundary of the sector in which it is defined. This function approaches infinity like r^α as $r \to \infty$ on every ray $\theta = $ constant. The Phragmèn-Lindelöf theorem asserts that the growth of the above function w as $r \to \infty$ is characteristic of harmonic functions which are unbounded in a sector. That is, any harmonic function which vanishes on the boundary and is not identically zero must grow as fast as r^α. Moreover, if a harmonic (or subharmonic) function is bounded along the entire boundary of a sector D of angle opening (π/α) and if it grows more slowly than r^α as $r \to \infty$, then it does not grow at all.

THEOREM 18. (Phragmèn-Lindelöf). Let u satisfy the inequality

$$\Delta u \geq 0$$

in a sector D of angle opening π/α. Assume that $u \leq M$ on the boundary $\theta = \pm \pi/2\alpha$ and suppose that*

$$\liminf_{R \to \infty} \{R^{-\alpha} \max_{r=R} u(r, \theta)\} \leq 0.$$

Then $u \leq M$ in D.

Proof. For a fixed number R, we consider the domain D_R bounded by

---

*The **limit inferior** $\liminf_{R \to A} F(R)$ is defined to be the smallest number a (possibly $\pm \infty$) such that there is a sequence $R_n \to A$ for which $F(R_n) \to a$. The **limit superior** is defined similarly, so that $\limsup_{R \to A} F(R) = -\liminf_{R \to A} [-F(R)]$.

the rays $\theta = \pi/2\alpha$, $\theta = -\pi/2\alpha$ and an arc of the circle $r = R$. The region D_R is contained in D (see Fig. 12).

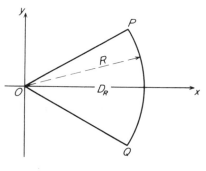

FIGURE 12

The main device of the proof consists in the determination of a harmonic function in D_R which remains bounded away from zero and has an appropriate growth as $R \to \infty$. This function is used as a comparison function in the maximum principle. A function having these properties is

$$w_R(r, \theta) = 1 + \frac{2}{\pi} R^\alpha \tan^{-1} \frac{2R^\alpha r^\alpha \cos \alpha\theta}{R^{2\alpha} - r^{2\alpha}}.$$

It is easily verified that this function is harmonic* in D_R, has the value 1 for $\theta = \pm\pi/2\alpha$, $0 \leq r < R$, and has the value $1 + R^\alpha$ for $r = R$, $-(\pi/2\alpha) < \theta < (\pi/2\alpha)$. The function

$$v(r, \theta) = \frac{u(r, \theta) - M}{w_R(r, \theta)}$$

satisfies the differential inequality

$$\Delta v + \frac{2}{w_R} \operatorname{grad} w_R \cdot \operatorname{grad} v \geq 0,$$

and so, by Theorem 5 in Section 3, v satisfies a maximum principle in D_R. By hypothesis, $v \leq 0$ on the rays $\theta = \pm\pi/2\alpha$, $0 \leq r < R$. Also, on the arc $\{r = R, -(\pi/2\alpha) < \theta < (\pi/2\alpha)\}$, we have

$$v(R, \theta) \leq \frac{\max_{(-\pi/2\alpha) \leq \theta \leq (\pi/2\alpha)} [u(R, \theta) - M]}{1 + R^\alpha}.$$

Theorem 5 shows that for all (r, θ) in D_R:

$$v(r, \theta) = \frac{u(r, \theta) - M}{w_R(r, \theta)} \leq \max \left\{ 0, \frac{\max_{r=R} [u(r, \theta) - M]}{1 + R^\alpha} \right\},$$

or

$$u(r, \theta) \leq M + w_R(r, \theta) \max \{0, (1 + R^\alpha)^{-1} \max_{r=R} [u(r, \theta) - M]\}. \qquad (1)$$

---

*This verification is made most simply by observing that the real and imaginary parts of an analytic function of the complex variable $z = re^{i\theta}$ are harmonic; the function w_R is the imaginary part of the analytic function

$$f(z) = i + \frac{2}{\pi} R^\alpha \log \frac{R^\alpha + iz^\alpha}{R^\alpha - iz^\alpha}.$$

We now keep (r, θ) fixed and let $R \to \infty$. It is an easy matter to verify that $w_R(r, \theta)$ remains bounded. In fact, l'Hôpital's rule yields

$$\lim_{R \to \infty} R^\alpha \tan^{-1} \frac{2R^\alpha r^\alpha \cos \alpha\theta}{R^{2\alpha} - r^{2\alpha}} = 2r^\alpha \cos \alpha\theta.$$

By hypothesis there is a sequence of radii R_n, $n = 1, 2, \ldots$, tending to infinity, with the property that for any $\epsilon > 0$ there is an integer N such that $(1 + R_n^\alpha)^{-1} \max_{r=R} [u(r, \theta) - M] < \epsilon$ for $n \geq N$. From (1) we conclude that

$$u(r, \theta) < M + \epsilon w_{R_n}(r, \theta)$$

in D_{R_n}. Since each point (r, θ) in D is in D_{R_n} when n is sufficiently large, we establish the theorem by letting ϵ tend to zero and n to infinity.

Remarks (i) We may now conclude from Theorem 2 in Section 1 that $u < M$ in D unless $u \equiv M$.

(ii) If D^* is any domain contained in a sector of angle opening π/α and if u is bounded by M on the boundary of D^*, then the above proof yields the same Phragmèn-Lindelöf theorem with the same exponent α, since $1 \leq w_R \leq 1 + R^\alpha$ throughout D_R.

(iii) The same argument yields the result for functions defined in a half-plane. We simply take $\alpha = 1$; the proof is also valid for $\alpha < 1$.

(iv) The result of Theorem 18 shows that if a subharmonic function defined in a sector or half-plane is unbounded from above, then the function must tend to infinity at least as fast as r^α on some sequence of points tending to infinity.

The basic argument given in Theorem 18 applies to much more general situations. We now extend the theorem to include not only more general elliptic operators but also bounded as well as unbounded domains.

Let L be a uniformly elliptic second-order operator defined in an n-dimensional domain D. Suppose u satisfies

$$(L + h(\mathbf{x}))[u] \geq 0 \text{ in } D, \tag{2}$$

where $h(\mathbf{x})$ may be positive. The domain D may be bounded or unbounded. Let Γ be a subset of ∂D which may be all of ∂D. Suppose we are given that

$$u \leq 0 \text{ on } \Gamma.$$

We wish to impose sufficient additional hypotheses to conclude that

$$u \leq 0 \text{ in } D. \tag{3}$$

For example, if D is the sector in Theorem 18 and $\Gamma \equiv \partial D$, then the growth condition $R^{-\alpha} \max u(R, \theta) \to 0$ yields (3) for the case $L = \Delta$ and

$h \equiv 0$. If $\Gamma \equiv \partial D$ and D is bounded, then (3) is the result given by Theorem 10 in Section 5. Therefore, the interesting cases are those where either Γ is a proper subset of D or where D is unbounded.

Given the domain D and the portion Γ of ∂D, we suppose that an increasing sequence of bounded regions $D_1 \subset D_2 \subset \cdots \subset D_k \cdots$ can be found with the properties:

(i) Each D_k is contained in D; for each point $\mathbf{x} \in D$ there is an integer N such that $\mathbf{x} \in D_N$ (and hence $\mathbf{x} \in D_k$ for all $k \geq N$).

(ii) The boundary of each D_k consists of two parts Γ_k and Γ'_k, where Γ_k is a subset of Γ and Γ'_k is a subset of D.

Suppose that on each domain D_k, there is a function $w_k(\mathbf{x})$ with the properties

$$\left. \begin{array}{l} w_k(\mathbf{x}) > 0 \text{ on } D_k \cup \partial D_k, \\ (L + h)[w_k] \leq 0 \text{ in } D_k. \end{array} \right\} \quad (4)$$

THEOREM 19. (Phragmèn-Lindelöf principle). **Let D be a domain, bounded or unbounded, and let u satisfy**

$$(L + h)[u] \geq 0 \text{ in } D,$$
$$u \leq 0 \text{ on } \Gamma,$$

where Γ is a subset of ∂D. Suppose that there is an increasing sequence of domains $\{D_k\}$ with properties (i) and (ii) above and that there exists a sequence $\{w_k\}$ which satisfies (4). Assume there is a function $w(\mathbf{x})$ with the property that at each point x of D the inequality

$$w_k(\mathbf{x}) \leq w(\mathbf{x})$$

holds for all k above a certain integer N_x. If $u(\mathbf{x})$ satisfies the growth condition

$$\liminf_{k \to \infty} \left[\sup_{\Gamma'_k} \frac{u(\mathbf{x})}{w_k(\mathbf{x})} \right] \leq 0, \quad (5)$$

then

$$u \leq 0 \text{ in } D.$$

Proof. We form the function

$$v_k = \frac{u}{w_k}$$

and observe that by Theorem 10, v_k satisfies a maximum principle in D_k. Since $\Gamma_k \subset \Gamma$ and since by hypothesis, $u \leq 0$ on Γ, we have

$$v_k \leq 0 \text{ on } \Gamma_k.$$

On Γ'_k, we have

$$v_k \leq \sup_{\mathbf{x} \in \Gamma'_k} \frac{u(\mathbf{x})}{w_k(\mathbf{x})}.$$

Hence, for $\mathbf{x} \in D_k$, we find

$$v_k(\mathbf{x}) = \frac{u(\mathbf{x})}{w_k(\mathbf{x})} \leq \max\left\{0, \sup_{\mathbf{x} \in \Gamma_{k}'} \frac{u(\mathbf{x})}{w_k(\mathbf{x})}\right\}$$

and, because $w_k(\mathbf{x}) \leq w(\mathbf{x})$,

$$u(\mathbf{x}) \leq \max\left\{0, \sup_{\Gamma_{k}'} \frac{u(\mathbf{x})}{w_k(\mathbf{x})}\right\} w(\mathbf{x}).$$

Inequality (5) states that there is a sequence $k_n \to \infty$ with the property that for any $\epsilon > 0$ there is an integer N such that $\sup_{\Gamma_{k_n}}[u/w_{k_n}] < \epsilon$ for $n \geq N$. It follows that $u(\mathbf{x}) \leq \epsilon w(\mathbf{x})$ in D_{k_n}. Letting ϵ tend to zero and n to infinity, we conclude that $u(\mathbf{x}) \leq 0$ in D.

Remarks. (i) If $h \equiv 0$, we can replace u by $u - M$, where M is any constant. Then the statement $u \leq 0$ may be replaced by $u \leq M$ both in the hypotheses and conclusion of Theorem 19. If $h(\mathbf{x}) \leq 0$, the same statement holds for M nonnegative.

(ii) The usefulness of the Phragmèn-Lindelöf principle hinges on the determination of the domains $\{D_k\}$ and the functions $\{w_k\}$ with the appropriate properties. In Theorem 18, the $\{D_k\}$ are the sectors of circles of radii R_k and the $\{w_k\}$ are given by explicit formulas. We give additional examples below.

Example. Let u satisfy

$$\Delta u \geq 0$$

in a plane domain D consisting of the exterior of one or more closed curves Γ. We choose a polar coordinate system with origin inside Γ and select a sequence of disks $\{K_n\}$ of radii R_n such that $R_n \to \infty$ and each ∂K_n is in D. We define

$$D_n = D \cap K_n$$

and

$$w_n(r, \theta) = \log\frac{r}{r_0} \text{ for } r_0 < r < R_n,$$

where the circle $r = r_0$ is so small that it lies entirely in the complement of $D \cup \Gamma$. Suppose that $u \leq M$ on Γ and that

$$\liminf_{R \to \infty}\left[\max_{0 \leq \theta \leq 2\pi} \frac{u(R, \theta)}{\log R}\right] = 0.$$

We observe that if the sequence $\{R_n\}$ is chosen appropriately, the $\{D_n\}$ and $\{w_n\}$ satisfy all the hypotheses of the theorem with $\Gamma_n = \Gamma$, $\Gamma_n' = \{r = R_n, 0 \leq \theta \leq 2\pi\}$, and the w_n harmonic and positive in D_n. We conclude that

$$u \leq M \text{ in } D.$$

In other words, if u is subharmonic in the exterior of Γ, is bounded on Γ, and if $u(R, \theta)$ does not grow as rapidly as $\log R$ as $R \to \infty$, then u must be bounded everywhere. As we shall see in Section 12, this result may also be obtained from the three-circles theorem of Hadamard.

The above application of the Phragmèn-Lindelöf principle yields a uniqueness theorem for an exterior boundary value problem. Suppose there are two solutions u_1 and u_2 of

$$\Delta u = f \text{ in } D,$$
$$u = g \text{ on } \partial D,$$

and suppose that both u_1 and u_2 do not grow as rapidly as $\log r$ as $r \to \infty$; that is, we assume

$$\lim_{r \to \infty} \frac{\max_{0 \leq \theta \leq 2\pi} |u_i(r, \theta)|}{\log r} = 0, \qquad i = 1, 2. \tag{6}$$

Then both $u_1 - u_2$ and $u_2 - u_1$ are harmonic in D and vanish on ∂D. The Phragmèn-Lindelöf principle, Theorem 19, applies so that $u_1 - u_2 \leq 0$ and $u_2 - u_1 \leq 0$ in D. We conclude that $u_1 \equiv u_2$. Note that the hypothesis (6) is essential since the function $\log r$, which vanishes on $r = 1$, may be added to any solution defined in the particular domain $D_1: \{r > 1, 0 \leq \theta \leq 2\pi\}$. There is no uniqueness of solutions if growth as rapid as $\log r$ is allowed.

In the above example, we selected $w_n = \log(r/r_0)$ for $r_0 < r < R_n$, and we see that w_n is merely the restriction of the function $w = \log(r/r_0)$ to the annulus $r_0 < r < R_n$. The function w has all the properties required of Theorem 19 and, whenever such a function is available, we obtain the conclusion of the theorem without intermediate steps. This special case is stated as a corollary.

COROLLARY. Suppose that u, D, and Γ are as in Theorem 19 and that there is a function $w(\mathbf{x})$ with the properties

$$\left. \begin{array}{l} w > 0 \text{ on } D \cup \partial D, \\ (L + h)[\mathbf{w}] \leq 0 \text{ in } D, \\ \lim_{\mathbf{x} \to \partial D - \Gamma} w(\mathbf{x}) = \infty \\ \lim_{|\mathbf{x}| \to \infty} w(\mathbf{x}) = \infty \text{ if } D \text{ is unbounded.} \end{array} \right\} \tag{7}$$

If $u \leq 0$ on ∂D and

$$\liminf_{A \to \infty} \left[\sup_{\substack{w(\mathbf{x}) = A \\ \mathbf{x} \in D}} \frac{u(\mathbf{x})}{w(\mathbf{x})} \right] \leq 0,$$

then

$$u \leq 0 \text{ in } D.$$

We now illustrate the application of the Corollary to a problem in a bounded domain.

Example. Let D be a bounded plane domain and suppose u satisfies

$$\left.\begin{array}{l}\Delta u \geq 0 \text{ in } D \\ u \leq 0 \text{ on } \partial D \text{ except at one point } P.\end{array}\right\} \quad (8)$$

We let P be the origin of coordinates and denote by d the diameter of D. We shall employ the harmonic function

$$w = \log \frac{2d^2}{x^2 + y^2}$$

in the Corollary. Clearly $w > 0$ on $D \cup \partial D$. The Corollary states that if

$$\liminf_{r \to 0} \left[\frac{\sup_{x^2+y^2=r^2} u(x,y)}{\log(1/r)} \right] \leq 0, \quad (9)$$

then $u \leq 0$.

For bounded plane domains, it might appear that conditions (8) alone would imply $u \leq 0$ everywhere since there is only one exceptional point on the boundary. However, the function

$$u(x, y) = \frac{1 - x^2 - y^2}{(x+1)^2 + y^2} \quad (10)$$

is harmonic in the unit disk $D: \{x^2 + y^2 < 1\}$, is zero everywhere on the boundary except at the one point $P(-1, 0)$, but is *positive* everywhere in D. Therefore, a condition such as (9) is necessary to conclude that $u \leq 0$ in the interior.

If $u \leq 0$ on the boundary of a bounded plane domain D except at a finite set of boundary points $P_1(x_1, y_1), P_2(x_2, y_2), \ldots, P_k(x_k, y_k)$, then we select

$$w = \sum_{i=1}^{k} \log \frac{2d^2}{(x-x_i)^2 + (y-y_i)^2}. \quad (11)$$

If the growth condition (9) is satisfied at each P_i, we conclude from the Corollary that $u \leq 0$ throughout D.

Once again we can deduce a uniqueness theorem for the first boundary value problem. Suppose we are given two solutions u_1 and u_2 of the boundary value problem

$$\Delta u = f \text{ in } D, \quad (12a)$$
$$u = g \text{ on } \partial D, \quad (12b)$$

where g is continuous on the boundary except at a finite number of points $P_1, P_2, \ldots, P_k$. At these points g is not defined. We shall determine conditions under which u_1 and u_2 coincide. Suppose u_1 and u_2 satisfy (12b) at all points of ∂D except at $P_1, P_2, \ldots, P_k$. We also assume that u_i/w, $i = 1, 2$ with w given by (11) approach zero as $w \to \infty$ at each of the points $P_1, P_2, \ldots, P_k$. In particular, it is sufficient to assume u_1 and u_2 are bounded. We conclude that $u_1 - u_2 \leq 0$ in D and $u_2 - u_1 \leq 0$ in D. Hence $u_1 \equiv u_2$.

The example of the function (10) shows that a growth condition is needed for uniqueness theorems. The function (10) grows as $[(x + 1)^2 + y^2]^{-1/2}$ which, of course, is faster than $\log[(x + 1)^2 + y^2]$. In fact, if the exceptional boundary point is taken as the origin and if the boundary is so smooth that the domain D lies outside some circle $(x - \bar{x})^2 + (y - \bar{y})^2 = \bar{x}^2 + \bar{y}^2$, we can use the harmonic function

$$w = 1 - \frac{x\bar{x} + y\bar{y}}{x^2 + y^2}$$

in the Corollary to replace $\log(1/r)$ by $(1/r)$ in the growth condition (9). If there are several such exceptional boundary points, we may use a sum of such functions w.

Growth conditions and uniqueness theorems for subharmonic functions in n dimensions, $n \geq 3$, can be obtained by the same method. Suppose u satisfies

$$\Delta u \geq 0 \text{ in } D,$$
$$u \leq 0 \text{ on } \partial D,$$

except at one point $P(x_1^0, x_2^0, \ldots, x_n^0)$. For D a bounded domain, the harmonic function

$$w(\mathbf{x}) = \left[\sum_{i=1}^{n}(x_i - x_i^0)^2\right]^{-(n-2)/2} \tag{13}$$

satisfies all the conditions of the Corollary. Therefore, if

$$\liminf_{R \to 0}\left[R^{n-2}\sup_{|\mathbf{x}-\mathbf{x}^0|=R}u(\mathbf{x})\right] \leq 0,$$

then $u \leq 0$ in D. If there is a finite number of exceptional boundary points, we choose for w a sum of functions of the type (13). As above, we may replace R^{n-2} by R^{n-1} in the growth condition if the boundary is sufficiently smooth at the exceptional points.

A uniqueness theorem for the boundary value problem

$$\left.\begin{array}{l}\Delta u = f \text{ in } D \\ u = g \text{ on } \partial D \text{ except at } P_1, \ldots, P_k\end{array}\right\} \tag{14}$$

is established exactly as in the two-dimensional case.

In general, the boundary values are prescribed on an $(n-1)$-dimensional surface, and the discontinuities will occur along $(n-2)$-dimensional surfaces. In particular, if $n = 3$, the discontinuities of the boundary values will occur along a one-dimensional curve C or a collection of such curves on ∂D. Suppose C is a smooth one-dimensional curve on ∂D, given parametrically in terms of its arc length by

$$C: x = \xi(s), \quad y = \eta(s), \quad z = \zeta(s), \quad 0 \leq s \leq l.$$

It is easy to verify that the function

$$w(x, y, z) = \int_0^l \{[x - \xi(s)]^2 + [y - \eta(s)]^2 + [z - \zeta(s)]^2\}^{-1/2}\, ds \quad (15)$$

is harmonic for (x, y, z) not on C. Also $w \to \infty$ as (x, y, z) approaches a point on C. Hence we may use this function w in the corollary. We conclude that in the class of bounded solutions of (14), any two solutions which have the same boundary values except on a curve C actually coincide everywhere in D. More generally, we reach the same conclusion for solutions which satisfy an appropriate growth condition.

Suppose now that the set $D \cup (\partial D - \Gamma)$ is itself a domain bounded by Γ. That is, we suppose that the points of the exceptional boundary $\partial D - \Gamma$, which we denote by Σ, are interior points of the closure of D. We assume that there is a bounded subdomain D' of $D \cup \Sigma$ which contains the exceptional boundary set Σ in its interior (see Fig. 13), and which has the property that the boundary value problem

$$(L + h)[v] \equiv \sum_{i,j=1}^n a_{ij}(\mathbf{x}) \frac{\partial^2 v}{\partial x_i \partial x_j} + \sum_{i=1}^n b_i(\mathbf{x}) \frac{\partial v}{\partial x_i} + h(\mathbf{x})v = 0 \text{ in } D'$$

$$v = \varphi \text{ on } \partial D'$$

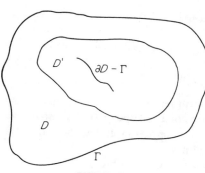

FIGURE 13

can be solved for arbitrary continuously differentiable boundary values φ. In addition, we assume that there exists a function w with the properties $(L + h)[w] \leq 0$ in $D' - \Sigma$, $w > 0$ in $D' \cup \partial D' - \Sigma$, $\lim_{\mathbf{x} \to \Sigma} w(\mathbf{x}) = \infty$.

Let u be a solution of $(L + h)[u] = 0$ in D for which $u/w \to 0$ as $w \to \infty$. Let v be the solution of the problem $(L + h)[v] = 0$ in D', $v = u$ on $\partial D'$. Then $(L + h)[u - v] = 0$ in $D' - \Sigma$, $(u - v)/w \to 0$ as $w \to \infty$, and $u - v = 0$ on $\partial D'$. Hence by the Corollary, $u - v \equiv 0$. Thus u is equal to the twice continuously differentiable function v near the set Σ. In other words, u may be extended as a solution of $(L + h)[u] = 0$ in the whole domain $D \cup \Sigma$ by defining it suitably (namely, by $u \equiv v$) on the set Σ. We then say that u has a **removable singularity** on Σ.

We have shown that if the boundary value problem in D' can be solved, any solution u of $(L + h)[u] = 0$ in D for which $u/w \to 0$ as $w \to \infty$ has a removable singularity on Σ.

For example, if $\Delta u = 0$ for $0 < x^2 + y^2 < R^2$, we let $w = -\log(x^2 + y^2)$, and $D'\colon x^2 + y^2 < (\tfrac{1}{2}R)^2$. Since the boundary value problem for Laplace's equation in a disk can be solved, we conclude that any function u which is harmonic in a punctured disk $0 < x^2 + y^2 < R^2$ and for which $u/\log(x^2 + y^2) \to 0$ as $x^2 + y^2 \to 0$ has a limit as $x^2 + y^2 \to 0$, and the function which results by defining $u(0, 0)$ to be this limit is harmonic in the whole disk $0 \leq x^2 + y^2 < R^2$.

In order to apply the Corollary and the above remarks to a general operator $L + h$ in n independent variables, we must construct an appropriate function $w(\mathbf{x})$. We now show how this can be done for a bounded domain D with a single exceptional point $\mathbf{O}$ on ∂D. We shall take $\mathbf{O}$ to be the origin of coordinates, and we restrict our attention to the case

$$h \leq 0 \text{ in } D.$$

Recalling that the coefficients of the second order terms in L form the elements of a symmetric matrix $(a_{ij}(\mathbf{x}))$, we let the constant matrix (A_{ij}) be the inverse of the coefficient matrix $(a_{ij}(\mathbf{O}))$. We define the quantity

$$\rho = \left\{ \sum_{i,j=1}^{n} A_{ij} x_i x_j \right\}^{1/2}.$$

We shall show that if $n \geq 3$ the function*

$$w = \rho^{2-n} + c\rho^{2-n+\epsilon} + K[e^{\alpha d} - e^{\alpha x_1}],$$

where ϵ is a fixed constant between zero and one, and c, d, α, and K are appropriate constants, has the desired properties whenever the coefficients of L are sufficiently smooth.

A computation yields

$$(L + h)[\rho^{2-n} + c\rho^{2-n+\epsilon}] =$$
$$\rho^{-n-2}[n(n - 2) + c(n - \epsilon)(n - 2 - \epsilon)\rho^{\epsilon}] \sum_{i,j,k,l} a_{kl}(\mathbf{x}) A_{ik} A_{jl} x_i x_j$$
$$- \rho^{-n}[n - 2 + c(n - 2 - \epsilon)\rho^{\epsilon}] \sum_{k,l} a_{kl}(\mathbf{x}) A_{kl}$$
$$- \rho^{-n}[n - 2 + c(n - 2 - \epsilon)\rho^{\epsilon}] \sum_{i,j} A_{ij} b_i x_j + h \rho^{2-n}[1 + c\rho^{\epsilon}].$$

We now write

$$a_{ij}(\mathbf{x}) = a_{ij}(\mathbf{O}) + [a_{ij}(\mathbf{x}) - a_{ij}(\mathbf{O})]$$

and make use of the identity

$$\sum_{p=1}^{n} a_{pq}(\mathbf{O}) A_{pr} = \delta_{qr} \equiv \begin{cases} 1 \text{ when } q = r \\ 0 \text{ when } q \neq r. \end{cases}$$

In this way, we obtain

$$(L + h)[\rho^{n-2} + c\rho^{n-2+\epsilon}] = -c\epsilon(n - 2 - \epsilon)\rho^{-n+\epsilon}$$
$$+ \rho^{-n}[n(n - 2) + c(n - \epsilon)(n - 2 - \epsilon)\rho^{\epsilon}]$$
$$\times \left\{ \sum_{i,j,k,l} [a_{kl}(\mathbf{x}) - a_{kl}(\mathbf{O})] A_{ik} A_{jl} \frac{x_i x_j}{\rho^2} \right\}$$
$$- \rho^{-n}[n - 2 + c(n - 2 - \epsilon)\rho^{\epsilon}] \sum_{k,l} [a_{kl}(\mathbf{x}) - a_{kl}(\mathbf{O})] A_{kl}$$
$$- \rho^{-n}[n - 2 + c(n - 2 - \epsilon)\rho^{\epsilon}] \sum_{i,j} A_{ij} b_j x_i + h(\mathbf{x})\rho^{2-n}[1 + c\rho^{\epsilon}].$$

---

*We observe that if (a_{ij}) is a constant matrix, then $L_1 [\rho^{2-n}] = 0$ except at $\rho = 0$, where L_1 is the principal part of L.

If $a_{ij}(\mathbf{x})$ is so smooth near $\mathbf{O}$ that the quantities $\rho^{-\epsilon}[a_{ij}(\mathbf{x}) - a_{ij}(\mathbf{O})]$ are bounded, then we can choose c so large that $(L+h)[\rho^{2-n} + c\rho^{2-n+\epsilon}] \to -\infty$ as $\rho \to 0$. The smoothness condition is satisfied if, for example, the elements of (a_{kl}) are differentiable.

We make a second computation to find that

$$L[e^{\alpha d} - e^{\alpha x_1}] = -(\alpha^2 a_{11} + \alpha b_1)e^{\alpha x_1}$$

and, since $a_{11} > 0$, we can select α so large that $-L[e^{\alpha d} - e^{\alpha x_1}]$ is positive on $D \cup \partial D$. We choose d so that $d \geq x$, and K so that

$$-KL[e^{\alpha d} - e^{\alpha x_1}] \geq \max_{D} (L+h)[\rho^{2-n} + c\rho^{2-n+\epsilon}].$$

Since $h \leq 0$, we conclude that w has the properties

$$(L+h)[w] \leq 0 \text{ in } D,$$

$$w > 0 \text{ on } D \cup \partial D.$$

We now apply the Corollary on page 99 to find that if

$$(L+h)[u] \geq 0 \text{ in } D,$$

$$\text{if } u \leq 0 \text{ on } \partial D \text{ except at } \mathbf{O},$$

and if u grows more slowly than $1/\rho^{n-2}$ as $\mathbf{x} \to \mathbf{O}$, then $u \leq 0$ in D.

By addition of several functions w of the same type, the same result may be obtained if there is a finite number of exceptional points on the boundary. Also, the function w may be integrated over an $(n-2)$-dimensional bounding hypersurface to show that such exceptional sets may be neglected when prescribing boundary data.

In particular, we see that if the coefficients a_{ij} are sufficiently smooth at a point $\mathbf{O}$, a bounded solution u of $(L+h)[u] \geq 0$ in a deleted neighborhood of $\mathbf{O}$ cannot attain a maximum at $\mathbf{O}$. Less restrictive smoothness conditions are given by Gilbarg and Serrin [1]. If the first boundary value problem can be solved in some neighborhood of $\mathbf{O}$, we obtain a removable singularity theorem. Since our construction with $K = 0$ makes $(L+h)[w] < 0$ in some neighborhood of the origin regardless of the sign of h, these results hold even when h is sometimes positive.

The case of two dimensions can be treated in the same way, after replacing the first two terms in the definition of w by $\log(1/\rho) - c_1\rho^\epsilon + c_2$.

The following example due to Gilbarg and Serrin shows that it is not sufficient for the coefficients to be just continuous at $\mathbf{O}$, no matter how smooth they are away from $\mathbf{O}$. The function

$$u(r) = 1 + \left(\frac{1}{\log r}\right)$$

($r = \sqrt{x^2 + y^2}$) is a solution of the uniformly elliptic equation
$$\left[1 + \frac{2x^2}{r^2(-\log r - 2)}\right]u_{xx} + \frac{4xy}{r^2(-\log r - 2)}u_{xy}$$
$$+ \left[1 + \frac{2y^2}{r^2(-\log r - 2)}\right]u_{yy} = 0$$
for $0 < r \leq 0.1$. The coefficients are infinitely differentiable except at the origin, and they are continuous at the origin. The solution u clearly attains a maximum at the origin.

The construction given above for $n > 2$ shows that if the coefficients are continuous at **O** then for any $\epsilon > 0$ there is a function w of the form $\rho^{2-n+\epsilon} + K[e^{\alpha d} - e^{\alpha x_1}]$ which satisfies the conditions of the Corollary. Thus any solution of $(L + h)[u] = 0$ with an unremovable singularity at **O** must grow as fast as $\rho^{2-n+\epsilon}$ for every $\epsilon > 0$. Gilbarg and Serrin [1] give an example of such a solution which grows more slowly than ρ^{2-n}.

We now consider the problem of mixed boundary conditions. Suppose that u is a solution of
$$(L + h)[u] = f \text{ in } D,$$
$$\frac{\partial u}{\partial v} + \alpha u = g_1 \text{ on } \Gamma_a,$$
$$u = g_2 \text{ on } \Gamma_b,$$
where Γ_a and Γ_b are disjoint, but $\Gamma_a \cup \Gamma_b$ may not be all of ∂D. The corresponding Phragmèn-Lindelöf theorem is obtained in precisely the same way as before. We state the result.

THEOREM 20. Suppose u satisfies
$$(L + h)[u] \geq 0 \text{ in } D$$
$$\frac{\partial u}{\partial v} + \alpha u \leq 0 \text{ on } \Gamma_a$$
$$u \leq 0 \text{ on } \Gamma_b,$$
and suppose that a sequence of bounded regions $\{D_k\}$ can be found as in Theorem 19 with $\Gamma = \Gamma_a \cup \Gamma_b$. Suppose a sequence of functions $\{w_k(x)\}$ can be found as in Theorem 19 which satisfy, in addition to the conditions (4), the condition
$$\frac{\partial w_k}{\partial v} + \alpha w_k \geq 0 \text{ on } \Gamma_k \cap \Gamma_a.$$
Then $u \leq 0$ in D.

Other forms of the Phragmèn-Lindelöf principle have been given by Hopf [3], Gilbarg [1, 4], Serrin [5], and Meyers and Serrin [1]. Further bibliographical references may be found in these papers.

For theorems concerning exceptional boundary sets for bounded domains, see Kryżański [2] who also treats related questions of the existence of solutions. Picone [5] and Leja [1] obtained such results for the special operator $\Delta + h$, which generalized the classical results of Zaremba [1] for the Laplace operator.

EXERCISES

1. Show that the equation considered by Gilbarg and Serrin:

$$\left[1 - \frac{2x^2}{r^2(2 + \log r)}\right]\frac{\partial^2 u}{\partial x^2} - \frac{4xy}{r^2(2 + \log r)}\frac{\partial^2 u}{\partial x\, \partial y} + \left[1 - \frac{2y^2}{r^2(2 + \log r)}\right]\frac{\partial^2 u}{\partial y^2} = 0$$

is uniformly elliptic for $r \leq 0.1$. Prove that the coefficients are not differentiable at $x = y = 0$.

2. The equation

$$[r^2(2 + \log r) - 2x^2]\frac{\partial^2 u}{\partial x^2} - 4xy\frac{\partial^2 u}{\partial x\, \partial y} + [r^2(2 + \log r) - 2y^2]\frac{\partial^2 u}{\partial y^2} = 0 \qquad (*)$$

is equivalent to the equation in Exercise 1 in that it is attained by the multiplication of that equation by $r^2(2 + \log r)$. Therefore $1 + (1/\log r)$ is a solution of $(*)$. Prove that the coefficients of $(*)$ are differentiable. Explain why the maximum principle fails.

3. Show that if $\Delta u = 0$ in a bounded two-dimensional domain D in the upper half-plane $y > 0$, if the origin **O** is a boundary point of D, and if

$$\lim_{(x^2+y^2) \to 0} (x^2 + y^2)^{1/2} u = 0,$$

then u is bounded by its maximum on $\partial D - \mathbf{O}$.

4. Show that if $u(x, y, z)$ satisfies $\Delta u = 0$ for $0 < x^2 + y^2 + z^2 < R_0^2$ and if

$$\lim_{(x^2+y^2+z^2) \to 0} (x^2 + y^2 + z^2)^{1/2} u = 0,$$

then u may be defined at the origin so that it is harmonic for $0 \leq x^2 + y^2 + z^2 < R_0^2$.

SECTION 10. THE HARNACK INEQUALITIES

The maximum principle enables us to estimate a solution of an elliptic equation at all interior points of a domain in terms of the extreme values of the solution on the boundary. It was discovered by Harnack that if u is a positive harmonic function in a domain D, it is possible to

Sec. 10 The Harnack Inequalities

bound the ratio of the values of u at any two points of a closed bounded subset S of D by a constant which depends only on the sets S and D. These estimates, known as the **Harnack inequalities**, have proved extremely useful in the development of the theory of harmonic functions.

The basis for Harnack's inequalities is the explicit formula for the solution of the Laplace equation in a disk. Recalling the results of Section 7, we see that if

$$\Delta u = 0$$

in the disk K_R of radius R with center at the origin, then at any point $P(\xi, \eta)$ in the interior of K_R, we have

$$u(\xi, \eta) = -\int_{\partial K_R} u \frac{\partial G}{\partial \mathbf{n}} \, ds, \tag{1}$$

where G is the Green's function for the disk K_R. An explicit expression for G is

$$G = \frac{1}{4\pi} \log \frac{R^4 - 2R^2(x\xi + y\eta) + (x^2 + y^2)(\xi^2 + \eta^2)}{R^2[(x - \xi)^2 + (y - \eta)^2]}.$$

The derivative in the direction normal to the boundary is the same as the radial derivative, and so $\partial G/\partial \mathbf{n}$ can be calculated directly. It is convenient to use polar coordinates. After making the change of variables $\xi = r \cos \phi$, $\eta = r \sin \phi$, $x = R \cos \theta$, $y = R \sin \theta$, we find (see Fig. 14)

$$u(r \cos \phi, r \sin \phi) = \frac{R^2 - r^2}{2\pi} \int_0^{2\pi} \frac{u(R \cos \theta, R \sin \theta) \, d\theta}{R^2 - 2Rr \cos (\theta - \phi) + r^2}. \tag{2}$$

Equation (2) is known as the **Poisson formula** for the solution of the Laplace equation in a disk. In fact, it is a simple matter to show by direct

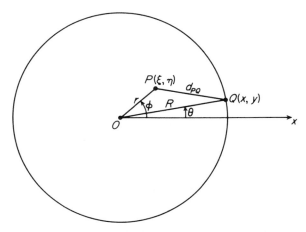

FIGURE 14

differentiation that the integral on the right in (2) satisfies the Laplace equation. So long as $P(\xi, \eta)$ is interior to K_R, differentiation under the integral sign is valid, and the verification is direct. If the boundary values are continuous, it can be shown that as P tends to the boundary of K_R, $u(P)$ tends to the boundary values. Under these circumstances the Poisson formula provides an existence theorem as well as a solution of the Dirichlet problem for a disk.

Referring to Fig. 14 where the distance between P and Q is denoted by d_{PQ}, we can rewrite (2) in the form

$$u(P) = \frac{R^2 - r^2}{2\pi R} \int_{\partial K_R} \frac{u(Q)\, ds_Q}{d_{PQ}^2}, \tag{3}$$

where Q is the variable point of integration and s_Q is arc length along ∂K_R. Now, if K_R denotes a ball of radius R in n-dimensional space, we can extend the Poisson formula to solutions u of the Laplace equation in n dimensions having prescribed values on the boundary ∂K_R. The result is

$$u(P) = \frac{R^2 - r^2}{R\omega_n} \int_{\partial K_R} \frac{u(Q)\, dS_Q}{d_{PQ}^n}, \tag{4}$$

where P is an interior point at distance r from the origin, Q is a boundary point, d_{PQ} is the distance $\overline{PQ}$, and ω_n is the surface area of the unit sphere in n dimensions. Once again, the fact that the integral (4) satisfies the Laplace equation in n dimensions may be verified by differentiation under the integral sign.

If P is the center O of the ball K_R, formula (4) yields the mean value theorem for harmonic functions (see Eq. 7 in Sec. 1):

$$u(0, 0, \ldots, 0) = \frac{1}{R^{n-1}\omega_n} \int_{\partial K_R} u(Q)\, dS_Q.$$

LEMMA. Let $u(x, y)$ be a **nonnegative harmonic function** in the disk K_R: $x^2 + y^2 < R^2$. If $P(x, y)$ is at distance $r < R$ from the origin, then

$$\frac{R - r}{R + r} u(0, 0) \leq u(x, y) \leq \frac{R + r}{R - r} u(0, 0). \tag{5}$$

Proof. Referring Fig. 14, we see that for fixed P, the distance d_{PQ} to any point on the boundary satisfies the inequalities

$$R - r \leq d_{PQ} \leq R + r.$$

Now formula (3) and the hypothesis that $u(Q)$ is nonnegative give the estimate

$$u(P) \leq \frac{R^2 - r^2}{2\pi(R - r)^2} \int_0^{2\pi} u(R\cos\theta, R\sin\theta)\, d\theta.$$

Employing the mean value theorem for harmonic functions, we obtain

$$u(x, y) \leq \frac{R+r}{R-r} u(0, 0).$$

In a similar way, we find

$$u(x, y) \geq \frac{R^2 - r^2}{2\pi(R+r)^2} \int_0^{2\pi} u(R\cos\theta, R\sin\theta)\, d\theta = \frac{R-r}{R+r} u(0, 0).$$

Remark. The extension of (5) to harmonic functions in n dimensions uses the Poisson formula (4). The result is

$$\frac{R-r}{(R+r)^{n-1}} R^{n-2} u(O) \leq u(P) \leq \frac{R+r}{(R-r)^{n-1}} R^{n-2} u(O). \tag{6}$$

Suppose P_1 and P_2 are two points which are at distance less than $\frac{1}{2}R$ apart, and suppose that u is harmonic in the disk $K_R(P_1)$ of radius R with center at P_1 (Fig. 15). We can use (5) to obtain a relation between $u(P_1)$ and $u(P_2)$.

Applying the above lemma to the disk $K_R(P_1)$, we find

$$\frac{R-r}{R+r} u(P_1) \leq u(P_2) \leq \frac{R+r}{R-r} u(P_1).$$

Since $r < \frac{1}{2}R$, we may write

$$\frac{R - \frac{1}{2}R}{\frac{3}{2}R} u(P_1) \leq u(P_2) \leq \frac{\frac{3}{2}R}{\frac{1}{2}R} u(P_1),$$

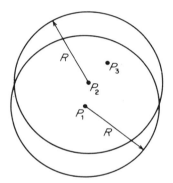

FIGURE 15

and so

$$\tfrac{1}{3} u(P_1) \leq u(P_2) \leq 3 u(P_1). \tag{7}$$

If P_3 is a third point at distance less than $\frac{1}{2}R$ from P_2, we may apply the lemma with P_2 as center and derive the inequalities

$$\tfrac{1}{3} u(P_2) \leq u(P_3) \leq 3 u(P_2),$$

from which we find

$$(\tfrac{1}{3})^2 u(P_1) \leq u(P_3) \leq (3)^2 u(P_1).$$

We may continue this process for any finite number of points.

If S is a closed, bounded, connected subset of a domain D and R is smaller than the distance of any point in S to the boundary of D, then, by the Heine-Borel theorem,* it is possible to cover S by a finite number k of disks of

---

*The Heine-Borel thorem states that if a closed, bounded set is contained in the union of an infinite collection of open sets, then it is already contained in the union of a finite number of open sets in the collection. In the present case, S is certainly in the union of all open disks of radius $\frac{1}{4}R$ having centers in S. Hence, S is in the union of a finite number of such disks.

radius $\frac{1}{2}R$, with centers in S. By choosing concentric disks of radius R, we see that if P and Q are any two points in S, they may be connected by a chain of k disks in D, each of radius R, such that the first disk has its center at P, the last its center at Q, and the centers of two successive disks are always at distance less than $R/2$ from each other (Fig. 16).

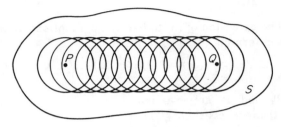

FIGURE 16

Employing (7) repeatedly, we obtain for any harmonic function D and any two points in S

$$(\tfrac{1}{3})^{k-1}u(P) \leq u(Q) \leq 3^{k-1}u(P).$$

Note that the number k depends only on the set S and the domain D. Since any closed bounded subset of D is contained in a closed, bounded, connected subset of D, the above development yields the next result.

THEOREM 21. (Harnack's Inequality). Let $u(x, y)$ be a nonnegative harmonic function defined in a domain D and let S be a closed bounded set contained in D. Then there is a positive constant A depending on S and D but not on u such that for every pair of points P and Q in S, we have

$$Au(P) \leq u(Q) \leq A^{-1}u(P). \tag{8}$$

Remarks. (i) An analogous result holds in any number of dimensions.

(ii) The selection of successive points closer together than $\frac{1}{2}R$ was one of convenience; any distance less than R is possible.

The utility of the Harnack inequalities is exhibited in the following convergence theorem which is a simple consequence.

THEOREM 22. Let $u_k(x, y)$, $k = 1, 2, \ldots$, be a nonincreasing sequence of functions, harmonic in a domain D. If the sequence $\{u_k\}$ converges at a single point $P(x_0, y_0)$ of D, it converges uniformly on every closed bounded subset of D.

Proof. We observe that for $i < j$ the function $v_{ij}(x, y) \equiv u_i(x, y) - u_j(x, y)$ is harmonic and nonnegative. If S is a closed bounded subset of D and $Q(x, y)$ is a point of S, we apply Harnack's inequality (8) to $v_{ij}(x, y)$, getting

$$0 \leq v_{i,j}(x, y) \leq A^{-1} v_{i,j}(x_0, y_0) = A^{-1}[u_i(x_0, y_0) - u_j(x_0, y_0)].$$

Since $\{u_k(x_0, y_0)\}$ converges by hypothesis, $\{v_{i,j}(x, y)\}$ approaches zero as $i, j \to \infty$, and the result follows.

Remarks. (i) The Harnack inequality and convergence theorems can be applied to harmonic functions which are bounded from above or below. If u is bounded from below by a constant m, then the harmonic function $v = u - m$ is nonnegative and Harnack's inequality is valid for it. Similarly, if u is bounded above by M, the function $w = M - u$ is harmonic and nonnegative.

(ii) If (r, θ) are polar coordinates, the function

$$u_\epsilon(r, \theta) = \begin{cases} \log(R/r) & \text{for } r \geq \epsilon \\ \log(R/\epsilon) + \dfrac{3}{4} - \dfrac{r^2}{\epsilon^2} + \dfrac{r^4}{4\epsilon^4} & \text{for } r \leq \epsilon \end{cases}$$

is superharmonic and nonnegative in the disk $r < R$, but the ratio $u_\epsilon(0, \theta)/u_\epsilon(\frac{1}{2}R, \theta)$ can be made arbitrarily large by choosing ϵ sufficiently small. Therefore, there is no Harnack inequality for the class of superharmonic functions. The example $(\frac{3}{4}) + \epsilon + \log(R/\epsilon) - u_\epsilon(r, \theta)$ shows that there is also no Harnack inequality for the class of subharmonic functions.

(iii) Suppose that $\Delta u = 0$ and $u \geq 0$ in all of n-space. Then if P and O are any points at distance r from each other, we have inequality (6) for any $R > r$. Letting $R \to \infty$, we find that

$$u(O) \leq u(P).$$

Interchanging the roles of P and O, we also find that

$$u(P) \leq u(O),$$

and hence that $u(P) = u(O)$. Therefore, any nonnegative function harmonic in all of n-space is constant. In the light of remark (i) above, we may conclude that **any harmonic function bounded either above or below in all of n-space is constant.** This statement is known as **Liouville's Theorem.**

Inequalities of Harnack's type have been obtained for general elliptic equations with smooth coefficients by Lichtenstein [1] and Feller [1]. We present here a modification of a proof due to Serrin [6] for the two-dimensional case. The advantages of Serrin's proof are twofold: (i) no special continuity or smoothness assumptions concerning the coefficients are made, and (ii) the considerations are entirely elementary in the sense that only the maximum principle is employed.

THEOREM 23. Let $u(x, y)$ satisfy the equation

$$(L + h)[u] \equiv au_{xx} + 2bu_{xy} + cu_{yy} + du_x + eu_y + hu = 0, \quad h \leq 0,$$

in a disk K_R of radius R centered at O, and let
$$u \geq 0 \text{ in } K_R.$$
Suppose that L is uniformly elliptic and that the coefficients of L are bounded in K_R. Then there is a constant $A > 0$ depending only on μ, the ellipticity constant, and on M, the bound for the coefficients of L, such that
$$u(P) \geq Au(O)$$
whenever P is at distance less than $\tfrac{1}{2}R$ from O.

Proof. Assume that $u \not\equiv 0$. Then by the maximum principle, $u(O) > 0$. Now consider the set of all points P such that $u(P) > \tfrac{1}{2}u(O)$. The particular component of this set which contains O must extend to the boundary of K_R. For otherwise its boundary would be a set where $u = \tfrac{1}{2}u(O)$, a number less than $u(O)$. Therefore u would have a positive relative maximum at an interior point, contradicting the maximum principle for elliptic operators. Therefore we can find a curve Γ connecting the point O to a boundary point Q such that
$$u(P) \geq \tfrac{1}{2}u(O) \text{ along } \Gamma.$$

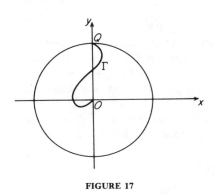

FIGURE 17

We translate and rotate the coordinate system so that the point O is at the origin and the point Q is on the $\bar{y}$-axis in the new $\bar{x}, \bar{y}$-coordinate system (Fig. 17). As we saw in Section 2, the transformation of coordinates takes L into a new elliptic operator of the same form with bounded coefficients $\bar{a}, \bar{b}, \bar{c}, \bar{d}, \bar{e}$ and with the same ellipticity constant μ. For convenience in notation, we shall drop the bars and use the letters x, y to denote the new coordinates; also, we shall designate the new coefficients of L by a, b, c, d, and e, as before.

The method of auxiliary functions, so useful in the previous applications, is employed here also. We introduce two functions z_+ and z_- defined by the formula
$$z_{\pm} = e^{-\alpha T_{\pm}} - 1,$$
where
$$T_{\pm} = [(y - \tfrac{1}{2}R)^2 \pm \tfrac{1}{3}Rx - \tfrac{1}{4}R^2]R^{-2}$$
and α is a positive constant to be determined appropriately. The function z_+ corresponds to T_+ and the function z_- to T_-. We compute $(L + h)[z_{\pm}]$ and find
$$(L + h)[z_{\pm}] = \alpha^2 R^{-2} e^{-\alpha T_{\pm}} [\tfrac{1}{9}a \pm \tfrac{4}{3}bR^{-1}(y - \tfrac{1}{2}R) + 4cR^{-2}(y - \tfrac{1}{2}R)^2]$$
$$- \alpha e^{-\alpha T_{\pm}} [2cR^{-2} \pm \tfrac{1}{3}dR^{-1} + 2R^{-2}e(y - \tfrac{1}{2}R)] + hz_{\pm}.$$

The uniform ellipticity of the operator L asserts that there is a positive constant μ such that for all real numbers ξ and η, the inequality
$$a\xi^2 + 2b\xi\eta + c\eta^2 \geq \mu(\xi^2 + \eta^2)$$
holds. Setting $\xi = \pm\frac{1}{3}$, $\eta = 2R^{-1}(y - \frac{1}{2}R)$, we find that
$$[\tfrac{1}{9}a \pm \tfrac{4}{3}bR^{-1}(y - \tfrac{1}{2}R) + 4cR^{-2}(y - \tfrac{1}{2}R)^2] \geq \mu[\tfrac{1}{9} + 4R^{-2}(y - \tfrac{1}{2}R)^2].$$
Inserting this inequality in the expression at the bottom of page 112, we obtain
$$(L + h)[z_\pm] \geq R^{-2}e^{-\alpha T_\pm}\{\alpha^2\mu[\tfrac{1}{9} + 4R^{-2}(y - \tfrac{1}{2}R)^2]$$
$$- \alpha[2c \pm \tfrac{1}{3}dR + 2e(y - \tfrac{1}{2}R)] + hR^2\} - h.$$
Since the coefficients $a, b, \ldots, e, h$ are bounded and h is nonpositive, we see that if α is selected sufficiently large, then
$$(L + h)[z_\pm] > 0$$
throughout the disk K_R. The size of α depends only on the value of μ and the bounds for the coefficients of $L + h$. The exponential quantities in $z_\pm$ play a special role. Figure 18 shows sketches of the parabolas
$$T_+ = 0: \quad (y - \tfrac{1}{2}R)^2 + \tfrac{1}{3}Rx - \tfrac{1}{4}R^2 = 0$$
and
$$T_- = 0: \quad (y - \tfrac{1}{2}R)^2 - \tfrac{1}{3}Rx - \tfrac{1}{4}R^2 = 0.$$
We note that
$$z_+ = 0 \text{ on } T_+ = 0$$
and
$$z_- = 0 \text{ on } T_- = 0.$$

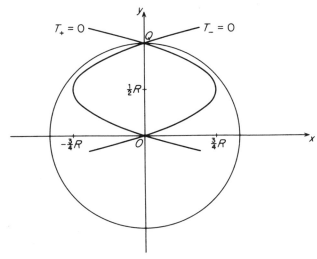

FIGURE 18

To the left of the parabola $T_+ = 0$, we have
$$R^2 T_+ \equiv (y - \tfrac{1}{2}R)^2 + \tfrac{1}{3}Rx - \tfrac{1}{4}R^2 < 0,$$
so that $z_+ > 0$. In the entire disk K_R, we obtain
$$-R^2 T_+ \equiv \tfrac{1}{4}R^2 - \tfrac{1}{3}Rx - (y - \tfrac{1}{2}R)^2 \leq \tfrac{1}{4}R^2 - \tfrac{1}{3}Rx \leq \tfrac{1}{4}R^2 + \tfrac{1}{3}R^2 = \tfrac{7}{12}R^2.$$
Consequently, inserting this last inequality in the expression for z_+, we find
$$0 < z_+ < e^{7\alpha/12}, \tag{9}$$
valid in the portion of K_R to the left of the parabola $T_+ = 0$. In a similar way, we derive the inequalities
$$0 < z_- < e^{7\alpha/12},$$
valid in the portion of K_R to the right of $T_- = 0$.

Let G^* denote the union of the closed sets $T_+ \leq 0$ and $T_- \leq 0$. Let Γ^* be the component which contains Q of the intersection of the curve Γ where $u \geq \tfrac{1}{2}u(O)$ with G^*. Let G denote the region between the parabolas, which is downward shaded in Fig. 19. Let V be any point of G and, for convenience,

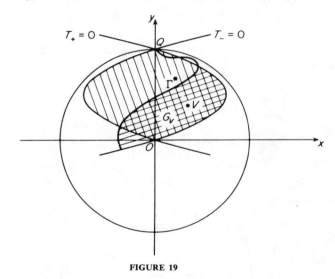

FIGURE 19

suppose it lies within the part of G^* to the right of the oriented curve Γ^*. Let G_V be the smallest subdomain of G^* containing V and bounded by arcs of Γ^* and of the parabola $T_+ = 0$ (shown upward shaded in Fig. 19). We recall that
$$u(P) \geq \tfrac{1}{2}u(O) \text{ for } P \text{ on } \Gamma^*$$
and hence, taking (9) into account,
$$u(P) \geq \tfrac{1}{2}e^{-7\alpha/12}u(O)z_+(P) \tag{10}$$

for P on the portion of Γ^* that lies on the boundary of G_V. Since $z_+ = 0$ on $T_+ = 0$, the inequality

$$u(P) \geq \tfrac{1}{2}e^{-7\alpha/12}u(O)z_+(P) \text{ for } P \text{ on } T_+ = 0$$

amounts to the assertion that $u(P) \geq 0$. Thus inequality (10) holds for all points on the boundary of G_V. Furthermore, we know that for $(x, y) \in G_V$

$$(L + h)[\tfrac{1}{2}e^{-7\alpha/12}u(O)z_+ - u] = \tfrac{1}{2}e^{-7\alpha/12}u(O)(L + h)[z_+] > 0.$$

Therefore, by the maximum principle, the inequality (10) which holds on the boundary of G_V must hold throughout the interior and, in particular, at the point V:

$$u(V) \geq \tfrac{1}{2}e^{-7\alpha/12}u(O)z_+(V).$$

Similarly, if V is any point of G to the left of Γ^* in G^*, we obtain

$$u(V) \geq \tfrac{1}{2}e^{-7\alpha/12}u(O)z_-(V).$$

We conclude that for any point of G

$$u(V) \geq \tfrac{1}{2}e^{-7\alpha/12}u(O) \min (z_+(V), z_-(V)). \tag{11}$$

The line segment I defined by

$$I: -\tfrac{3}{4}R < x < \tfrac{3}{4}R, \qquad y = \tfrac{1}{2}R$$

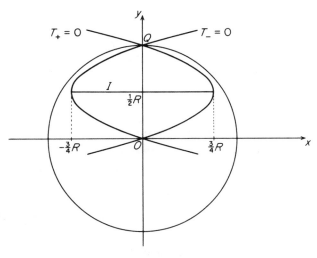

FIGURE 20

is in G (see Fig. 20). Along I, inequality (11) yields

$$u(x, \tfrac{1}{2}R) \geq \tfrac{1}{2}e^{-7\alpha/12}u(O)[e^{\alpha[(1/4) - (1/3)|x|R^{-1}]} - 1]. \tag{12}$$

We now form the parabola

$$\bar{P}: x^2 - \tfrac{3}{8}R(y + R) = 0.$$

It is easily checked that the function

$$z \equiv e^{-(2/9)\alpha R^{-2}[x^2 - (3/8)R(y+R)]} - 1$$

is zero on $\bar{P}$ and, for sufficiently large α, satisfies

$$(L + h)[z] > 0$$

in the part of K_R where $z \geq 0$. We now choose α so large that both $(L + h)[z_\pm] > 0$ and $(L + h)[z] > 0$ are satisfied. As before, the quantity α depends only on μ, R and bounds for the coefficients of $L + h$. The parabola $\bar{P}$ passes through the endpoints of I and has its vertex at $(0, -R)$ (see

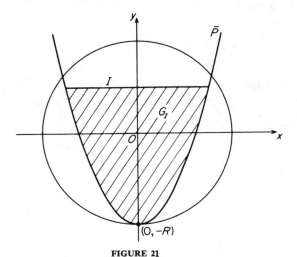

FIGURE 21

Fig. 21). We now argue for the region G_I bounded by I and $\bar{P}$ as we did for G_V. Along I we see by computation that

$$z(x, \tfrac{1}{2}R) = e^{-(2/9)\alpha R^{-2}[x^2 - (9/16)R^2]} - 1,$$
$$= e^{(1/8)\alpha - (2/9)\alpha x^2 R^{-2}} - 1,$$
$$\leq e^{\alpha[(1/4) - (1/3)|x|R^{-1}]} - 1.$$

Hence by using the vanishing of z on $\bar{P}$, the nonnegativity of u, and the inequality (12), we find

$$u(P) \geq \tfrac{1}{2} e^{-7\alpha/12} u(O) z(P)$$

for P on the boundary of G_I. The maximum principle yields the same inequality at interior points of G_I. In particular, this is true for the disk

$$K_{R/2}: x^2 + y^2 \leq \tfrac{1}{4}R^2$$

which lies entirely within G_I. A computation shows that

$$z \geq e^{\alpha/51} - 1 \text{ for } (x, y) \text{ in } K_{R/2}.$$

As a final conclusion, we get

$$u(P) \geq \tfrac{1}{2}e^{-7\alpha/12}(e^{\alpha/51} - 1)u(O) \text{ for } P \text{ in } K_{R/2} \tag{13}$$

which is the desired Harnack inequality with

$$A = \tfrac{1}{2}e^{-7\alpha/12}(e^{\alpha/51} - 1).$$

Remarks. (i) Instead of selecting a curve Γ joining O to the boundary with $u > \tfrac{1}{2}u(O)$ on Γ, we can use a curve Γ_η with $u > \eta u(O)$ along Γ_η with any value for η such that $0 < \eta < 1$. The remainder of the argument for the proof of the theorem is unchanged, and in this way the factor $\tfrac{1}{2}$ may be eliminated from the constant A.

(ii) Once inequality (13) is at hand we can argue exactly as in the case of harmonic functions to obtain bounds for a positive solution of $(L + h)[u] = 0$ on any closed bounded subset S of a domain D.

(iii) If the function h is not nonpositive, but if R is so small that there is a positive w which satisfies $(L + h)[w] \leq 0$ in K_R, we can obtain the Harnack inequality $u(P) \geq Au(O)$ for P in $K_{R/2}$ by applying the above theorem to the function u/w. We can then still obtain a Harnack inequality for any closed bounded subset S of a domain D where $(L + h)[u] = 0$ by covering it with sufficiently small disks.

(iv) Suppose a Harnack inequality $u(P) \geq Ku(Q)$ relates the values at two points P and Q of all positive solutions of $(L + h)[u] = 0$ in a domain D. If D is so small that there exists a positive function w which satisfies $(L + h)[w] \leq -1$, we can obtain a corresponding inequality for nonnegative solutions v of the inhomogeneous equation

$$(L + h)[v] = f$$

which are continuous in the closure of D. Let ϕ be the solution of $(L + h)[\phi] = 0$ which has the same (nonnegative) boundary values as v. Then we see by the methods of Section 6 that

$$|\phi - v| \leq w \sup_D |f|.$$

Combining this estimate with the Harnack inequality applied to ϕ, we obtain

$$v(P) \geq Kv(Q) - [Kw(Q) + w(P)] \sup_D |(L + h)[v]|.$$

This inequality holds for every function v which is positive and twice continuously differentiable in D. Serrin [6] has shown that an inequality of this form can be derived as in the proof of Theorem 23 without the hypothesis that the function ϕ exists.

(v) It may be seen by computing $(L + h)[z_+]$ and $(L + h)[z]$ that the constant α in (13) can be chosen as

$$\alpha = \sup_{x^2+y^2 \leq R^2} \{2(288)^2 \mu^{-2}(a^2 + 2b^2 + c^2 + R^2 d^2 + R^2 e^2) - 288\mu^{-1} R^2 h\}^{1/2} \tag{14}$$

in terms of the coefficients and the ellipticity constant of the operator L in the original coordinate system. We observe that this quantity α is a nondecreasing function of the radius R.

We shall now use the Harnack inequality to obtain information about the behavior of solutions of the equation $L[u] = 0$ at an isolated singularity. We follow the treatment of Gilbarg and Serrin [1].

Let u be a solution of the equation

$$L[u] \equiv au_{xx} + 2bu_{xy} + cu_{yy} + du_x + eu_y = 0,$$

in the punctured disk $0 < x^2 + y^2 < R_0^2$. We assume there is a constant B such that the inequality

$$a^2 + 2b^2 + c^2 + (x^2 + y^2)(d^2 + e^2) \leq B \tag{15}$$

is valid for all x, y. It is not difficult to verify that any two points P and Q on a circle $C_{2R}(O)$: $x^2 + y^2 = (2R)^2$ can be connected by a chain of at most 13 points on the circle at distance at most $\frac{1}{2}R$ from each other. If $R < \frac{1}{3}R_0$, then disks of radius R centered on $C_{2R}(O)$ lie in the domain $0 < x^2 + y^2 < R_0^2$. Applying the Harnack inequality repeatedly, we find that $u(Q) \geq A^{13} u(P)$, where the constant A depends only on the quantity α in (14). Now in a disk of radius R centered on the circle C_{2R}, we have $R^2 \leq x^2 + y^2$. Hence, by the boundedness hypothesis (15) on the coefficients, A can be chosen so that it is independent of R.

We now suppose that $u(x, y)$ has a finite limit superior M as $x^2 + y^2 \to 0$. Then for any $\epsilon > 0$ there is a radius R_ϵ such that $M - u + \epsilon > 0$ for $x^2 + y^2 < R_\epsilon^2$, and there is a sequence of points $\{P_\nu\}$ approaching the origin such that $u(P_\nu) > M - \epsilon$. Let the distance from the origin to P_ν be $2R_\nu$. Then $R_\nu \to 0$, and we may choose a subsequence so that R_ν is monotone decreasing and each R_ν is less than $\frac{1}{3}R_0$ as well as R_ϵ. We now apply the above Harnack inequality on the circle $C_{2R_\nu}(O)$ to the function $M - u + \epsilon$, observing that $L[M - u + \epsilon] = 0$. We find that $M - u + \epsilon < 2A^{-13}\epsilon$ on C_{2R_ν}. Then, by the maximum principle, $u > M - (1 + 2A^{-13})\epsilon$ in the annular regions between these circles. Thus we find that for any $\epsilon > 0$ there is an R_1 such that

$$|u - M| < (1 + 2A^{-13})\epsilon$$

for $x^2 + y^2 < (2R_1)^2$. In other words, $u \to M$ as $x^2 + y^2 \to 0$.

If the limit superior of u is $-\infty$, then by definition u has the limit $-\infty$. Thus if u is bounded above, it has a limit. Since $L[-u] = 0$, we reach the same conclusion if u is bounded below.

We have established the following theorem:

THEOREM 24. If u is a solution of the uniformly elliptic differential equation
$$L[u] \equiv au_{xx} + 2bu_{xy} + cu_{yy} + du_x + eu_y = 0$$
in a punctured disk $0 < x^2 + y^2 < R_0^2$, if u is bounded either from above or from below, and if the quantity
$$a^2 + 2b^2 + c^2 + (x^2 + y^2)(d^2 + e^2)$$
is bounded, then $\lim_{x^2+y^2 \to 0} u(x, y)$ exists. (It may be $\pm \infty$.)

Remarks. (i) We observe that in the above theorem the coefficients d and e are permitted to be unbounded.

(ii) Suppose that u is a solution of $(L+h)u = f$ in the punctured disk, and that, in addition to the hypotheses of Theorem 24,
$$\lim_{x^2+y^2 \to 0} (x^2 + y^2)(d^2 + e^2 + |h| + |f|) = 0.$$
It is then easily verified that on the disk $x^2 + y^2 \leq (3R)^2$ the function $w = \mu^{-1}(2(3R)^2 - x^2 - y^2)$ satisfies the inequalities $\mu^{-1}(3R)^2 \leq w \leq 2\mu^{-1}(3R)^2$, and $(L+h)w \leq -1$ when R is sufficiently small. Then the proof of Theorem 24 with Theorem 23 replaced by its modification given in Remarks (iii) and (iv) on page 117 and with R chosen sufficiently small gives the existence of the limit of a solution of $(L+h)u = f$. The fact that $u = \cos \theta$ satisfies both the equations $\Delta u + r^{-2}u = 0$ and $\Delta u = -r^{-2} \cos \theta$ shows that a growth condition on f and h is needed.

We now show how this result may be used to establish a decomposition theorem for solutions of elliptic equations in a punctured domain.

THEOREM 25. Suppose that w is a solution of $L[w] = 0$ in the punctured disk $D: 0 < x^2 + y^2 < R_0^2$, and that $w \to +\infty$ as $x^2 + y^2 \to 0$. Also suppose that L is uniformly elliptic in D and that the quantity
$$a^2 + 2b^2 + c^2 + (x^2 + y^2)(d^2 + e^2)$$
$$+ (x^2 + y^2)w^{-2}\{(2aw_x + 2bw_y)^2 + (2bw_x + 2cw_y)^2\}$$
is bounded. Then any solution u of $L[u] = 0$ in D for which u/w is bounded either above or below may be decomposed into a sum of a multiple of w and a solution q of $L[q] = 0$ which attains its maximum on the circle $x^2 + y^2 = R_0^2$ and has a limit at $(0, 0)$.

Proof. We first note that u/w satisfies in D an equation of the form $\tilde{L}[u/w] = 0$ to which Theorem 24 can be applied. If u/w is bounded above or below, then $\lim (u/w)$ exists as $x^2 + y^2 \to 0$. If $u/w \to +\infty$ then $u \to +\infty$ and $w/u \to 0$ as $x^2 + y^2 \to 0$. In this case we can apply the Phragmèn-

Lindelöf principle (Corollary to Theorem 19) and conclude that w is bounded in D, contrary to hypothesis. Similarly, the limit cannot be $-\infty$. Thus $u/w \to c$, a finite number. We define $q = u - cw$ and observe that q is a solution of $(L + h)[q] = 0$ which satisfies $q/w \to 0$ and hence is bounded by the Corollary to Theorem 19.

In particular, if a removable singularity theorem holds, then u is equal to cw plus a solution in the whole disk $x^2 + y^2 < R_0^2$.

Remark. By Remark (ii) after Theorem 24, the statement of Theorem 25 can be extended to solutions of $(L+h)[u] = f$ when also $(x^2 + y^2) \cdot (d^2 + e^2 + |h| + |f|) \to 0$ as $x^2 + y^2 \to 0$.

If u is a solution of $L[u] = 0$ in an exterior domain $x^2 + y^2 > R_0^2$, the change of variables $\xi = x/(x^2 + y^2)$, $\eta = y/(x^2 + y^2)$ makes u a solution of a uniformly elliptic equation of the same form in the domain $0 < x^2 + y^2 < R_0^{-2}$. By applying Theorem 24 to the transformed problem, we get the next result.

THEOREM 26. *If u is a solution of the uniformly elliptic equation*

$$L[u] \equiv au_{xx} + 2bu_{xy} + cu_{yy} + du_x + eu_y = 0$$

in the exterior domain $x^2 + y^2 > R_0^2$, if u is bounded either above or below, and if the quantity $a^2 + 2b^2 + c^2 + (x^2 + y^2)(d^2 + e^2)$ is bounded, then $\lim\limits_{x^2+y^2\to\infty} u(x, y)$ exists. (It may be $\pm \infty$.)

Remark. We observe that the condition on the coefficients is such that d and e must approach zero as $x^2 + y^2 \to \infty$. The necessity of such a restriction is demonstrated by the function $u = e^x$ which is a solution of

$$u_{xx} + u_{yy} - u_x = 0.$$

If u satisfies $L[u] = 0$ in the entire x, y-plane, we see from the maximum principle that u cannot be larger or smaller than its limit at infinity. Thus we obtain the Liouville theorem:

THEOREM 27. (Liouville). *If u is a solution of the uniformly elliptic equation*

$$L[u] \equiv au_{xx} + 2bu_{xy} + cu_{yy} + du_x + eu_y = 0,$$

which is bounded either above or below in the entire x, y-plane, and if the quantity

$$a^2 + 2b^2 + c^2 + (x^2 + y^2)(d^2 + e^2)$$

is bounded, then u is a constant.

Serrin [6, 7] has also established a Harnack inequality for solutions of elliptic equations in n dimensions with $n > 2$. However, in contrast with the two-dimensional case, he has to assume either some smoothness conditions on the coefficients or a bound for the gradient of the solution.

By using the latter result, Gilbarg and Serrin [1] have established results analogous to Theorems 24, 25, and 26 in n dimensions, provided the coefficients have a limit at the isolated singular point.

Moser [1] has obtained a Harnack inequality for a uniformly elliptic equation in the divergence form

$$(L + h)[u] \equiv \sum_{i,j=1}^{n} \frac{\partial}{\partial x_i}\left(a_{ij} \frac{\partial u}{\partial x_j}\right) + \sum_{i=1}^{n} b_i \frac{\partial u}{\partial x_i} + hu = 0, \qquad h \leq 0$$

with few restrictions on the coefficients. Such a result is useful in the study of nonlinear equations.

EXERCISES

1. Verify that Poisson's formula (2) on page 107 gives a solution to the Laplace equation in the plane. Verify the same result in n dimensions, $n > 2$, for formula (4) on page 108.

2. Define $u_k(x, y) = a_k r^k \cos k\theta$, $k = 1, 2, \ldots$, where the $\{a_k\}$ form a bounded sequence. Show that $\{u_k\}$ converges uniformly on any closed set contained in D: $x^2 + y^2 < 1$. To what extent can the boundedness hypothesis on $\{a_k\}$ be weakened? Is the limit function harmonic?

3. Verify formula (14) on page 118.

4. Suppose that in the operator $L + h$ the quantity
$$a^2 + 2b^2 + c^2 + (x^2 + y^2)(d^2 + e^2 - h)$$
is bounded. Let u be a nonnegative solution of $(L + h)[u] = 0$ in a disk of radius R with center at a point P at distance $2R$ from the origin. Assume that $h \leq 0$. Show that there is a constant A independent of R such that $u(Q) > Au(P)$ for any point Q within distance $\frac{1}{2}R$ of P.

5. Using the result of Exercise 4, show that if $(L + h)[u] = 0$ for $0 < x^2 + y^2 < R_0^2$, then the quantity $u(x, y)(x^2 + y^2)^{\log(1/A)/2\log(5/4)}$ is bounded.

6. Show that if $|\Delta u| \leq M$ and $u \geq 0$ in the disk $x^2 + y^2 < R^2$, then
$$\frac{R-r}{R+r}\{u(0, 0) - \tfrac{1}{4}M[R^2 + (R + r)^2]\} \leq u(x, y)$$
$$\leq \frac{R+r}{R-r}\{u(0, 0) + \tfrac{1}{4}M[R^2 + (R - r)^2]\}$$

SECTION 11. CAPACITY

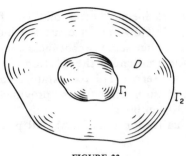

FIGURE 22

Let D be a three-dimensional domain whose boundary ∂D consists of two smooth, simple, closed surfaces Γ_1 and Γ_2, with Γ_1 lying entirely inside Γ_2 (Fig. 22). It can be shown that there is a function $u(x, y, z)$ in D with the properties:

$$\Delta u = 0 \text{ in } D,$$
$$u = 1 \text{ on } \Gamma_1,$$
$$u = 0 \text{ on } \Gamma_2.$$

Theorem 9 of Section 4 establishes the uniqueness of such a function. In appropriate units, the quantity u may be interpreted physically as the electrostatic potential at any point $P(x, y, z)$ of D when a potential difference of exactly one unit is maintained between the perfect conductors Γ_1 and Γ_2.

The *total charge* $C = C(\Gamma_1, \Gamma_2)$ induced by this potential difference is given by the formula

$$C(\Gamma_1, \Gamma_2) = -\frac{1}{4\pi} \int_{\Gamma_2} \frac{\partial u}{\partial \mathbf{n}} dS.$$

The quantity C is called the **electrostatic capacity** or simply the **capacity of Γ_1 with respect to Γ_2**.

Theorem 15 of Section 7 shows that if u is any function harmonic in D and possessing a continuous normal derivative on the boundary, the relation

$$\int_{\partial D} \frac{\partial u}{\partial \mathbf{n}} dS = 0$$

holds. For the annular domain under consideration, this identity shows that

$$-\frac{1}{4\pi} \int_{\Gamma_2} \frac{\partial u}{\partial \mathbf{n}} dS = \frac{1}{4\pi} \int_{\Gamma_1} \frac{\partial u}{\partial \mathbf{n}} dS.$$

Since the function $u_1 = 1 - u$ is harmonic in D, equal to 1 on Γ_2, and equal to 0 on Γ_1, we see that the relation

$$-\frac{1}{4\pi} \int_{\Gamma_1} \frac{\partial u_1}{\partial \mathbf{n}} dS = \frac{1}{4\pi} \int_{\Gamma_1} \frac{\partial u}{\partial \mathbf{n}} dS$$

implies that the capacity of Γ_1 with respect to Γ_2 is the same as that of Γ_2 with respect to Γ_1; that is,

$$C(\Gamma_1, \Gamma_2) = C(\Gamma_2, \Gamma_1).$$

Moreover, if Γ' is any simple closed surface situated between Γ_1 and Γ_2 and separating them, we find that $\int_{\Gamma'} (\partial u/\partial \mathbf{n})\, dS = \int_{\Gamma_1} (\partial u/\partial \mathbf{n})\, dS$, so that the capacity may be computed by integrating the normal derivative of u along the arbitrary surface Γ'.

We shall now show how the maximum principle may be employed to obtain comparisons between the capacities of conductors of various sizes. We consider the effect of replacing the conductor Γ_1 by a conductor $\bar{\Gamma}_1$ which is larger in the sense that it lies in the domain D and separates Γ_1 and Γ_2 (see Fig. 23). We denote by $\bar{D}$ the domain bounded by $\bar{\Gamma}_1$ and Γ_2. The function $\bar{u}$ is the electrostatic potential obtained when a unit potential difference is maintained between $\bar{\Gamma}_1$ and Γ_2. That is, $\bar{u}$ is the solution of the problem

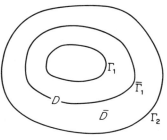

FIGURE 23

$$\Delta \bar{u} = 0 \text{ in } \bar{D},$$
$$\bar{u} = 1 \text{ on } \bar{\Gamma}_1,$$
$$\bar{u} = 0 \text{ on } \Gamma_2.$$

Calling $\bar{C} = C(\bar{\Gamma}_1, \Gamma_2)$ the capacity of $\bar{\Gamma}_1$ with respect to Γ_2, we have

$$\bar{C} = -\frac{1}{4\pi} \int_{\Gamma_2} \frac{\partial \bar{u}}{\partial \mathbf{n}}\, dS.$$

According to the maximum principle, the values of the function u lie between zero and one in D. Defining (in $\bar{D}$) the function $v = u - \bar{u}$, we see that

$$\Delta v = 0 \text{ in } \bar{D},$$
$$v \leq 0 \text{ on } \bar{\Gamma}_1,$$
$$v = 0 \text{ on } \Gamma_2.$$

The maximum of v is zero and it is attained everywhere on Γ_2. According to Theorem 7 we know that

$$\frac{\partial v}{\partial \mathbf{n}} > 0 \text{ on } \Gamma_2.$$

It follows that

$$C - \bar{C} = -\frac{1}{4\pi} \int_{\Gamma_2} \left(\frac{\partial u}{\partial \mathbf{n}} - \frac{\partial \bar{u}}{\partial \mathbf{n}} \right) dS$$

$$= -\frac{1}{4\pi} \int_{\Gamma_2} \frac{\partial v}{\partial \mathbf{n}}\, dS < 0.$$

Thus we have the inequality
$$C < \bar{C}.$$
In other words, **for fixed Γ_2 the electrostatic capacity increases as Γ_1 moves outward.** Similarly, it can be shown that for fixed Γ_1 the capacity increases as Γ_2 moves inward.

The above property of capacity may be used to obtain bounds for the capacity of any pair of conductors. Let K_1 and K_2 be concentric spheres of radii R_1 and R_2, respectively ($R_2 > R_1$). The capacity of this pair of conductors may be found explicitly. Taking the common center of the spheres as the origin of coordinates and letting $r^2 = x^2 + y^2 + z^2$, we easily verify that

$$u(x, y, z) = \frac{R_1 R_2}{R_2 - R_1}\left(\frac{1}{r} - \frac{1}{R_2}\right)$$

is harmonic in the region between K_1 and K_2. Furthermore, $u = 0$ when $r = R_2$ and $u = 1$ when $r = R_1$. Thus u is the required electrostatic potential function. Since $\partial/\partial \mathbf{n} = \partial/\partial r$ on K_2, we have

$$\left.\frac{\partial u}{\partial \mathbf{n}}\right|_{r=R_2} = -\frac{R_1}{R_2(R_2 - R_1)},$$

and the capacity $C(K_1, K_2)$ of the two concentric spheres is

$$C(K_1, K_2) = \frac{1}{4\pi}\frac{R_1}{R_2(R_2 - R_1)}\int_{K_2} dS = \frac{R_1 R_2}{R_2 - R_1}.$$

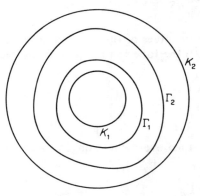

FIGURE 24

If K_1 is interior to a fixed surface Γ_1 and K_2 encloses a surface Γ_2 containing Γ_1 (see Fig. 24), then the capacity $C(\Gamma_1, \Gamma_2)$ satisfies the inequality

$$C(\Gamma_1, \Gamma_2) > C(K_1, K_2) = \frac{R_1 R_2}{R_2 - R_1}.$$

Similarly, this method gives an upper bound for $C(\Gamma_1, \Gamma_2)$ whenever it is possible to find concentric spheres lying in the domain between Γ_1 and Γ_2 which also separate Γ_1 and Γ_2.

We have shown that the capacity of a surface Γ_1 with respect to Γ_2 decreases as Γ_2 increases in size. Suppose we select for Γ_2 a sphere K_R of radius R. Then the capacity of Γ_1 with respect to K_R decreases steadily as R tends to infinity. Since capacity is always nonnegative, it must tend to a limit as $R \to \infty$. We are now in a position to define the capacity of a single conductor Γ_1.

DEFINITION. The capacity of a conductor Γ_1 is the limit as $R \to \infty$ of the capacity of Γ_1 with respect to the sphere K_R. We denote this quantity by $C(\Gamma_1)$.

Let u be the function which satisfies
$$\Delta u = 0 \text{ in the exterior of } \Gamma_1$$
and the additional conditions
$$u = 1 \text{ on } \Gamma_1,$$
$$\lim_{r \to \infty} u(x, y, z) = 0.$$

By the last condition, we see that for any $\epsilon > 0$ there is an R such that $0 < u < \epsilon$ on the sphere K_R. By the argument used in deriving the monotonicity of the capacity, we conclude that
$$\int_{\Gamma_1} \frac{\partial u}{\partial \mathbf{n}} dS \leq C(\Gamma_1, K_R).$$

On the other hand, if we apply the same argument to the function $(u - \epsilon)/(1 - \epsilon)$ which is nonpositive on K_R and 1 on Γ_1, we find that
$$\frac{1}{1 - \epsilon} \int_{\Gamma_1} \frac{\partial u}{\partial \mathbf{n}} dS \geq C(\Gamma_1, K_R).$$

Thus
$$(1 - \epsilon) C(\Gamma_1, K_R) \leq \int_{\Gamma_1} \frac{\partial u}{\partial \mathbf{n}} dS \leq C(\Gamma_1, K_R).$$

Letting $R \to \infty$ and $\epsilon \to 0$, we obtain
$$C(\Gamma_1) = \frac{1}{4\pi} \int_{\Gamma_1} \frac{\partial u}{\partial \mathbf{n}} dS.$$

Again, by Theorem 15 of Section 7, we can write
$$C(\Gamma_1) = -\frac{1}{4\pi} \int_{K_R} \frac{\partial u}{\partial r} dS,$$
where K_R is *any* sphere containing Γ_1.

If Γ_1 is a sphere K_1 of radius R_1, we find
$$C(K_1) = \lim_{R \to \infty} C(K_1, K_R) = \lim_{R \to \infty} \frac{R R_1}{R - R_1} = R_1.$$

The capacity of a sphere is given by its radius.

By a limiting argument it can be shown that **the capacity of a conductor increases as the conductor moves outward.**

We shall now use the maximum principle to obtain comparisons between two conductors, neither one of which includes the other. Suppose

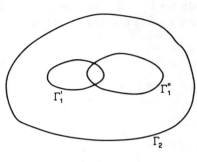

FIGURE 25

Γ_1' and Γ_1'' are two simple closed surfaces contained in the interior of the closed surface Γ_2. The surfaces Γ_1' and Γ_1'' may or may not intersect (Fig. 25). We denote by Γ_1 the union of the surfaces Γ_1' and Γ_1'', and we define three potential functions:

(i) The function u' has the properties

$\Delta u' = 0$ between Γ_1' and Γ_2,

$u' = 1$ on Γ_1', $\qquad u' = 0$ on Γ_2.

(ii) The function u'' has the properties

$\Delta u'' = 0$ between Γ_1'' and Γ_2, $\qquad u'' = 1$ on Γ_1'', $\qquad u'' = 0$ on Γ_2.

(iii) The function u has the properties

$\Delta u = 0$ between Γ_1 and Γ_2, $\qquad u = 1$ on Γ_1, $\qquad u = 0$ on Γ_2.

We now define the function

$$v = u' + u'' - u$$

in the domain between Γ_1 and Γ_2, and observe that $\Delta v = 0$ and that

$$v > 0 \text{ on } \Gamma_1,$$
$$v = 0 \text{ on } \Gamma_2.$$

Since the minimum of v occurs at every point of Γ_2, the maximum principle yields

$$\frac{\partial v}{\partial \mathbf{n}} < 0 \text{ on } \Gamma_2.$$

Therefore, in obvious notation,

$$C' + C'' - C = -\frac{1}{4\pi} \int_{\Gamma_2} \left(\frac{\partial u'}{\partial \mathbf{n}} + \frac{\partial u''}{\partial \mathbf{n}} - \frac{\partial u}{\partial \mathbf{n}} \right) dS = -\frac{1}{4\pi} \int_{\Gamma_2} \frac{\partial v}{\partial \mathbf{n}} dS > 0,$$

or

$$C < C' + C''. \qquad (1)$$

In other words, the **capacity with respect to a given surface Γ_2 is a subadditive functional of Γ_1**. In particular, the inequality holds if Γ_2 is a sphere K_R of radius R. We let $R \to \infty$ and conclude that if $C(\Gamma')$, $C(\Gamma'')$, and $C(\Gamma)$ are the capacities of single conductors with $\Gamma = \Gamma' \cup \Gamma''$, then

$$C(\Gamma) \leq C(\Gamma') + C(\Gamma''). \qquad (2)$$

Inequalities (1) and (2) can be improved if we use the fact that u' and u'' are bounded away from zero on Γ_1. Let m' and m'' be positive constants such that

$$u' \geq m' \text{ on } \Gamma_1'',$$
$$u'' \geq m'' \text{ on } \Gamma_1'.$$

We introduce the harmonic function

$$v = u - \frac{(1-m'')u' + (1-m')u''}{1 - m'm''}$$

and observe that $v = 0$ on Γ_2 and $v \leq 0$ on Γ_1. Then by the maximum principle $(\partial v/\partial \mathbf{n}) > 0$ on Γ_2 and, using the same reasoning as before, we find

$$C < \frac{(1-m'')C' + (1-m')C''}{1 - m'm''}.$$

We can write this inequality in the form

$$C < C' + C'' - \frac{m''(1-m')C' + m'(1-m'')C''}{1 - m'm''}. \tag{3}$$

The bound (3) is an improvement over (1).

In order for (3) to be useful it is essential to be able to estimate m' and m'' without actual knowledge of the potential functions u' and u''. We show how this can be done in terms of the geometry of the conductors.

Applying Green's third identity, as given in equation (12) of Section 7, to the potential u' at a point (ξ, η, ζ) in the region between Γ_1' and Γ_2, we find

$$u'(\xi, \eta, \zeta) = \frac{1}{4\pi} \int_{\Gamma_2} \frac{1}{r} \frac{\partial u'}{\partial \mathbf{n}} dS - \frac{1}{4\pi} \int_{\Gamma_1'} \frac{1}{r} \frac{\partial u'}{\partial \mathbf{n}} dS, \tag{4}$$

where r denotes distance from (ξ, η, ζ). (Note that $\int_{\Gamma_1'} (\partial/\partial \mathbf{n})(1/r)\, dS = 0$ by Green's first identity.) The derivatives $\partial/\partial \mathbf{n}$ are taken in the direction from the interior to the exterior of each surface. We observe that $\partial u'/\partial \mathbf{n} < 0$ on both Γ_1' and Γ_2 and that

$$C' = -\frac{1}{4\pi} \int_{\Gamma_1'} \frac{\partial u'}{\partial \mathbf{n}} dS = -\frac{1}{4\pi} \int_{\Gamma_2} \frac{\partial u'}{\partial \mathbf{n}} dS. \tag{5}$$

We define d_1' and δ_1' as the maximum and minimum distances, respectively, from (ξ, η, ζ) to any point of Γ_1'. Similarly, we define d_2 and δ_2 to be maximum and minimum distances to Γ_2. Then, inserting these maxima and minima, in turn, in (4) and making use of (5), we get

$$C'\left(\frac{1}{d_1'} - \frac{1}{\delta_2}\right) \leq u'(\xi, \eta, \zeta) \leq C'\left(\frac{1}{\delta_1'} - \frac{1}{d_2}\right). \tag{6}$$

Similar bounds hold for u''. In this way values for m' and m'' may be obtained for use in (3).

In the case of the capacity of a single conductor, Γ_2 is at infinity, and so $1/\delta_2 = 1/d_2 = 0$. Denoting by d the greatest distance between any point of Γ_1' and any point of Γ_1'', we obtain from (6) the inequality $u' \geq C'/d$ on Γ_2'. Thus in (3), we may take $m' = C'/d$. Similarly, we may choose $m'' = C''/d$. Then (3) becomes

$$C < C' + C'' - \frac{C'C''(2d - C' - C'')}{d^2 - C'C''},$$

an estimate which depends only on the geometry of the conductors.

EXERCISES

1. Find upper and lower bounds for the capacity of a cube with edge of length 1.

2. Find upper and lower bounds for the capacity of a regular tetrahedron with edge of length 1.

3. A tetrahedron has vertices at $(0, 0, 0)$, $(1, 0, 0)$, $(0, 1, 0)$, and $(0, 0, 1)$. A cube with edge of length 10 is situated so that its center is at the origin and its sides are parallel to the coordinate planes. Find upper and lower bounds for the capacity of the tetrahedron with respect to the cube.

4. Find upper and lower bounds for the capacity of an array of three spheres each of unit radius and with centers at $(0, 0, 0)$, $(-2, 0, 0)$, $(2, 0, 0)$.

5. The capacity of a disk of radius a is $(2/\pi)a$. Find bounds for the capacity of the set consisting of two parallel disks each of radius a with the line joining the centers of the disks of length $2a$ and perpendicular to the planes containing the disks.

SECTION 12. THE HADAMARD THREE-CIRCLES THEOREM

Consider the ring-shaped plane domain D formed by two concentric circles with radii R_1 and R_2, $(R_2 > R_1)$. Let u be a subharmonic function defined in D; that is,

$$\Delta u \geq 0 \text{ in } D.$$

Setting $r^2 = x^2 + y^2$ and taking the common center of the circles as the origin of coordinates (see Fig. 26), we define

$$M(r) = \max_{x^2+y^2=r^2} u(x, y).$$

In other words, the function $M(r)$ is the maximum of u on the concentric circle of radius r.

Sec. 12   The Hadamard Three-Circles Theorem

We observe that any function of the form

$$\varphi(r) = a + b \log r, \qquad (1)$$

with a and b constant, is harmonic for $r \neq 0$. If r_1 and r_2 are any two numbers between R_1 and R_2, we may choose a and b so that

$$\varphi(r_1) = M(r_1),$$
$$\varphi(r_2) = M(r_2).$$

A simple computation shows that

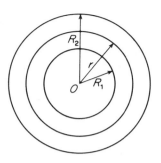

FIGURE 26

$$\varphi(r) = \frac{M(r_1) \log (r_2/r) + M(r_2) \log (r/r_1)}{\log (r_2/r_1)}. \qquad (2)$$

Defining $v(x, y) = u(x, y) - \varphi(\sqrt{x^2 + y^2})$, we have

$$\Delta v \geq 0,$$
$$v \leq 0 \text{ on } r = r_1 \text{ and } r = r_2.$$

Hence, by the maximum principle,

$$v \leq 0 \text{ for } r_1 < r < r_2$$

with equality if and only if $u \equiv \phi(r)$. Thus we find that when $x^2 + y^2 = r^2$,

$$u \leq \varphi(r), \qquad r_1 < r < r_2,$$

and therefore that

$$M(r) \leq \varphi(r), \qquad r_1 < r < r_2.$$

This argument establishes the following result.

THEOREM 28. (Hadamard Three-Circles Theorem). *Let $u(x, y)$ be subharmonic in a domain D containing concentric circles of radii r_1 and r_2 and the region between them. If $M(r)$ denotes the maximum of u on any concentric circle of radius r, then for $r_1 < r < r_2$*

$$M(r) \leq \frac{M(r_1) \log (r_2/r) + M(r_2) \log (r/r_1)}{\log (r_2/r_1)}. \qquad (3)$$

Equality occurs if and only if $u \equiv \varphi$, where φ is given by (2).

A function $f(x)$ is a **convex function** of x if for any two numbers x_1 and x_2, we have

$$f(x) \leq f(x_1) \frac{x_2 - x}{x_2 - x_1} + f(x_2) \frac{x - x_1}{x_2 - x_1} \text{ for } x_1 \leq x \leq x_2.$$

That is, the graph of f lies below the straight line connecting the points $(x_1, f(x_1))$ and $(x_2, f(x_2))$. The Hadamard three-circles theorem states that $M(r)$ **is a convex function of** $\log r$.

Suppose that u is subharmonic in the whole x, y-plane, except possibly at the origin, and is bounded above by a constant M. If we let $r_2 \to \infty$ in inequality (3), we obtain

$$M(r) \leq M(r_1) \lim_{r_2 \to \infty} \frac{\log(r_2/r)}{\log(r_2/r_1)} + \lim_{r_2 \to \infty} \left(M(r_2) \frac{\log(r/r_1)}{\log(r_2/r_1)} \right).$$

Since $M(r_2) \leq M$, the second limit on the right is zero; a simple application of l'Hôpital's rule shows that the first limit is 1. Thus we find

$$M(r) \leq M(r_1) \text{ for } r \geq r_1.$$

Similarly, letting $r_1 \to 0$ in inequality (3) yields the result

$$M(r) \leq M(r_2) \text{ for } r \leq r_2.$$

Since r_1 and r_2 are arbitrary, we conclude that $M(r)$ is constant. Therefore by the strong maximum principle u must be a constant. We have proved the following theorem.

THEOREM 29. (Liouville's Theorem). **If u is subharmonic in the whole x, y-plane except possibly at the origin and if u is uniformly bounded above, then u is a constant.**

Remarks. (i) From the method of proof it is clear that the boundedness hypothesis of u may be weakened. By taking limits over sequences of values of r_1 and r_2, we can show that u is constant if

$$\liminf \frac{M(r)}{|\log r|} \leq 0$$

as $r \to 0$ and as $r \to \infty$.

(ii) The preceding remark shows that for a nonconstant subharmonic function defined in the entire plane except at the origin, $M(r)$ must approach infinity at least as rapidly as $|\log r|$, either as r tends to zero or as r tends to infinity.

The two subharmonic functions:

$$u = \begin{cases} -(\tfrac{3}{4} - r^2 + \tfrac{1}{4}r^4), & \text{for } r \leq 1, \\ \log r, & \text{for } r \geq 1, \end{cases}$$

and

$$u = \begin{cases} \log(1/r) - (\tfrac{3}{4} - r^2 + \tfrac{1}{4}r^4), & \text{for } r \leq 1, \\ 0, & \text{for } r \geq 1, \end{cases}$$

grow logarithmically, one at infinity, the other at the origin. Therefore we have obtained the best possible result concerning the growth of a subharmonic function.

(iii) It is clear that the analogue of Liouville's theorem holds for superharmonic functions which are bounded from below. In particular, a function which is harmonic in the whole x, y-plane except possibly at one point and which is bounded either above or below must be constant.

The three-circles theorem depends on the special behavior of the fundamental solution $\log r$. In extending this theorem to functions of n variables, $n \geq 3$, we make use of the fundamental solution $r^{-(n-2)}$. Let D be the domain between two concentric n-dimensional spheres of radii R_1 and R_2 with $R_2 > R_1$. Suppose that u satisfies

$$\Delta u \geq 0 \text{ in } D.$$

We define

$$M(r) = \max_{\sum_{i=1}^{n} x_i^2 = r^2} u(x_1, x_2, \ldots, x_n)$$

and observe that

$$\varphi(r) \equiv a + br^{-(n-2)}$$

is a harmonic function for $r \neq 0$. Setting

$$\varphi(r_1) = M(r_1),$$
$$\varphi(r_2) = M(r_2),$$

where $R_1 < r_1 < r_2 < R_2$, we get a three-spheres theorem by an argument similar to that used for the three-circles theorem.

THEOREM 30. (Three-Spheres Theorem). Suppose $\Delta u \geq 0$ in a domain D containing two concentric spheres of radii r_1 and r_2 and the region between them. If $r_1 < r < r_2$, then

$$M(r) \leq \frac{M(r_1)(r^{2-n} - r_2^{2-n}) + M(r_2)(r_1^{2-n} - r^{2-n})}{r_1^{2-n} - r_2^{2-n}}.$$

Equality holds if and only if $u = a + br^{-(n-2)}$.

Remarks. (i) The inequality in this theorem states that $M(r)$ is a convex function of r^{2-n}.

(ii) Letting $r_1 \to 0$, we see that if

$$\liminf_{r_1 \to 0} r_1^{n-2} M(r_1) \leq 0,$$

then $M(r)$ is a nondecreasing function of r. It follows that u is bounded above near the origin. In other words, if a subharmonic function is not bounded above at a point O, it must go to $+\infty$ as rapidly as r^{2-n} on some sequence of points tending to O.

(iii) Letting $r_2 \to \infty$, we see that if

$$\liminf_{r_2 \to \infty} M(r_2) \leq 0,$$

then $r^{n-2}M(r)$ is a nonincreasing function of r. Hence if $\limsup_{r \to \infty} u \leq 0$, then $r^{n-2}u$ is bounded above.

(iv) Theorem 30 does not lead to a Liouville theorem. In fact, subharmonic functions in three or more variables which are bounded above in the entire space are not necessarily constant. For example, the function

$$u(r) = \begin{cases} -\tfrac{1}{8}(15 - 10r^2 + 3r^4), & \text{for } r \leq 1 \\ -1/r, & \text{for } r \geq 1 \end{cases}$$

is subharmonic in all of Euclidean three-space and is bounded everywhere. The same function $u(r)$ considered in Euclidean two-space satisfies the inequality $\Delta u + \min(1, r^{-2})(xu_x + yu_y) \geq 0$, which shows that the Liouville theorem as given in Theorem 27 cannot be extended to functions which satisfy an elliptic differential inequality instead of a differential equation.

The three-circles (or three-spheres) theorem may be extended to more general elliptic differential inequalities of the form

$$L[u] \equiv \sum_{i,j=1}^{n} a_{ij} \frac{\partial^2 u}{\partial x_i \, \partial x_j} + \sum_{i=1}^{n} b_i \frac{\partial u}{\partial x_i} \geq 0. \tag{4}$$

We begin by observing that the proofs of Theorems 28 and 30 do not require the auxiliary function $\varphi(r)$ to be harmonic. To employ the maximum principle in these proofs it is only necessary that $\varphi(r)$ be superharmonic; that is, φ must satisfy the inequality $\Delta \varphi \leq 0$.

If u is any nonconstant solution of $L[u] \geq 0$ and $M(r)$ is defined as before, it follows from the maximum principle that $M(r)$ cannot be constant in any interval and, in fact, cannot have an interior maximum in any interval. Furthermore, it cannot have a local maximum, and so may have at most one minimum. Therefore $M(r)$ may first decrease and then increase, or it may always increase, or it may always decrease.

Suppose we can find an auxiliary function $\psi(r)$ such that $L[\psi] \leq 0$. Furthermore, assume that $M(r)$ and $\psi(r)$ both increase or both decrease for $R_1 < r < R_2$. We define the function

$$\varphi(r) = \frac{\psi(r_2)M(r_1) - \psi(r_1)M(r_2)}{\psi(r_2) - \psi(r_1)} + \frac{M(r_2) - M(r_1)}{\psi(r_2) - \psi(r_1)} \psi(r), \tag{5}$$

for $R_1 < r_1 \leq r \leq r_2 < R_2$, and note that

$$\varphi(r_1) = M(r_1), \qquad \varphi(r_2) = M(r_2).$$

Hence we obtain
$$u - \varphi \leq 0 \text{ for } r = r_1, \quad r = r_2$$
and
$$L[u - \varphi] \geq 0 \text{ in } D.$$

The last inequality depends on the fact that the coefficient of $\psi(r)$ in (5) is nonnegative; that is, it depends on the hypothesis that M and ψ are either both increasing or both decreasing functions of r.

The maximum principle applied to the function $u - \varphi$ gives $u \leq \varphi$ for $r_1 < r < r_2$; hence

$$M(r) \leq \varphi(r), \quad r_1 < r < r_2,$$

with equality if and only if $u \equiv \varphi$. Again this can be stated by saying that $M(r)$ **is a convex function of** $\psi(r)$. The applicability of such a result hinges on our ability to find a function $\psi(r)$ which is increasing and satisfies $L[\psi] \leq 0$. Similarly, we must be able to find a decreasing function satisfying the same inequality.

We now exhibit a method for finding such functions. Let $c_1(r)$ and $c_2(r)$ be functions such that

$$c_1(r) \leq \frac{\sum_{i=1}^{n} a_{ii} - \sum_{i,j=1}^{n} a_{ij} \frac{x_i}{r} \frac{x_j}{r} + \sum_{i=1}^{n} b_i x_i}{\sum_{i,j=1}^{n} a_{ij} \frac{x_i}{r} \frac{x_j}{r}} \leq c_2(r), \quad R_1 \leq r \leq R_2.$$

Functions of this kind can always be found when L is uniformly elliptic and (a_{ii}), (b_i) are uniformly bounded in D.

Define the increasing function

$$\psi_+(r) = \int_a^r e^{-\int_b^s [c_2(\rho)/\rho]d\rho} \, ds$$

and the decreasing function

$$\psi_-(r) = \int_r^a e^{-\int_b^s [c_1(\rho)/\rho]d\rho} \, ds,$$

where a and b are constants to be prescribed. It is easily verified that both ψ_+ and ψ_- satisfy $L[\psi] \leq 0$. In fact, a computation shows that

$$L[\psi(r)] = \left(\sum_{i,j=1}^{n} a_{ij} \frac{x_i}{r} \frac{x_j}{r}\right) \psi_{rr} + \left(\sum_{i=1}^{n} a_{ii} - \sum_{i,j=1}^{n} a_{ij} \frac{x_i}{r} \frac{x_j}{r} + \sum_{i=1}^{n} b_i x_i\right) \frac{\psi_r}{r}.$$

The definition of $c_2(r)$ and the facts that $\psi_{+rr} = -(c_2(r)/r)\psi_{+r}$ and $\psi_{+r} > 0$ yield the inequality $L[\psi_+] \leq 0$. Similarly, we find $L[\psi_-] \leq 0$.

If L is the Laplace operator, we can select $c_1(r) \equiv c_2(r) = n - 1$ and both functions reduce to linear combinations of 1 and $\log r$ in two dimensions, and of 1 and $r^{-(n-2)}$ in n dimensions, $n \geq 3$.

The result of Hadamard can be extended even further by considering a one-parameter family of surfaces instead of a family of concentric spheres. Let $f(x_1, x_2, \ldots, x_n)$ be a twice continuously differentiable function whose gradient never vanishes in a domain D. We suppose that for $\rho_1 \leq \rho \leq \rho_2$ the one-parameter family of equations

$$S_\rho: f(\mathbf{x}) = \rho$$

represents a family of smooth closed $(n-1)$-dimensional surfaces lying in D. In addition, we suppose that a surface corresponding to a larger value of ρ contains in its interior that for a smaller ρ. For example, if

$$f(\mathbf{x}) = (x_1^2 + x_2^2 + \ldots + x_n^2)^{1/2},$$

then S_ρ is a sphere of radius ρ.

Let $u(\mathbf{x})$ satisfy the inequality

$$L[u] \geq 0,$$

where L is the operator given by (4). We define

$$M(\rho) = \max_{f(\mathbf{x})=\rho} u(\mathbf{x}).$$

As before, we find that $M(\rho)$ can have no local maximum and at most one minimum. Now we seek auxiliary functions $\psi_+(\rho)$ and $\psi_-(\rho)$, both of which satisfy $L[\psi] \leq 0$ and with the properties that ψ_+ is an increasing function of ρ while ψ_- is a decreasing function of ρ. Then on an interval where $M(\rho)$ is increasing, it is a convex function of ψ_+; on an interval where $M(\rho)$ is decreasing, it is a convex function of ψ_-.

A computation shows that

$$L[\psi] = \psi''(\rho) \sum_{i,j=1}^n a_{ij} \frac{\partial f}{\partial x_i} \frac{\partial f}{\partial x_j} + \psi'(\rho) L[f].$$

We now define the functions $d_1(\rho)$ and $d_2(\rho)$ so that

$$d_1(\rho) \leq \frac{L[f]}{\sum_{i,j=1}^n a_{ij} \frac{\partial f}{\partial x_i} \frac{\partial f}{\partial x_j}} \leq d_2(\rho) \text{ for } f(\mathbf{x}) = \rho.$$

The uniform ellipticity of the operator L assures us that the functions d_1 and d_2 can always be found. Then with a and b arbitrary numbers, the increasing function

$$\psi_+(\rho) = \int_a^\rho e^{-\int_b^s d_2(\sigma)d\sigma} ds$$

and the decreasing function

$$\psi_-(\rho) = \int_\rho^a e^{-\int_b^s d_1(\sigma)d\sigma} ds$$

satisfy the required conditions. In the case $f(\mathbf{x}) = (x_1^2 + x_2^2 + \ldots + x_n^2)^{1/2}$, we easily see that $d_1(\rho) = c_1(\rho)/\rho$, $d_2(\rho) = c_2(\rho)/\rho$ and we recapture the three-spheres result.

Example. Let $L = \Delta$, the Laplace operator in the plane, and select for f the function
$$f(x, y) = (\alpha x^2 + \beta y^2)^{1/2}, \quad \alpha > \beta > 0.$$
Then the curves S_ρ form a family of ellipses as shown in Fig. 27. A simple computation yields
$$\frac{\Delta f}{|\text{grad } f|^2} = \frac{\alpha\beta(x^2 + y^2)}{[\alpha^2 x^2 + \beta^2 y^2][\alpha x^2 + \beta y^2]^{1/2}}.$$

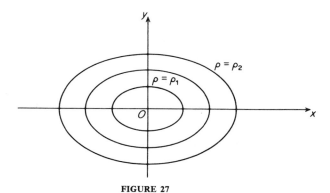

FIGURE 27

Hence we may take
$$d_1(\rho) = \frac{\beta}{\alpha\rho}, \quad d_2(\rho) = \frac{\alpha}{\beta\rho}.$$
Integrating the expressions for ψ_+ and ψ_-, we find
$$\psi_-(\rho) = -\rho^{1-(\beta/\alpha)}, \quad \psi_+(\rho) = -\rho^{-[(\alpha/\beta)-1]}.$$
Thus if $M(\rho)$ is increasing, it is a convex function of $-\rho^{-[(\alpha/\beta)-1]}$. That is, $M(\rho)$ satisfies the inequality
$$M(\rho) \leq \frac{M(\rho_1)[\rho^{(\beta-\alpha)/\beta} - \rho_2^{(\beta-\alpha)/\beta}] + M(\rho_2)[\rho_1^{(\beta-\alpha)/\beta} - \rho^{(\beta-\alpha)/\beta}]}{\rho_1^{(\beta-\alpha)/\beta} - \rho_2^{(\beta-\alpha)/\beta}},$$
$$\rho_1 < \rho < \rho_2.$$

The extended three-surfaces theorem allows us to obtain convexity theorems for functions defined in an unbounded set which is not the entire space. We illustrate this situation with an example in the plane.

Example. Let $L = \Delta$ and select for f the function
$$f(x, y) = \frac{1 + x^2 + y^2}{2y}, \quad y > 0.$$

The function f is positive in the upper half-plane and the curves $f(x, y) = \rho$ with ρ a constant, are circles as shown in Fig. 28. The circles corresponding to larger values of ρ contain those for smaller values. The one-parameter family

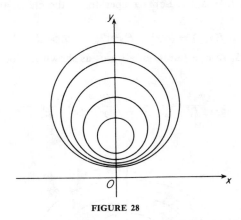

FIGURE 28

$f = \rho$ for $1 \leq \rho < \infty$ fills the upper half-plane $y > 0$. We select for $\psi(\rho)$ the function

$$\psi(\rho) = \log \frac{\rho + 1}{\rho - 1}, \qquad \rho > 1,$$

and a computation shows that $\Delta \psi = 0$. Because ψ is harmonic, $M(\rho)$ is a convex function of ψ regardless of whether it is increasing or decreasing, and we have

$$M(\rho) \leq \frac{M(\rho_2)[\psi(\rho) - \psi(\rho_1)] + M(\rho_1)[\psi(\rho_2) - \psi(\rho)]}{\psi(\rho_2) - \psi(\rho_1)}, \qquad \rho_1 < \rho < \rho_2$$

for any function which is subharmonic in the upper half-plane.

If $h(x) \leq 0$, the three-surfaces theorem can be extended to functions which satisfy
$$(L + h)[u] \geq 0.$$
We merely replace $M(\rho)$ by the function
$$N(\rho) = \max (M(\rho), 0)$$
in all the formulas.

A different generalization of the three-circles theorem has been given by E. M. Landis [2, 3].

EXERCISES

1. Suppose that u is harmonic in the annular region $1 < r < 2$ and that
$$u(1, \theta) = \cos \theta, \qquad u(2, \theta) = 3 \sin \theta, \qquad 0 \leq \theta \leq 2\pi.$$
Find upper and lower bounds for $u(\tfrac{3}{2}, 0)$.

2. Suppose that $u(x, y, z)$ is a subharmonic function in all of three-space. If $u \to 0$ as $r^2 \equiv x^2 + y^2 + z^2 \to +\infty$ and $u(0, 0, 0) = 0$, show that $u \equiv 0$. Use the three-spheres theorem to show how the hypotheses at $(0, 0, 0)$ and at ∞ may be weakened.

3. Let $(\alpha x^2 + \beta y^2 + \gamma z^2)^{1/2} = \rho$ be a family of ellipsoids. If u is subharmonic in all of three-space, find the appropriate exponents δ such that $M(\rho)$ is a convex function of ρ^δ.

4. Obtain a three-spheres theorem for a one-parameter family of spheres lying in the half-space $z > 0$, given that $u(x, y, z)$ is subharmonic for $z > 0$.

SECTION 13. DERIVATIVES OF HARMONIC FUNCTIONS

If $u(x, y)$ is a harmonic function in D, then each of its derivatives is also harmonic and therefore satisfies the maximum principle.

If C_r is the circle with center at (x_0, y_0) and radius r, we saw in Section 1 that for any harmonic function the mean value theorem holds:

$$u(x_0, y_0) = \frac{1}{2\pi r} \oint_{C_r} u \, ds = \frac{1}{2\pi r} \int_0^{2\pi} ur \, d\theta.$$

We multiply both sides by r and integrate with respect to r from 0 to a fixed number R:

$$\int_0^R r u(x_0, y_0) dr = \frac{1}{2\pi} \int_0^R \int_0^{2\pi} ur \, dr \, d\theta,$$

or

$$u(x_0, y_0) = \frac{1}{\pi R^2} \iint_K u \, dA, \tag{1}$$

where K is the interior of the circle of radius R centered at (x_0, y_0) and contained in D. Equation (1) is a statement of the "area mean value theorem," which asserts that **the value of a harmonic function at any point P is the mean of its values taken over the area of any disk K (in D) of which P is the center.**

Since equation (1) holds for any harmonic function and since $\partial u/\partial x$ is again harmonic, we have

$$\frac{\partial u(x_0, y_0)}{\partial x} = \frac{1}{\pi R^2} \iint_K \frac{\partial u}{\partial x} dA.$$

Applying the divergence theorem to the right side, we obtain

$$\frac{\partial u(x_0, y_0)}{\partial x} = \frac{1}{\pi R^2} \int_0^{2\pi} u(x_0 + R \cos \theta, y_0 + R \sin \theta) R \cos \theta \, d\theta. \tag{2}$$

Now if u satisfies the inequalities

$$m \leq u \leq M \qquad (3)$$

on ∂D, the boundary of D, then by the maximum principle the same inequalities hold throughout D. If c is any constant, $u + c$ is harmonic whenever u is. We apply (2) to the function $u - \frac{1}{2}(M + m)$, getting

$$\frac{\partial u(x_0, y_0)}{\partial x} = \frac{1}{\pi R^2} \int_0^{2\pi} \left[u - \frac{1}{2}(M + m) \right] R \cos \theta \, d\theta.$$

Thus we have the estimate

$$\left| \frac{\partial u(x_0, y_0)}{\partial x} \right| \leq \frac{1}{\pi R} \int_0^{2\pi} \left| u - \frac{1}{2}(M + m) \right| \left| \cos \theta \right| d\theta$$

$$\leq \frac{1}{\pi R} \frac{M - m}{2} \int_0^{2\pi} \left| \cos \theta \right| d\theta$$

$$= \frac{2(M - m)}{\pi R}.$$

In this way we obtain a bound for $\partial u/\partial x$ in terms of R and the maximum and minimum values of u. Since we can rotate the coordinates to make any direction the x-direction (or carry out the analogous argument), we get an estimate for the directional derivative in any direction. In particular, we may take the direction to coincide with the gradient of u. Hence, if u satisfies (3) on the boundary of D, then

$$\left| \text{grad } u(x, y) \right| \leq \frac{2(M - m)}{\pi d}, \qquad (4)$$

where d is the minimum distance from (x, y) to any point on the boundary. That is, d is the radius of the largest disk having (x, y) as center and lying entirely in D. Inequality (4) is the best possible inequality of this type, since there are harmonic functions for which the equality in (4) holds. This fact is exhibited by the function

$$u = \tan^{-1} \frac{2y}{1 - (x^2 + y^2)} \qquad (5)$$

which is harmonic in the unit disk $K: x^2 + y^2 < 1$. At the origin we have

$$\frac{\partial u(0, 0)}{\partial y} = 2$$

On the other hand,

$$u \to \tfrac{1}{2}\pi$$

as (x, y) tends to the upper half of the boundary of K, while

$$u \to -\tfrac{1}{2}\pi$$

as (x, y) tends to the lower half of the boundary of K. Therefore $M - m = \pi$, and the right side of (4) has the value 2.

The estimate (4) gets worse as (x, y) approaches ∂D, since the quantity d tends to zero. The gradient of the harmonic function (5) behaves like $1/d$ as (x, y) approaches ∂K, and so the denominator in (4) cannot be replaced by a quantity which goes to zero more slowly than d, the distance of the point from the boundary.

The harmonic function (5) has discontinuous boundary values, and it may be argued that these discontinuities produce the growth of $|\operatorname{grad} u|$ as (x, y) approaches the boundary. However, the function

$$u = y \log [(x-1)^2 + y^2] + 2(1-x)\tan^{-1} \frac{y}{1-x}$$

is harmonic in the unit disk K and has continuous boundary values. Nevertheless $\operatorname{grad} u$ is not bounded in K. A simple calculation shows that $|\operatorname{grad} u|$ behaves like

$$|\log [(x-1)^2 + y^2]| \text{ as } (x, y) \to (1, 0).$$

The above method for obtaining bounds of the derivatives of harmonic functions may be employed in any number of dimensions. Let ω_{n-1} and ω_n denote the surface area of the unit sphere in $(n-1)$ and n dimensions, respectively. We can derive the inequality

$$|\operatorname{grad} u(x_1, x_2, \ldots, x_n)| \leq \frac{n\omega_{n-1}}{(n-1)\omega_n d}(M - m) \tag{6}$$

for the derivatives of any harmonic function u defined in a domain D where

$$m \leq u \leq M \text{ on } \partial D,$$

and where d is the minimum distance of $(x_1, x_2, \ldots, x_n)$ to ∂D. Inequalities (4) and (6) are valid for harmonic functions but false for subharmonic or superharmonic functions.

Examples. (i) The simple subharmonic function $u = (x^2 + y^2) - 1$ vanishes identically on the boundary of the unit disk; hence the gradient of u in the unit disk K cannot be bounded by the boundary values of u on ∂K.

(ii) The superharmonic function

$$u = \begin{cases} 1 + \left[\dfrac{(3 - e^{-2}r^2)(1 - e^{-2}r^2)}{4 \log (1/\epsilon)}\right] & \text{for } r \leq \epsilon \\ \dfrac{\log (1/r)}{\log (1/\epsilon)} & \text{for } r \geq \epsilon \end{cases}$$

satisfies $0 < u \leq 1 + (3/4 \log 2)$ when $0 < \epsilon \leq \frac{1}{2}$ and $0 \leq r \leq 1$. However,

$$|\operatorname{grad} u| = 1/[\epsilon \log(1/\epsilon)] \text{ at } r = \epsilon,$$

and hence is arbitrarily large for ϵ sufficiently small, even though the distance to the boundary is $1 - \epsilon$. Therefore a bounded superharmonic function may have an arbitrarily large gradient arbitrarily far from the boundary. The function $2 - u$ leads to the same conclusion for subharmonic functions.

Estimates of the form (6) have many applications. As an example, we obtain a weak form of Liouville's theorem for harmonic functions. **If u is harmonic for all values of $\mathbf{x} = (x_1, x_2, \ldots, x_n)$ and is bounded above and below, then u is constant.** To prove this result we merely let $d \to \infty$ in (6) to conclude that the gradient must be zero. As we saw in Section 12, it is sufficient in two dimensions to assume that u is only subharmonic and bounded above to conclude that u is constant. However, the example given in Remark (iv) on page 132 shows that such a theorem for subharmonic functions cannot hold in three dimensions. Similar examples can be constructed in higher dimensions.

The first derivatives of solutions of general second-order elliptic equations satisfy inequalities of the same character as (6). The techniques required to establish these estimates are less elementary, since the mean value theorem is not available and, furthermore, derivatives of solutions of elliptic equations in general are not themselves solutions of the same equation. For a treatment of this topic see Bers, John, and Schechter [1, pp. 231–237] or Miranda [3, pp. 113–126 and pp.135–137].

The method for obtaining bounds for the first derivatives can be adapted to obtain bounds for derivatives of any order. We illustrate the technique for second derivatives of harmonic functions in the plane.

Since $\partial u/\partial y$ is harmonic when u is, the representation formula (2) gives

$$\frac{\partial^2 u(x_0, y_0)}{\partial x \, \partial y} = \frac{1}{\pi r^2} \int_0^{2\pi} \frac{\partial u(x_0 + r \cos\theta, y_0 + r \sin\theta)}{\partial y} r \cos\theta \, d\theta.$$

We multiply this equation by πr^3 and integrate with respect to r from 0 to R. Noting that $x = r\cos\theta$, we find

$$\frac{1}{4}\pi R^4 \frac{\partial^2 u(x_0, y_0)}{\partial x \, \partial y} = \int_0^R \int_0^{2\pi} x \frac{\partial u}{\partial y} r \, dr \, d\theta$$

$$= \iint_{(x-x_0)^2 + (y-y_0)^2 \leq R^2} x \frac{\partial u}{\partial y} dx \, dy.$$

Applying the divergence theorem to this last integral, we obtain

$$\frac{\partial^2 u(x_0, y_0)}{\partial x \, \partial y} = \frac{4}{\pi R^4} \int_0^{2\pi} R^2 u \sin\theta \cos\theta \, d\theta$$

$$= \frac{2}{\pi R^2} \int_0^{2\pi} u \sin 2\theta \, d\theta.$$

Similarly, we find at the center (x_0, y_0) of a circle of radius R,

$$\frac{\partial^2 u(x_0, y_0)}{\partial x^2} = -\frac{\partial^2 u}{\partial y^2} = \frac{2}{\pi R^2} \int_0^{2\pi} u \cos 2\theta \, d\theta.$$

If u is harmonic in a region S and u satisfies $m \leq u \leq M$ on ∂S, we obtain for any second derivative of u at (x_0, y_0), denoted $D^2 u$, the estimate

$$|D^2 u| \leq \frac{4(M-m)}{\pi d^2},$$

where d is the minimum distance of (x_0, y_0) to any point of the boundary set ∂S. We see that the bounds for the second derivatives of harmonic functions tend to infinity like d^{-2} as (x_0, y_0) approaches the boundary. More generally, the estimate for a partial derivative of order k of a harmonic function has the pointwise bound

$$|D^k u| \leq \frac{A(M-m)}{d^k},$$

where A is a constant depending only on k and the number of independent variables, n.

EXERCISES

1. Suppose that $u(x, y)$ is harmonic in $D: |x| < 1$, $|y| < 1$ and that u has the boundary values $u = |x|$ for $|y| = 1$, $u = |y|$ for $|x| = 1$. Find upper and lower bounds for $\partial u/\partial x$ at $(\frac{1}{2}, \frac{1}{2})$.

2. Suppose that $\Delta u - u^3 = 0$ in $D: |x| < 1$, $|y| < 1$. If $u \equiv 1$ on ∂D, find upper and lower bounds for $\partial u/\partial x$ at $(0, 0)$.

3. Suppose that $u(x, y)$ is harmonic in the unit disk $D: x^2 + y^2 < 1$. Find the smallest number A such that

$$\left|\frac{\partial^3 u(P)}{\partial x^3}\right| \leq \frac{A(M-m)}{d^3},$$

where $m \leq u \leq M$ in D and d is the distance from P to ∂D.

SECTION 14. BOUNDARY ESTIMATES FOR THE DERIVATIVES

In the preceding section we obtained bounds for the gradient of a harmonic function at a point in a domain D in terms of the maximum and minimum values of the function on the boundary ∂D and in terms of the distance of the point in question from the boundary. In this section we shall derive uniform estimates for the gradient of a harmonic function in terms of the derivatives of the boundary values.

We shall find the following lemma useful.

LEMMA. If v is harmonic, then v^2 is subharmonic.

Proof. A direct computation yields the result:

$$\Delta(v^2) = 2v \Delta v + 2 \sum_{i=1}^{n} \left(\frac{\partial v}{\partial x_i}\right)^2$$

$$= 2 \sum_{i=1}^{n} \left(\frac{\partial v}{\partial x_i}\right)^2 \geq 0.$$

If u is harmonic, so is each of its derivatives $\partial u/\partial x_i$, $i = 1, 2, \ldots, n$. From the above lemma, we conclude that $(\partial u/\partial x_i)^2$ is subharmonic. Since the sum of subharmonic functions is subharmonic, we find that

$$|\text{grad } u|^2 = \sum_{i=1}^{n} \left(\frac{\partial u}{\partial x_i}\right)^2$$

is also subharmonic. We shall apply the maximum principle to $|\text{grad } u|^2$ to get bounds for the derivatives of a harmonic function.

We consider first the problem of estimating the first derivatives of a harmonic function of two variables. Suppose $u(x, y)$ is a solution of

$$\Delta u = 0$$

in a bounded plane domain D and that

$$u = g(x, y) \text{ on } \partial D.$$

We wish to get a bound for $\partial u/\partial \mathbf{n}$ at a point P on the boundary. However, in the preceding section we gave examples of harmonic functions with continuous boundary values whose gradient becomes infinite on the boundary. Therefore we know that at the very least the function g must have a certain degree of smoothness. We shall assume that g can be extended as a twice continuously differentiable function $g(x, y)$ on $D \cup \partial D$. Then the Laplacian of g is bounded on $D \cup \partial D$, and there is a number $A > 0$ such that

$$|\Delta g| \leq A \text{ in } D.$$

In addition, we make assumptions about the smoothness of the boundary of D. We shall suppose that there is a number ρ such that for each point P of ∂D a circle of radius ρ can be drawn which passes through P and whose interior is entirely exterior to D. This means, for example, that D cannot have a reentrant corner of the type shown in Fig. 29.

We fix a point $P(x_0, y_0)$ on ∂D and construct a circle of radius ρ which passes through P and whose interior K_ρ is outside D. We suppose D is bounded so that it is contained in a disk K_R of radius R and concentric with K_ρ (Fig. 30). For convenience we choose the origin of a polar coordinate system at the center of K_ρ and define the functions

$$z_1(x, y) = g(x, y) - \frac{1}{4}A\left[(R^2 - \rho^2)\frac{\log(r/\rho)}{\log(R/\rho)} - r^2 + \rho^2\right],$$

$$z_2(x, y) = g(x, y) + \frac{1}{4}A\left[(R^2 - \rho^2)\frac{\log(r/\rho)}{\log(R/\rho)} - r^2 + \rho^2\right].$$

A simple computation shows that

$$\Delta z_1 \geq 0 \text{ in } D$$
$$z_1 \leq g \text{ on } \partial D$$

and

$$\Delta z_2 \leq 0 \text{ in } D$$
$$z_2 \geq g \text{ on } \partial D.$$

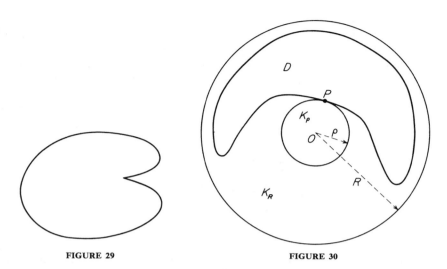

FIGURE 29   **FIGURE 30**

Moreover,
$$z_1(x_0, y_0) = z_2(x_0, y_0) = g(x_0, y_0).$$

Therefore the maximum principle as given in Theorem 7 applied to the functions $z_1 - u$ and $u - z_2$ gives the bounds at $P(x_0, y_0)$:

$$\frac{\partial z_2}{\partial \mathbf{n}} \leq \frac{\partial u}{\partial \mathbf{n}} \leq \frac{\partial z_1}{\partial \mathbf{n}}$$

or

$$\frac{\partial g}{\partial \mathbf{n}} - A\frac{R^2 - \rho^2 - 2\rho^2 \log (R/\rho)}{4\rho \log (R/\rho)} \leq \frac{\partial u}{\partial \mathbf{n}} \leq \frac{\partial g}{\partial \mathbf{n}} + A\frac{R^2 - \rho^2 - 2\rho^2 \log (R/\rho)}{4\rho \log (R/\rho)}. \tag{1}$$

By hypothesis, we can choose the same constants ρ and R for every point of ∂D. Therefore the inequalities (1) are valid for all boundary points.

We now observe that since $u = g$ on ∂D, the tangential derivatives of u and g coincide at each boundary point. Hence $|\text{grad}\,(u - g)|$ is precisely $|(\partial/\partial \mathbf{n})(u - g)|$ at each point P on ∂D. We conclude from (1) that

$$|\text{grad}\, u| \leq |\text{grad}\, g| + A\frac{R^2 - \rho^2 - 2\rho^2 \log (R/\rho)}{4\rho \log (R/\rho)} \text{ on } \partial D.$$

Since we have shown in the lemma at the beginning of this section that $|\text{grad}\, u|^2$ is subharmonic, the maximum of $|\text{grad}\, u|$ on $D \cup \partial D$ must occur on the boundary. Therefore, for any (x, y) in D, we have

$$|\text{grad}\, u(x, y)| \leq \max_{\partial D} |\text{grad}\, g| + \sup_{D} |\Delta g|\frac{R^2 - \rho^2 - 2\rho^2 \log (R/\rho)}{4\rho \log (R/\rho)}.$$

A similar bound is easily obtained for harmonic functions of three or more variables. If D is bounded and ∂D is so smooth that through each point of ∂D a sphere of radius ρ can be constructed passing through P whose interior lies outside D, then the same argument as the one in two dimensions may be used. If n is the number of dimensions, the comparison functions z_1 and z_2 are

$$z_1(\mathbf{x}) = g(\mathbf{x}) - \frac{A}{2n}\left[(R^2 - \rho^2)\frac{\rho^{2-n} - r^{2-n}}{\rho^{2-n} - R^{2-n}} - r^2 + \rho^2\right],$$

$$z_2(\mathbf{x}) = g(\mathbf{x}) + \frac{A}{2n}\left[(R^2 - \rho^2)\frac{r^{2-n} - \rho^{2-n}}{R^{2-n} - \rho^{2-n}} - r^2 + \rho^2\right].$$

The bound for $|\operatorname{grad} u|$ at any point of D is

$$|\operatorname{grad} u(\mathbf{x})| \leq \max_{\partial D} |\operatorname{grad} g| + \sup_{D} |\Delta g| \cdot \frac{\rho}{2n}\left\{\frac{(n-2)(R^2 - \rho^2)}{\rho^2\{1 - (\rho/R)^{n-2}\}} - 2\right\}. \tag{2}$$

The **diameter** of a domain D is the least upper bound of the distances between any two points of D. If d denotes the diameter of D, then we can select $R = d + \rho$ provided that every boundary point has a sphere of radius ρ of the appropriate type passing through it. In particular, if D is convex, we can let $\rho \to \infty$ and obtain the estimate

$$|\operatorname{grad} u| \leq \max_{\partial D} |\operatorname{grad} g| + \frac{d}{2} \sup_{D} |\Delta g|,$$

valid in any number of dimensions.

For solutions of more general elliptic equations it is still possible to get bounds similar to (2). In fact, if Theorem 13 on approximation applies, we have the bounds $z_2(\mathbf{x}) \leq u(\mathbf{x}) \leq z_1(\mathbf{x})$ for a solution of the boundary value problem with mixed boundary conditions. If in addition to the inequalities needed for Theorem 13, we have $z_2(\mathbf{x}_0) = u(\mathbf{x}_0) = z_1(\mathbf{x}_0)$, where $\mathbf{x}_0$ is a point on the part Γ_2 of the boundary ∂D where u itself is prescribed, then clearly

$$\frac{\partial z_1}{\partial \mathbf{n}} \leq \frac{\partial u}{\partial \mathbf{n}} \leq \frac{\partial z_2}{\partial \mathbf{n}} \text{ at } x_0.$$

Because the tangential derivatives are known at x_0, we get a bound for $|\operatorname{grad} u|$ at this point.

Since the gradient of a solution of a general elliptic equation does not, in general, satisfy an elliptic differential inequality, we *cannot* obtain bounds for the gradient at interior points by a simple application of the maximum principle. An exception to this last statement occurs in the case of an elliptic equation in two dimensions of the form

$$L[u] \equiv a(x, y)\frac{\partial^2 u}{\partial x^2} + 2b(x, y)\frac{\partial^2 u}{\partial x \partial y} + c(x, y)\frac{\partial^2 u}{\partial y^2} = 0.$$

If a, b, and c have continuous first derivatives, then it is easy to see that $\partial u/\partial x$ satisfies an elliptic equation. Writing $v = \partial u/\partial x$, we merely differentiate the above equation, getting

$$a\frac{\partial^2 v}{\partial x^2} + 2b\frac{\partial^2 v}{\partial x\,\partial y} + c\frac{\partial^2 v}{\partial y^2} + \frac{\partial a}{\partial x}\frac{\partial v}{\partial x} + 2\frac{\partial b}{\partial x}\frac{\partial v}{\partial y} + \frac{\partial c}{\partial x}\frac{\partial^2 u}{\partial y^2} = 0.$$

We substitute for $\partial^2 u/\partial y^2$ from the original equation $L[u] = 0$ and find that

$$L[v] + \left(\frac{\partial a}{\partial x} - \frac{a}{c}\frac{\partial c}{\partial x}\right)\frac{\partial v}{\partial x} + 2\left(\frac{\partial b}{\partial x} - \frac{b}{c}\frac{\partial c}{\partial x}\right)\frac{\partial v}{\partial y} = 0.$$

Thus $v = \partial u/\partial x$ satisfies the maximum principle. Similarly, we can show that $\partial u/\partial y$ satisfies the maximum principle. Each of these functions is therefore bounded by the maximum value of $|\text{grad } u|$ on the boundary.

EXERCISES

1. Suppose that $u(x, y)$ is harmonic in the square D: $|x| < 1$, $|y| < 1$ and that u has the boundary values $u = |x|$ for $|y| = 1$, $u = |y|$ for $|x| = 1$. Find upper and lower bounds for $\partial u/\partial x$ at the point $(-1, 0)$.

2. Suppose u is harmonic in the disk D: $r < 1$, and that u has the boundary values $u(1, \theta) = e^{\cos\theta}$, $0 \leq \theta \leq 2\pi$. Find bounds for $\partial u/\partial r$ at the point $(1, 0)$.

3. Suppose that $u(x, y)$ satisfies an elliptic differential equation. Show that u^2 satisfies an elliptic differential inequality with the same principal part.

SECTION 15. APPLICATIONS OF BOUNDS FOR DERIVATIVES

In Section 11 we discussed the electrostatic potential in a domain bounded by two perfect conductors when a potential difference of exactly one unit is maintained between them. This potential function u is the harmonic function which has boundary value 1 on one of the conductors and boundary value 0 on the other. Suppose a unit electrical charge is placed at some point P between the conductors. Then the components of the force acting on this unit charge are given by the first derivatives of the potential function u. The magnitude of this force vector is precisely $|\text{grad } u|$, which may be estimated by the methods of Sections 13 and 14. For example, inequality (6) of Section 13 which, for harmonic functions, states that

$$|\text{grad } u(P)| \leq \frac{3\omega_2}{2\omega_3 d}(M - m) = \frac{3(M - m)}{4d}$$

may be applied to the special case of the electrostatic potential with $M = 1$, $m = 0$. We find

$$|\text{grad } u(P)| \leq \frac{3}{4d}, \tag{1}$$

where d is the distance of P from the boundary of the domain. The inequality (1) above, while useful, becomes weaker as the point P tends to the boundary. In such cases, the ideas used in Section 14 are often helpful. A specific example illustrates the technique.

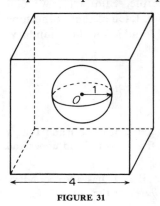

FIGURE 31

Suppose D is the domain between a sphere of radius 1 and a cube of side 4 whose center coincides with the center of the sphere (Fig. 31). We select the center of the sphere as the origin of coordinates. Let $u(x, y, z)$ be the harmonic function having the value 1 on the surface of the sphere and 0 on the surface of the cube. We wish to estimate $|\text{grad } u|$.

By the maximum principle we have $0 \leq u \leq 1$. We observe that the harmonic function $2 - x - u$ is nonnegative on the boundary of D. Therefore the maximum principle yields the inequality

$$0 < u < 2 - x \text{ in } D,$$

and by Theorem 7 we find that on the face $x = 2$ of the cube

$$-1 < \frac{\partial u}{\partial x} < 0.$$

Since $\partial u/\partial y = \partial u/\partial z = 0$ there, $|\text{grad } u| < 1$ on this face of the cube. By employing the functions $2 + x - u$, $2 \pm y - u$, and $2 \pm z - u$, we find in the same way that

$$|\text{grad } u| < 1$$

on all the faces of the cube.

To obtain a bound for the gradient of u on the spherical boundary, we observe that the harmonic function $u + 1 - 2(x^2 + y^2 + z^2)^{-1/2}$ is nonnegative on the boundary of D. Hence by Theorem 6

$$-1 + 2(x^2 + y^2 + z^2)^{-1/2} < u < 1$$

in D, and by Theorem 7

$$0 < \frac{\partial u}{\partial \mathbf{n}} < 2$$

on the sphere $x^2 + y^2 + z^2 = 1$. The tangential derivatives are zero on this sphere. Thus we find that

$$|\text{grad } u| < 2 \tag{2}$$

on the boundary of D. Hence, again by the maximum principle, the bound (2) holds throughout the domain D. Notice that this bound is valid up to the boundary and so is better than the inequality (1) when P is near the boundary. However, for points far from the boundary, (2) is a weaker inequality than (1).

Estimates for the derivatives of harmonic functions are important in problems concerning fluid flow. Under certain ideal conditions, the motion of an incompressible fluid is governed by a potential function. This function satisfies the Laplace equation and has the property that its first derivatives are the components of the velocity of the fluid particles. If u denotes the potential function, then the magnitude of the velocity, called the **speed**, is given by $|\text{grad } u|$. Since $|\text{grad } u|^2$ is subharmonic, we can employ the maximum principle to gain information about the motion. In particular, we consider the two-dimensional motion of an object through an ideal fluid. (Under appropriate conditions, air and water may be considered ideal fluids.) A change of coordinates yields the equivalent problem of a fluid moving past a stationary object (Fig. 32), called an obstacle. The potential function is defined in the entire

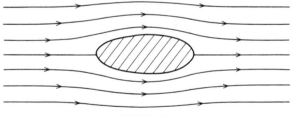

FIGURE 32

plane exterior to the fixed obstacle, the surface of which is the boundary of the flow. The speed of the flow is given at each point in space by $|\text{grad } u|$, and the maximum principle yields the important conclusion that **the maximum speed must occur at a point on the surface of the obstacle.**

If a long cylindrical rod with a simply connected, two-dimensional cross-section D is twisted at its ends, shear stresses are produced in each cross-sectional plane. Using x, y-coordinates in the plane of D we can express the stresses as derivatives of a function v called the **stress function**. We write

$$\tau_1 \equiv \frac{\partial v}{\partial y},$$

$$\tau_2 \equiv -\frac{\partial v}{\partial x}.$$

where τ_1 and τ_2 are the (x, z) and (y, z) components of the stress. The stress function v is a solution (using suitable units of measurement) of the equation

$$\Delta v = -2 \text{ in } D$$

with

$$v = 0 \text{ on } \partial D.$$

[Note that the function

$$u = v + \tfrac{1}{2}(x^2 + y^2)$$

satisfies Laplace's equation with $u = \tfrac{1}{2}(x^2 + y^2)$ on ∂D.] The **magnitude of the stress** is given by $\sqrt{\tau_1^2 + \tau_2^2} = |\text{grad } v|$. Since $\partial v/\partial x$ and $\partial v/\partial y$ are harmonic, it follows that

$$\Delta(|\text{grad } v|^2) \geq 0.$$

Therefore, the maximum stress must occur on the boundary of D. The question of when the elastic range ends and the plastic region of deformation starts is intimately connected with the size and location of the maximum stress.

Employing the methods of Section 14, we can get estimates for the size of the maximum stress. For example, if D is convex and has maximum width B, then for any point P on ∂D, we may choose $z_1(Q) \equiv 0$, $z_2(Q) = \tfrac{1}{4}B^2 - (\xi - \tfrac{1}{2}B)^2$, where ξ is the component of the line PQ along the normal vector to the boundary at P. Then

$$-B \leq \frac{\partial v(P)}{\partial \mathbf{n}} \leq 0.$$

Since $v = 0$ on ∂D, we know that its tangential derivative vanishes on the boundary. Hence

$$|\text{grad } v| \leq B$$

in D.

We can also use the maximum principle to find the effect on the stress of varying the size and shape of the domain. Let D_1 be a subdomain of D such that ∂D and ∂D_1 have points in common and suppose $F \in \partial D \cap \partial D_1$ (see Fig. 33).

We denote by v and v_1 the stress functions for D and D_1, respectively. Because v and v_1 are superharmonic, we find from the maximum principle that $v > 0$ in D and $v_1 > 0$ in D_1. We define $w = v - v_1$ in $D_1 \cup \partial D_1$ and observe

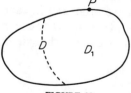

FIGURE 33

that $w \geq 0$ on ∂D_1, $\Delta w = 0$ in D_1. Therefore $w > 0$ in D_1. At the point P we have $v(P) = v_1(P) = 0$. By Theorem 7, we conclude that

$$\frac{\partial v(P)}{\partial \mathbf{n}} < \frac{\partial v_1(P)}{\partial \mathbf{n}}.$$

Applying Theorem 7 to the superharmonic function v_1, we also have $\partial v_1(P)/\partial \mathbf{n} < 0$. Since the tangential derivatives of v and of v_1 vanish at P, we conclude that

$$|\text{grad } v| > |\text{grad } v_1| \text{ at } P.$$

We interpret this inequality by stating that removal of a part of a beam (so that its cross section is diminished) decreases the magnitude of the stress required to produce a given amount of twist per unit length. We know that thin beams are easier to twist than thick ones, so the above inequality agrees with our intuition.

This type of comparison theorem can be used to obtain bounds for the stress on a given domain in terms of those for a domain in which the stress can be determined explicitly or estimated easily. See Weinberger [1].

SECTION 16. NONLINEAR OPERATORS

In Section 9 of Chapter 1 we showed how the maximum principle can be used to obtain a variety of results for nonlinear ordinary differential equations. We now employ the same methods to obtain information about solutions of nonlinear partial differential equations.

Let $F(x, y, u, p, q, r, s, t)$ be a continuously differentiable function of its eight variables. We say that a function $u(x, y)$ of two variables is a solution of the partial differential equation

$$F\left(x, y, u, \frac{\partial u}{\partial x}, \frac{\partial u}{\partial y}, \frac{\partial^2 u}{\partial x^2}, \frac{\partial^2 u}{\partial x \, \partial y}, \frac{\partial^2 u}{\partial y^2}\right) = f(x, y) \tag{1}$$

in a domain D if at each point (x, y) of D, the value obtained by inserting

$$\left. \begin{array}{l} u = u(x, y), \quad p = \dfrac{\partial u(x, y)}{\partial x}, \quad q = \dfrac{\partial u(x, y)}{\partial y}, \, r = \dfrac{\partial^2 u(x, y)}{\partial x^2}, \\[6pt] s = \dfrac{\partial^2 u(x, y)}{\partial x \, \partial y}, \quad t = \dfrac{\partial^2 u(x, y)}{\partial y^2} \end{array} \right\} \tag{2}$$

into F is equal to the given function $f(x, y)$.

Example: If we select

$$F = (1 - p^2)r + (1 + q^2)t + 1,$$

the corresponding differential equation is

$$\left[1 - \left(\frac{\partial u}{\partial x}\right)^2\right]\frac{\partial^2 u}{\partial x^2} + \left[1 + \left(\frac{\partial u}{\partial y}\right)^2\right]\frac{\partial^2 u}{\partial y^2} + 1 = f(x, y) \tag{3}$$

We say that the equation (1) is **elliptic with respect to a particular function u at a point** (x, y) if for all pairs of real numbers (ξ, η) with $\xi^2 + \eta^2 > 0$, we have

$$\frac{\partial F}{\partial r}\xi^2 + \frac{\partial F}{\partial s}\xi\eta + \frac{\partial F}{\partial t}\eta^2 > 0 \tag{4}$$

when the values given by (2) are inserted in F. We say that (1) is **elliptic in a domain** D if it is elliptic at each point of D. A nonlinear equation may be elliptic for some functions u but not for others. For example, equation (3) is elliptic for those functions u such that $|\partial u/\partial x| < 1$, and not otherwise. Of course, if (1) is linear, then the left-hand side of (4) is independent of u and ellipticity depends only on the character of F at each point (x, y) of D.

We now wish to establish conditions under which nonlinear elliptic equations satisfy a maximum principle. Suppose that $u(x, y)$ is a solution of (1) and that $w(x, y)$ satisfies the differential inequality

$$F\left(x, y, w, \frac{\partial w}{\partial x}, \frac{\partial w}{\partial y}, \frac{\partial^2 w}{\partial x^2}, \frac{\partial^2 w}{\partial x\,\partial y}, \frac{\partial^2 w}{\partial y^2}\right) \leq f(x, y).$$

We form the function

$$v(x, y) = u(x, y) - w(x, y)$$

and, using subscripts for partial derivatives, we consider the inequality

$$F(x, y, u, u_x, u_y, u_{xx}, u_{xy}, u_{yy}) - F(x, y, w, w_x, w_y, w_{xx}, w_{xy}, w_{yy}) \geq 0.$$

We apply the mean-value theorem of multi-dimensional calculus to the left side of the above inequality. The result is

$$\left(\frac{\partial F}{\partial r}\right)_0 \frac{\partial^2 v}{\partial x^2} + \left(\frac{\partial F}{\partial s}\right)_0 \frac{\partial^2 v}{\partial x\,\partial y} + \left(\frac{\partial F}{\partial t}\right)_0 \frac{\partial^2 v}{\partial y^2}$$
$$+ \left(\frac{\partial F}{\partial p}\right)_0 \frac{\partial v}{\partial x} + \left(\frac{\partial F}{\partial q}\right)_0 \frac{\partial v}{\partial y} + \left(\frac{\partial F}{\partial u}\right)_0 v \geq 0, \tag{5}$$

where the subscript zero indicates that the derivative is evaluated at the argument $(x, y, u_0, p_0, q_0, r_0, s_0, t_0)$ with

$$u_0 = w + \theta(u - w), \qquad p_0 = w_x + \theta(u_x - w_x), \qquad q_0 = w_y + \theta(u_y - w_y),$$
$$r_0 = w_{xx} + \theta(u_{xx} - w_{xx}), \qquad s_0 = w_{xy} + \theta(u_{xy} - w_{xy}),$$
$$t_0 = w_{yy} + \theta(u_{yy} - w_{yy}),$$

the quantity $\theta = \theta(x, y)$ being some number between zero and one at each point (x, y).

For each point (x, y) in D we assume that F is elliptic for all functions of the form $\theta w + (1 - \theta)u$, with $0 \leq \theta \leq 1$. Then (5) is a *linear* elliptic differential inequality for the function v. We may apply Theorem 6 to conclude that **if v is not constant and**

$$\left(\frac{\partial F}{\partial u}\right)_0 \leq 0,$$

then v cannot have a nonnegative maximum at any point of D. This maximum principle yields at once the following theorem on approximation.

THEOREM 31. Let $u(x, y)$ be a solution of

$$F(x, y, u, u_x, u_y, u_{xx}, u_{xy}, u_{yy}) = f(x, y) \text{ in } D,$$
$$u = g \text{ on } \partial D.$$

Let z and Z satisfy the inequalities

$$F(x, y, Z, Z_x, Z_y, Z_{xx}, Z_{xy}, Z_{yy}) \leq f(x, y) \leq F(x, y, z, z_x, z_y, z_{xx}, z_{xy}, z_{yy}) \tag{6}$$

in D and

$$z(x, y) \leq g(x, y) \leq Z(x, y) \text{ on } \partial D.$$

We assume that for each constant θ such that $0 \leq \theta \leq 1$, the function F is elliptic with respect to $u + \theta(z - u)$ and $u + \theta(Z - u)$ in D, and that $\partial F/\partial u \leq 0$ in D. Then we have

$$z(x, y) \leq u(x, y) \leq Z(x, y) \text{ in } D.$$

The difficulty in applying Theorem 31 occurs in trying to establish the ellipticity of F within the required class of functions $\theta z + (1 - \theta) u$, since u is generally unknown. Nevertheless, there are situations when this difficulty can be resolved. We give several examples.

Suppose Γ is a simple closed curve in three-space which intersects each line in the z-direction at most once, and whose projection on the x, y-plane is convex. It can be shown that there is a surface of minimum area which spans the space curve Γ. This surface has the equation $z = u(x, y)$, and u satisfies the nonlinear equation

$$(1 + u_y^2)u_{xx} - 2u_x u_y u_{xy} + (1 + u_x^2)u_{yy} = 0. \tag{7}$$

In the notation at the beginning of the section, this **minimal surface equation** (7) has the form

$$F \equiv (1 + q^2)r - 2pqs + (1 + p^2)t = 0.$$

The condition for ellipticity (4) becomes

$$F_r \xi^2 + F_s \xi \eta + F_t \eta^2 \equiv (1 + q^2)\xi^2 - 2pq\xi\eta + (1 + p^2)\eta^2 \tag{8}$$
$$= \xi^2 + \eta^2 + (q\xi - p\eta)^2 > 0,$$

and so (4) holds whenever $\xi^2 + \eta^2 > 0$. Therefore the minimal surface equation is *always* elliptic.

Since any constant satisfies (7), we can apply Theorem 31 to the minimal surface equation by letting z and Z be constants. Noting that $\partial F/\partial u \equiv 0$, we conclude that any minimal surface must attain its maximum and minimum on the boundary. That is, a minimal surface cannot have a vertical bulge, a result which is intuitively clear if the surface area is to be as small as possible.

Since any linear function is also a solution of (7), we find that a minimal surface must lie entirely above any plane which is below the boundary curve Γ. Similarly, it must remain entirely below any plane situated above Γ. That is, any plane which intersects the interior of a minimal surface must also intersect the boundary.

The minimal surface equation and its solutions have been studied intensively for a long period of time. For properties which are obtained by means of the maximum principle see Bernstein [1], Bers [2], Finn [1], Nitsche [1], Serrin [7] and the references listed in these papers.

As a second example we consider the equation

$$\frac{\partial^2 u}{\partial x^2}\frac{\partial^2 u}{\partial y^2} - \left(\frac{\partial^2 u}{\partial x \partial y}\right)^2 = f(x, y). \tag{9}$$

This equation is a special case of the **Monge-Ampere equation,** one which arises in a number of problems in differential geometry. Applying the criterion of ellipticity as given by (4), we obtain

$$t\xi^2 - 2s\xi\eta + r\eta^2 > 0.$$

Thus either (9) or the equation obtained from (9) by multiplying it by (-1) is elliptic whenever $rt - s^2$ is positive. If we assume that

$$f(x, y) > 0 \text{ in } D,$$

then (9) is elliptic for all solutions in D. It is often difficult to establish the ellipticity of (9) with respect to a whole family of functions $u + \theta(z - u)$. For example, the functions

$$Z = \tfrac{1}{2} + \tfrac{1}{2}(x^2 + y^2), \qquad z = \tfrac{3}{2} - \tfrac{1}{2}(x^2 + y^2), \tag{10}$$

both satisfy the equation

$$\frac{\partial^2 u}{\partial x^2}\frac{\partial^2 u}{\partial y^2} - \left(\frac{\partial^2 u}{\partial x \partial y}\right)^2 = 1$$

in the disk $K: x^2 + y^2 < 1$, and both z and Z have the boundary values 1 on ∂K. Thus we see that the Dirichlet problem for the Monge-Ampere equation does not possess a unique solution.* Furthermore, Theorem 31

---

*It can be shown that for the general Monge-Ampere equation, $A_1(rt - s^2) + A_2 r + A_3 s + A_4 t + A_5 = 0$, the Dirichlet problem has at most *two* solutions when u, A_1, $A_2, \ldots, A_5$ are such that the equation is elliptic. See Courant and Hilbert [1, p.324].

is violated when $u = z$. Although the equation is elliptic with respect to z and Z, the conditions of the theorem are not met because it is not elliptic with respect to $\frac{1}{2}(z + Z)$.

As a final example we consider the equation

$$[\mu(|\text{grad } \psi|) - \psi_y^2]\psi_{xx} + 2\psi_x\psi_y\psi_{xy} + [\mu(|\text{grad } \psi|) - \psi_x^2]\psi_{yy} = 0, \tag{11}$$

which arises in the study of the flow of compressible fluids. The x and y components of the velocity of the flow are given by

$$\left(\frac{1}{\rho}\psi_y, -\frac{1}{\rho}\psi_x\right), \tag{12}$$

and the *speed*, q, at any point is $|\text{grad } \psi|/\rho$. We assume that the pressure p is a known function of the density ρ. Then μ is equal to $\rho^2 dp/d\rho$, expressed as a function of $|\text{grad } \psi|$ by solving the Bernoulli relation $|\text{grad } \psi|^2 = 2\rho^2 \int_\rho^{\rho_0} \frac{1}{\rho}\frac{dp}{d\rho} d\rho$ for ρ. The constant ρ_0 is determined by prescribing conditions at infinity. Since $(d/d\rho)\left\{2\rho^2 \int_\rho^{\rho_0} \frac{1}{\rho}\frac{dp}{d\rho} d\rho\right\} = 4\rho \int_{\rho_0}^\rho \frac{1}{\rho}\frac{dp}{d\rho} d\rho - 2\rho\frac{dp}{d\rho} = 2\rho\left(q^2 - \frac{dp}{d\rho}\right)$, we can solve the Bernoulli relation for ρ as a single-valued function of $|\text{grad } \psi|$ as long as $q^2 < dp/d\rho$. Thus the function $\mu(|\text{grad } \psi|)$ is defined when q is small enough to make $q^2 < dp/d\rho$. It can be shown that the local speed of sound is $\sqrt{dp/d\rho}$. For this reason a flow for which $q^2 < dp/d\rho$ is said to be **subsonic**. Recalling that $|\text{grad } \psi|^2 = \rho^2 q^2$ and that $\mu(|\text{grad } \psi|) = \rho^2 dp/d\rho$, we can write the inequality $q^2 < dp/d\rho$ in the form

$$|\text{grad } \psi|^2 < \mu(|\text{grad } \psi|). \tag{13}$$

The velocity vector (12) is tangent to the curves $\psi = $ constant, which are called **streamlines** of the flow. The function $\psi = \psi(x, y)$ is called the **stream function**. The mass per unit time flowing between two streamlines $\psi = a$ and $\psi = b$ is $b - a$.

We observe that

$$[\mu(|\text{grad } \psi|) - \psi_y^2]\xi^2 + 2\psi_x\psi_y\xi\eta + [\mu(|\text{grad } \psi|) - \psi_x^2]\eta^2$$
$$= [\mu(|\text{grad } \psi|) - |\text{grad } \psi|^2](\xi^2 + \eta^2) + (\psi_x\xi + \psi_y\eta)^2$$
$$\geq [\mu(|\text{grad } \psi|) - |\text{grad } \psi|^2](\xi^2 + \eta^2),$$

so that by (13) the equation (11) is elliptic with respect to any subsonic flow.

We now give an application of the maximum principle as it applies to equation (11). We consider a subsonic flow in an infinite channel bounded by two rigid walls. For sufficiently large distances to the right and left, i.e., for sufficiently large values of $|x|$, we suppose that the walls are straight lines parallel to the x-axis (see Fig. 34). Calling the total

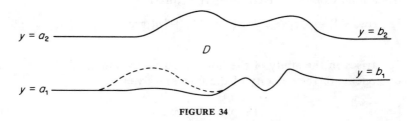

FIGURE 34

mass flux A and the stream function ψ, we choose $\psi = 0$ on the lower wall and $\psi = A$ on the upper wall. We assume that the flow becomes a uniform horizontal flow as $x \to \pm \infty$. More specifically, denoting the equations of the straight portions of the wall $y = a_1, y = a_2, y = b_1, y = b_2$ as shown in Fig. 34, we assume that

$$\lim_{x \to -\infty} \psi(x,y) = A \frac{y - a_1}{a_2 - a_1}, \qquad \lim_{x \to +\infty} \psi(x,y) = A \frac{y - b_1}{b_2 - b_1}, \qquad (14)$$

uniformly in y.

We observe that the expression on the left of equation (11) is linear in the second derivatives of ψ. An equation of this kind is said to be **quasilinear.** If we consider the coefficients of these second derivatives as fixed functions of x and y, we see that ψ satisfies a linear elliptic second order differential equation with no lower order terms. Therefore Theorems 5 and 7 apply. In particular, $0 < \psi < A$ in the entire channel.

Denoting the channel by D, we consider a second channel D_1 which is a subdomain of D obtained by increasing the height of a finite portion of the lower boundary wall of D. This changed portion is shown as a dotted line in Figure 34.

Let ψ_1 be the stream function which corresponds to a subsonic flow through the channel D_1 which carries the same mass flux A as the original flow described by ψ. Then ψ_1 satisfies the differential equation (11) in D_1, with the boundary conditions $\psi_1 = 0$ on the lower wall and $\psi_1 = A$ on the upper wall of D_1. The function ψ_1 also satisfies the conditions (14).

Because the equation (11) is quasilinear, we find that the difference between the equation for ψ and that for ψ_1 can be written in the form

$$[\mu(|\operatorname{grad} \psi|) - \psi_y^2](\psi - \psi_1)_{xx} + 2\psi_x \psi_y (\psi - \psi_1)_{xy}$$
$$+ [\mu(|\operatorname{grad} \psi|) - \psi_x^2](\psi - \psi_1)_{yy}$$

$$+ [(\psi + \psi_1)_y \psi_{1xy} - (\psi + \psi_1)_x \psi_{1yy}](\psi - \psi_1)_x$$
$$+ [(\psi + \psi_1)_x \psi_{1xy} - (\psi + \psi_1)_y \psi_{1xx}](\psi - \psi_1)_y$$
$$+ \Delta \psi_1 [\mu(|\text{grad } \psi|) - \mu(|\text{grad } \psi_1|)] = 0.$$

If we apply the mean value theorem to the last term, we obtain a linear combination of $(\psi - \psi_1)_x$ and $(\psi - \psi_1)_y$. In this way we find that $\psi - \psi_1$ satisfies a linear elliptic equation. We can now apply Theorems 5 and 7 to the difference $\psi - \psi_1$. We have $\psi - \psi_1 = 0$ on the upper wall of D_1 and $\psi - \psi_1 \geq 0$ on the lower wall of D_1. Also,

$$\lim_{x \to \pm\infty} (\psi - \psi_1) = 0.$$

We conclude that

$$\psi > \psi_1 \text{ in } D_1.$$

Since $\psi = \psi_1$ on the common boundary walls, we find that

$$\frac{\partial \psi}{\partial \mathbf{n}} < \frac{\partial \psi_1}{\partial \mathbf{n}} \text{ on } \partial D \cap \partial D_1. \tag{15}$$

Now $0 < \psi < A$ in D and $\psi \equiv A$ on the upper wall. Hence $\partial \psi / \partial \mathbf{n} > 0$ on the upper wall; similarly $\partial \psi_1 / \partial \mathbf{n} > 0$ there. Hence (15) implies that

$$|\text{grad } \psi_1| > |\text{grad } \psi| \text{ on the upper wall.}$$

It is not difficult to show that the speed q is an increasing function of $|\text{grad } \psi|$. Therefore, denoting by q_1 the speed corresponding to ψ_1, we have

$$q_1 > q$$

on the upper wall. At a portion of the lower wall common to D and D_1, we find $\partial \psi / \partial \mathbf{n} < \partial \psi_1 / \partial \mathbf{n} < 0$. Consequently, $q_1 < q$ on the common portion of the lower wall.

We have shown that **narrowing a channel by indenting a portion of one wall increases the speed on the opposite wall but decreases the speed on the unchanged portion of the indented wall.**

Remarks. (i) In the above comparison we have used only the facts that the equation (11) is quasilinear and that it is elliptic with respect to ψ. The result would still be valid if the function $\mu(|\text{grad } \psi|)$ were such that the equation (11) is not elliptic with respect to all functions for which it is defined. In particular, it need not be elliptic with respect to the second solution ψ_1.

(ii) Other comparison theorems of this type have been found by Gilbarg [2, 4] and Serrin [4].

BIBLIOGRAPHICAL NOTES

The maximum principle for harmonic (and subharmonic) functions goes back to C. F. Gauss [1] and S. Earnshaw [1]. The first proof of a maximum principle for elliptic operators more general than the Laplacian was given by A. Paraf [1] for $h < 0$ in two dimensions. It was extended to the case $h \leqslant 0$ by E. Picard [1, pp. 29–30] and L. Lichtenstein [1, 2].

The Paraf result was extended to more than two dimension by T. Moutard [1]. This result was used by M. Picone [1] to obtain a generalized maximum principle. Theorems 5 and 6 of Section 3, in which no continuity hypothesis is made about the coefficients, are due to E. Hopf [1].

The fact that the Green's function corresponding to the first boundary value problem for the Laplace operator satisfies $\partial G/\partial \mathbf{n} < 0$ was proved by C. Neumann [1] and A. Korn [1]. It was extended to more general operators by L. Lichtenstein [2]. Theorems 7 and 8 of Section 3 were proved under some continuity hypothesis on the coefficients by G. Giraud [1, 2]. In the general form given here they were discovered independently by E. Hopf [2] and O. Oleĭnik [1]. A similar result for an elliptic equation in divergence form in two variables was given by D. Gilbarg [4].

A very general form of the maximum principle with minimal hypotheses on the coefficients has been given by A. D. Aleksandrov [1, 2, 3] and by C. Pucci [8]. Extensions to partial differential equations which cease to be elliptic at some points have been found by C. Pucci [2, 3, 5] and M. Picone [1, 4].

The maximum principle has been extended in various ways. A generalization to weak solutions of elliptic inequalities was given by W. Littman [1, 2]. Extensions to nonlinear elliptic operators have been obtained by S. S. Dymkov [1], O. A. Ladyzhenskaya [1], A. McNabb [2], L. Nirenberg [2], M. Nagumo and S. Simoda [1], and R. M. Redheffer [1, 2, 3, 4]. An extension and an application to geometry of the maximum principle has been given by E. Calabi [1]. An extension to weakly coupled systems is due to P. Szeptycki [1]. M. Schechter [1] has given a maximum principle in the case when the coefficients become singular on the boundary. A strong maximum principle for equations whose second order derivatives appear in divergence form with minimal hypotheses on the coefficients was given by G. Stampacchia [1].

The concept of maximum principle is sometimes used in a generalized sense to indicate a certain norm (not necessarily the maximum norm) of the solution of a boundary value problem that is bounded in terms of a norm of its boundary values. A. D. Aleksandrov [1, 2] obtained such maximum principles for elliptic equations in a variety of integral norms. Results on such maximum principles involving maximum norms for higher order elliptic equations and systems may be found in the works of C. Miranda [1, 2], S. Agmon [1], G. Fichera [4], and R. J. Duffin [2].

Many textbooks, for example those of R. Courant and D. Hilbert [1], P. Garabedian [1], and H. F. Weinberger [4], contain expositions of the maximum principle.

The use of the maximum principle as a tool in approximation theory for both linear and nonlinear equations, and as an aid in obtaining bounds has occurred in the works of many authors. For example, estimates for the gradient of a solution of an elliptic equation were obtained by S. N. Bernstein [1] as long ago as 1910. More recent estimates for derivatives have been obtained by S. N. Bernstein [2], O. A. Ladyzhenskaya [1], and L. Nirenberg [2]. Results on approximation in which the maximum principle is an essential ingredient are to be found in the works of L. Collatz [1, 3, 4], L. Bers [1], C. Clark and C. A. Swanson [1], A. G. Meyer [1], and W. Walter [2].

The material in Section 7 is based on Green's identities, and these may be found in standard texts such as O. D. Kellogg [1] and H. F. Weinberger [4]. The corresponding formulas for general second order elliptic operators have been given by W. Feller [1].

The bounds on the eigenvalues of elliptic operators given in Section 8 are taken from M. H. Protter and H. F. Weinberger [2]. J. Barta [1] obtained the corresponding result for the Laplace operator in 1937. Since then a variety of lower bounds for eigenvalues and related comparison theorems was developed by R. J. Duffin [1], C. Clark and C. A. Swanson [1], P. Hartman and A. Wintner [1], J. Hersch [1, 2], W. Hooker [1], K. Kreith [1], and M. H. Protter [2].

The classical Phragmèn-Lindelöf theorem may be found in texts on function theory. For example E. C. Titchmarsh [1] states the theorem for analytic functions. See also M. H. Heins [1]. H. F. Weinberger [4] gives the result for harmonic functions. Various forms of the Phragmèn-Lindelöf principle for elliptic operators have been obtained by G. N. Blohina [1], A. Friedman [1], D. Gilbarg [1, 4], E. Hopf [3], E. M. Landis [1], A. Pfluger [1], J. B. Serrin [5], N. Meyers and J. B. Serrin [1], and P. C. Fife [1]. Questions concerning isolated singularities of solutions of elliptic equations have been treated by Huber [1], Gilbarg and Serrin [1], and C. Pucci [7].

Inequalities of Harnack type for second order elliptic equations with smooth coefficients have been obtained by L. Lichtenstein [1] and W. Feller [1]. The proof of Theorem 23 in Section 10 follows that of J. B. Serrin [6]. Another proof of this theorem and another Harnack inequality in two dimensions were given by L. Bers and L. Nirenberg [1]. Harnack inequalities for linear equations in n variables have been obtained by J. Moser [1], J. B. Serrin [7] and G. Stampacchia [1]. The nonlinear case was treated by E. Bohn and L. K. Jackson [1], J. B. Serrin [8] and N. S. Trudinger [1]. Theorems on isolated singularities have been obtained by D. Gilbarg and J. B. Serrin [1]. A two-dimensional Liouville theorem without the uniform ellipticity hypothesis was given by S. N. Bernstein [2].

For the definition and elementary properties of the capacity of a set, see

G. Polya and G. Szegö [1] or O. D. Kellogg [1]. The inequalities derived in Section 11 may be found in M. H. Protter and H. F. Weinberger [1].

The classical Hadamard three-circles theorem is usually stated in terms of the maximum modulus of an analytic function. This form of the theorem may be found in E. C. Titchmarsh [1] and M. H. Heins [2]. Various extensions and generalizations to solutions of general second order elliptic equations and in a variety of norms have been obtained by S. Agmon [2], E. M. Landis [2, 3], and K. Miller [1].

The maximum principle, especially in the form of the Phragmèn-Lindelöf theorem, has been used with great success in problems of fluid flow. For results involving the comparison of two flows and the use of the maximum principle at infinity, see L. Bers [2], R. Finn and D. Gilbarg [1, 2], D. Gilbarg [2, 3, 4], D. Gilbarg and M. Shiffman [1], M. Lavrentiev [1], and J. B. Serrin [1, 2, 3, 4].

Extensive summaries of related results are to be found in the books of L. Bers, F. John, and M. Schechter [1], A. Friedman [4] and C. Miranda [3], and in the survey articles of O. A. Ladyzhenskaya and N. N. Ural'tzeva [1] and of E. M. Landis [1, 3].

CHAPTER 3

PARABOLIC EQUATIONS

SECTION 1. THE HEAT EQUATION

Suppose a long, thin rod of length l is situated on the interval $(0, l)$ along the x-axis. We shall assume that the material of the rod is homogeneous. Heat may be put into or removed from the rod, and we assume that the temperature u at any point in the rod is a function only of x, the location of a particular cross section, and of t, the time. We write $u = u(x, t)$. Under certain assumptions on the physical properties of the rod, the differential equation governing the flow of heat (in appropriate units) in the rod is given by

$$\frac{\partial^2 u}{\partial x^2} - \frac{\partial u}{\partial t} = f(x, t).$$

The function f is the rate of heat removal in the bar. The temperature function $u(x, t)$ satisfies a maximum principle somewhat different from the one which was established for elliptic equations and inequalities.

Suppose $u(x, t)$ satisfies the strict inequality

$$L[u] \equiv \frac{\partial^2 u}{\partial x^2} - \frac{\partial u}{\partial t} > 0$$

in a region E of the x, t-plane (Fig. 1). It is clear that u cannot have a (local) maximum at any interior point. For at such a point

$$\frac{\partial^2 u}{\partial x^2} \leq 0 \text{ and } \frac{\partial u}{\partial t} = 0,$$

thereby violating $L[u] > 0$. We shall not only extend this statement to solutions u of $L[u] \geq 0$, but we shall also show that for operators of this type, the maximum principle takes a stronger form.

To illustrate a typical problem,

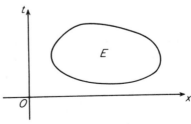

FIGURE 1

we shall suppose that the rod described above has its temperature prescribed initially (i.e., at time $t = 0$) and that the temperatures at the ends of the rod are known functions of time. The *principle of causality* states that the temperature distribution at any fixed time T is unaffected by any changes in the rod which happen at a time $t > T$. Thus it is natural to consider the rectangular region

$$E: \{0 < x < l, 0 < t \leq T\} \tag{1}$$

in the x, t-plane. We suppose that the temperature $u(x, t)$ is known on three sides of E:

$$S_1: \{x = 0, 0 \leq t \leq T\}, \qquad S_2: \{0 \leq x \leq l, t = 0\},$$
$$S_3: \{x = l, 0 \leq t \leq T\}.$$

On physical grounds, we expect that this information and the fact that the temperature u satisfies the equation

$$L[u] \equiv \frac{\partial^2 u}{\partial x^2} - \frac{\partial u}{\partial t} = 0$$

in E suffice to determine the temperature uniquely throughout E (Fig. 2).

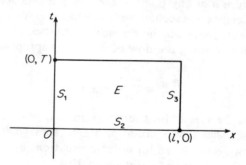

FIGURE 2

The uniqueness of the solution is easily established as a corollary of the following maximum principle.

THEOREM 1. Suppose $u(x, t)$ satisfies the inequality

$$L[u] \equiv \frac{\partial^2 u}{\partial x^2} - \frac{\partial u}{\partial t} \geq 0 \tag{2}$$

in the rectangular region E given by (1). Then the maximum of u on the closure $E \cup \partial E$ must occur on one of the three sides S_1, S_2 or S_3 (Fig. 2).

Proof. Suppose that M is the maximum of the values of u which occur on S_1, S_2, and S_3. We shall assume there is a point $P(x_0, t_0)$ of E where u has a value $M_1 > M$ and establish a contradiction. We define the auxiliary function

$$w(x) = \frac{M_1 - M}{2l^2}(x - x_0)^2.$$

Then, since $u \leq M$ on S_1, S_2, and S_3, we have

$$v(x, t) \equiv u(x, t) + w(x) \leq M + \frac{M_1 - M}{2} < M_1 \tag{3}$$

for all points of S_1, S_2, and S_3. Furthermore,

$$v(x_0, t_0) = u(x_0, t_0) = M_1 \tag{4}$$

and

$$L[v] \equiv L[u] + L[w] = L[u] + \frac{M_1 - M}{l^2} > 0 \tag{5}$$

throughout E. Conditions (3) and (4) show that v must assume its maximum either at an interior point of E or along the open interval

$$S_4: \{0 < x < l, t = T\}.$$

The inequality (5) shows that v cannot have an interior maximum. At a maximum along S_4, we have $\partial^2 v/\partial x^2 \leq 0$, implying that $\partial v/\partial t$ is strictly negative. Thus v must be larger at an earlier time so that the maximum in E cannot be on S_4. We see in this way that the assumption $u(x_0, t_0) > M$ leads to a contradiction.

Remarks. (i) The theorem states that the maximum not only cannot occur at an interior point of E, but also cannot occur at the "latest" time, except possibly at the ends of the rod unless $u \equiv$ const.

(ii) The maximum principle of Theorem 1 is not one of the strong form, since this theorem permits the maximum of u to occur both on the boundary and at interior points. Later we shall see that if the maximum occurs in E, then the solution must be constant in a certain region, a result which contains Theorem 1 as a special case.

(iii) For solutions of $L[u] = 0$ we obtain an associated minimum principle when we replace u by $-u$. The uniqueness theorem alluded to earlier then follows easily.

(iv) For elliptic differential inequalities the maximum of a solution could occur anywhere on the boundary. In the case of the heat equation, we have a stronger result—namely, the maximum can occur only on a specified portion of the boundary (unless $u \equiv$ const.). This fact is true both for more general equations of which the heat equation is the prototype, and for more general domains.

The equation of heat propagation in a three-dimensional, homogeneous object D is

$$L[u] \equiv \Delta u - \frac{\partial u}{\partial t} = f(x, y, z, t),$$

where $u = u(x, y, z, t)$ is the temperature at a point $P(x, y, z)$ of D at time t, and f is the rate of heat removal. As in the one-dimensional case, the function u, considered as a function of four variables, cannot have a local maximum at a point where

$$L[u] \equiv \Delta u - \frac{\partial u}{\partial t} > 0.$$

This fact follows when we recall that at a maximum $\Delta u \leq 0$ and $\partial u/\partial t = 0$. We wish to extend the maximum principle to functions u which satisfy the nonstrict inequality

$$L[u] \geq 0.$$

The simplest three-dimensional problem of physical interest is that of a fixed, bounded, homogeneous solid filling a domain D. We suppose that the problem begins at time $t = 0$ and that initially the temperature $u(x, y, z, 0)$ is a prescribed function of (x, y, z). Furthermore, the temperature on the boundary ∂D of D is prescribed for all times $t \geq 0$. The problem of heat flow concerns the determination of the temperature function $u(x, y, z, t)$ for all points $P(x, y, z)$ in D and for all time $t > 0$.

The domain of interest in four-dimensional space-time consists of the infinite cylinder $D \times (0, \infty)$. However, the principle of causality, as stated in the one-dimensional case, permits us to restrict the domain under consideration. The temperature distribution in D at some fixed positive time T is determined by what happens during the interval $0 \leq t \leq T$. Thus, the natural four-dimensional region to consider is the finite cylinder $D \times (0, T]$. The temperature u in this cylinder is to be determined by $L[u]$, by the values of u in D at $t = 0$, and by the values of u on the cylindrical wall, $\partial D \times (0, T]$.

Denoting the cylinder $D \times (0, T]$ by E, we expect the maximum principle to state that u takes its maximum value on the portion of the boundary of E which is either at the bottom of E or along the sides $\partial D \times [0, T]$ (see Fig. 3). Indeed, if

$$L[u] \equiv \Delta u - \frac{\partial u}{\partial t} > 0,$$

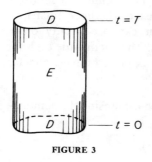

FIGURE 3

then we already have seen that a maximum cannot occur at an interior point of E. If the maximum of u is attained at time $t = T$ at an interior point of D, we reach a contradiction. To see this, we note that $\Delta u \leq 0$ at this maximum point; therefore the inequality $L[u] > 0$ implies that

$$\frac{\partial u}{\partial t} < 0$$

at this same point. Then u is larger at a slightly earlier time, and so the maximum must occur either along the sides or at the bottom of E.

We shall extend this maximum principle not only to the case $L[u] \geq 0$ but also to more general equations and domains.

EXERCISES

1. Verify that the function $u(x, t) = (1/\sqrt{t})e^{-x^2/4t}$ is a solution of the heat equation $u_{xx} - u_t = 0$ for $t > 0$. Show that for fixed $x \neq 0$, $\lim_{t \to 0^+} u(x, t) = 0$. Also show that u is unbounded in any neighborhood of $(0, 0)$.

2. Prove the analogue of Theorem 1 for the heat operator in n dimensions:
$$L[u] \equiv \frac{\partial^2 u}{\partial x_1^2} + \frac{\partial^2 u}{\partial x_2^2} + \cdots + \frac{\partial^2 u}{\partial x_n^2} - \frac{\partial u}{\partial t}.$$

SECTION 2. THE ONE-DIMENSIONAL PARABOLIC OPERATOR

The differential operator
$$L[u] \equiv a(x, t) \frac{\partial^2 u}{\partial x^2} + b(x, t) \frac{\partial u}{\partial x} - \frac{\partial u}{\partial t} \tag{1}$$
is said to be **parabolic** at a point (x, t) if
$$a(x, t) > 0.$$
The operator L is **uniformly parabolic** in a domain D of the x, t-plane if there is a positive constant μ such that
$$a(x, t) \geq \mu \text{ for all } (x, t) \text{ in } D.$$

The one-dimensional heat operator discussed in Section 1 is uniformly parabolic in the entire x, t-plane, since it is obtained from (1) by setting $a(x, t) \equiv 1$ and $b(x, t) \equiv 0$.

Let E be the rectangular region
$$E: \{0 < x < A, 0 < t \leq T\}$$
(see Fig. 4). It is clear that if u satisfies the strict inequality
$$L[u] > 0 \text{ in } E, \tag{2}$$

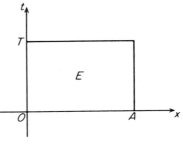

FIGURE 4

then u cannot have a local maximum at any interior point. For, at an interior maximum point, $\partial^2 u/\partial x^2 \leq 0$ and $\partial u/\partial x = \partial u/\partial t = 0$, in violation of (2). Moreover, the maximum of u cannot occur along the open segment forming the upper boundary of E—that is, along $0 < x < A$, $t = T$. To see this, we observe that at such a maximum point, $\partial u/\partial x = 0$, $\partial u/\partial t \geq 0$, and $\partial^2 u/\partial x^2 \leq 0$.

The maximum principle for the operator L will now be extended to solutions of the differential inequality $L[u] \geq 0$. The proof we present is due to Nirenberg [1] and uses a suitable variation of the method employed by Hopf for elliptic operators. As we shall see in the next section, this proof is easily modified to include more general equations and domains. The basic result depends on the following three lemmas.

LEMMA 1. Let u satisfy the differential inequality

$$L[u] \equiv a(x, t)\frac{\partial^2 u}{\partial x^2} + b(x, t)\frac{\partial u}{\partial x} - \frac{\partial u}{\partial t} \geq 0 \tag{3}$$

in a domain E of the x, t-plane where a and b are bounded and L is uniformly parabolic. Let K be a disk such that it and its boundary ∂K are contained in E. Suppose the maximum of u in E is M, that $u < M$ in the interior of K, and that $u = M$ at some point P on the boundary of K. Then the tangent to K at P is parallel to the x-axis. (That is, P is either the point at the top or the point at the bottom of the disk K.)

Proof. Let the disk K have its center at $(\bar{x}, \bar{t})$, and let R be the radius of K. We shall assume that the point P on ∂K is not at the top or bottom and reach a contradiction.

We may assume without loss of generality that P is the only boundary point where $u = M$. For, if not, we may replace K by a slightly smaller disk K' whose boundary is interior to K except at the one point P where $\partial K'$ and ∂K are tangent. Then K' has exactly one point P on its boundary where $u = M$ and the argument may be continued with K', if necessary.

Suppose P has coordinates (x_1, t_1) with $x_1 \neq \bar{x}$ (Fig. 5). We construct a disk K_1 with center at P and radius R_1 so small that

$$R_1 < |x_1 - \bar{x}|,$$

and also such that K_1 lies completely in E. The boundary ∂K_1 consists of two arcs: C' (which includes its endpoints) is the intersection of ∂K_1 with the closed disk $K \cup \partial K$, and C'' is the complement of C' with respect to ∂K_1 (Fig. 5). Since u is less than M on the closed arc C', a positive constant η can be found so that

$$u \leq M - \eta \text{ on } C'.$$

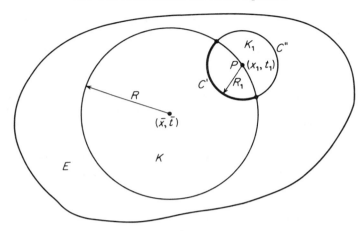

FIGURE 5

Moreover, since $u \leq M$ throughout E, we have
$$u \leq M \text{ on } C''.$$
We define the auxiliary function
$$v(x, t) = e^{-\alpha[(x-\bar{x})^2+(t-\bar{t})^2]} - e^{-\alpha R^2}.$$
Then for positive values of α, v is positive in K, zero on ∂K, and negative in the exterior of K. We compute
$$L[v] = 2\alpha e^{-\alpha[(x-\bar{x})^2+(t-\bar{t})^2]}[2\alpha a(x - \bar{x})^2 - a - b(x - \bar{x}) + (t - \bar{t})].$$
In the disk K_1 and on its boundary we have $|x - \bar{x}| \geq |x_1 - \bar{x}| - R_1 > 0$, and so it is possible to choose α so large that
$$L[v] > 0 \text{ for } (x, t) \text{ in } K_1 \cup \partial K_1.$$
We now form the function
$$w(x, t) = u(x, t) + \epsilon v(x, t)$$
where ϵ is a positive constant to be chosen. We observe that
$$L[w] = L[u] + \epsilon L[v] > 0 \text{ in } K_1. \tag{4}$$
Since $u \leq M - \eta$ on C', we can select ϵ so small that
$$w = u + \epsilon v < M \text{ on } C'.$$
Furthermore, since v is negative on C'' and $u \leq M$, we have
$$w = u + \epsilon v < M \text{ on } C''.$$
Thus $w < M$ on the entire boundary $\partial K_1 = C' \cup C''$. On the other hand, since v vanishes on ∂K, we find
$$w(x_1, t_1) = u(x_1, t_1) + \epsilon v(x_1, t_1) = u(x_1, t_1) = M.$$

Hence the maximum of w in K_1 must occur at an interior point. This fact contradicts (4) and the lemma is established. Notice that the argument fails if P is at the top or the bottom of K. For if $x_1 = \bar{x}$, we cannot choose $R_1 < |x_1 - \bar{x}|$.

Remark. It is essential that the inequality $u \leq M$ hold in a domain E containing $K \cup \partial K$, so that it is valid on the disk $K_1 \cup \partial K_1$ which is partly outside ∂K. For example, the function $u = x^2 + (t - 2)^2$ satisfies the inequality $u_{xx} - u_t \geq 0$ for $t \leq 4$ and $u < 1$ in the disk $x^2 + (t - 2)^2 < 1$, but it takes on the maximum value one everywhere on the boundary $x^2 + (t - 2)^2 = 1$. Lemma 1 does not apply because $u > 1$ outside this circle. Theorem 3 will show that the inequality (3) need only hold in K.

LEMMA 2. Suppose that in a domain E of the x, t-plane, u satisfies the inequality $L[u] \geq 0$ with L as in Lemma 1. Suppose that $u < M$ at some interior point (x_0, t_0) of E and that $u \leq M$ throughout E. If l is any horizontal line segment in the interior of E which contains (x_0, t_0), then $u < M$ on l (Fig. 6).

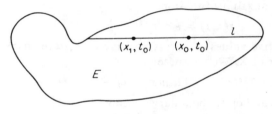

FIGURE 6

Proof. We suppose that $u = M$ at some interior point (x_1, t_0) on l, and that $u < M$ at (x_0, t_0). We shall reach a contradiction. For convenience we suppose that $x_1 < x_0$ and move x_1 to the right, if necessary, so that $u < M$ for $x_1 < x \leq x_0$. Let d_0 be either $x_0 - x_1$ or the minimum of the distances from any point of the line segment $x_1 \leq x \leq x_0$, $t = t_0$ to ∂E, whichever is smaller.

For $x_1 < x < x_1 + d_0$ we define $d(x)$ to be the distance from (x, t_0) to the nearest point in E where $u = M$. Since $u(x_1, t_0) = M$, $d(x) \leq x - x_1$. By Lemma 1, the nearest point is directly above or below (x, t_0). That is, either $u(x, t_0 + d(x)) = M$ or $u(x, t_0 - d(x)) = M$. Since the distance from a point $(x + \delta, t_0)$ to $(x, t_0 \pm d(x))$ is $\sqrt{d(x)^2 + \delta^2}$, we see that

$$d(x+\delta) \leq \sqrt{d(x)^2 + \delta^2} < d(x) + \frac{\delta^2}{2d(x)}. \tag{5}$$

Replacing x by $x + \delta$ and δ by $-\delta$, we see that also

$$d(x+\delta) > \sqrt{d(x)^2 - \delta^2}. \qquad (6)$$

Suppose now that $d(x) > 0$, and choose $0 < \delta < d(x)$. We subdivide the interval $(x, x+\delta)$ into n equal parts and apply the inequalities (5) and (6) to find

$$d\left(x + \frac{j+1}{n}\delta\right) - d\left(x + \frac{j}{n}\delta\right) \le \frac{\delta^2}{2n^2 d(x + (j/n)\delta)} \le \frac{\delta^2}{2n^2\sqrt{d(x)^2 - \delta^2}},$$
$$j = 0, 1, \ldots, n-1.$$

Summing from $j = 0$ to $n-1$ gives

$$d(x+\delta) - d(x) \le \frac{\delta^2}{2n\sqrt{d(x)^2 - \delta^2}}$$

for any integer n. Letting $n \to \infty$, we see that

$$d(x+\delta) \le d(x)$$

for $\delta > 0$. In other words, $d(x)$ is a nonincreasing function of x. Since $d(x) \le x - x_1$, which is arbitrarily small for x sufficiently close to x_1, we see that $d(x) \equiv 0$ for $x_1 < x < x_1 + d_0$. In other words, $u(x, t_0) \equiv M$ on this interval, contrary to our hypothesis that $u < M$ for $x_1 < x \le x_0$. Thus we have reached a contradiction of the statement that $u(x_0, t_0) < M$, $u(x_1, t_0) = M$.

Remark. Lemma 2 states that if there is a single interior point where $u = M$, then $u \equiv M$ along the largest horizontal segment containing this point whose interior lies in E.

In what follows, we shall have occasion to consider the differential inequality $L[u] \ge 0$ in a region of the form $E_T = \{(x, t) \in E : t \le T\}$ where E is a domain. We shall assume that u is continuously differentiable in x and t and twice differentiable in x throughout E_T, with $\partial u(x,t)/\partial T$ defined as a one-sided derivative

LEMMA 3. Suppose that in the lower half $K_{t_1} = \{(x, t) : (x - x_1)^2 + (t - t_1)^2 < R^2, t \le t_1\}$ of a disk K centered at $P(x_1, t_1)$ u satisfies the inequality $L[u] \ge 0$, with L as in Lemma 1. Suppose that $u < M$ in the portion of K where $t < t_1$. Then $u(P) < M$.

Proof. We define the function

$$v(x, t) = e^{-[(x-x_1)^2 + \alpha(t-t_1)]} - 1.$$

A simple computation shows that

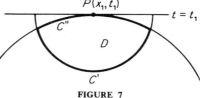

FIGURE 7

$$L[v] = e^{-[(x-x_1)^2 + \alpha(t-t_1)]}\left[4a(x-x_1)^2 - 2a - 2b(x-x_1) + \alpha\right].$$

We choose α positive and so large that
$$L[v] > 0 \text{ in } K \text{ for } t \leq t_1.$$
The parabola
$$(x - x_1)^2 + \alpha(t - t_1) = 0 \tag{7}$$
is tangent to the line $t = t_1$ at the point P. We denote by C' the portion (including the endpoints) of ∂K which is below the parabola (7), and we denote by C'' the portion of the parabola located within the disk K (Fig. 7). The region enclosed by C' and C'' is designated D. By hypothesis, $u < M$ on the closed arc C', and so there is an $\eta > 0$ such that
$$u \leq M - \eta \text{ on } C'.$$
We form the function
$$w(x, t) = u(x, t) + \epsilon v(x, t),$$
where ϵ is a positive constant to be chosen. We observe that $v = 0$ on C''. Therefore, we may choose ϵ so small that w has the properties:
(i) $L[w] = L[u] + \epsilon L[v] > 0$ in D,
(ii) $w = u + \epsilon v < M$ on C',
(iii) $w = u + \epsilon v \leq M$ on C''.
Condition (i) shows that w cannot attain its maximum in D; therefore the maximum of w is M, and it occurs at the point P. We conclude that
$$\frac{\partial w}{\partial t} \geq 0 \text{ at } P. \tag{8}$$
A simple computation shows that at P
$$\frac{\partial v}{\partial t} = -\alpha < 0. \tag{9}$$
Thus, we find from (8) and (9) that
$$\frac{\partial u}{\partial t} \geq -\epsilon \frac{\partial v}{\partial t} > 0 \text{ at } P. \tag{10}$$
On the other hand, since the maximum of u on $t = t_1$ occurs at P,
$$\frac{\partial u}{\partial x} = 0, \quad \frac{\partial^2 u}{\partial x^2} \leq 0 \text{ at } P.$$
These inequalities contradict the hypothesis that $L[u] \geq 0$, and the lemma is proved.

On the basis of the preceding lemmas, we can now establish the following result.

THEOREM 2. Let E be a domain and suppose that in $E_{t_1} = \{(x, t) \in E : t \leq t_1\}$ the inequality
$$L[u] \equiv a \frac{\partial^2 u}{\partial x^2} + b \frac{\partial u}{\partial x} - \frac{\partial u}{\partial t} \geq 0$$

holds, that a and b are bounded, and that L is uniformly parabolic in E_{t_1}. If $u \le M$ in E_{t_1} and $u(x_1, t_1) = M$, then $u = M$ at every point (x, t) in E_{t_1} which can be connected with (x_1, t_1) by a horizontal and a vertical line segment, both of which lie in E_{t_1}.

Proof. Suppose that $u(x_1, t_0) < M$ and that the line segment $l = \{(x, t) : x = t_1, t_0 \le t \le t_1\}$ lies in E. Let τ be the least upper bound of values of u on l such that $u(x_1, t) < M$. By continuity, $u(x_1, \tau) = M$ while Lemma 2 shows that there is an $R > 0$ such that $u < M$ for $|x - x_1| < R$, $t_0 \le t < \tau$. This leads to a contradiction of Lemma 3.

Remarks. (i) It is possible for a solution u of (3) to attain its maximum in a region E without being identically constant. For example, this will actually occur in a heat flow problem along a rod, if the rod is initially at a uniform temperature M and if this same temperature is maintained at the ends up to a time $t = t_1$. Then, if the temperature at the ends is decreased, thereafter the solution is no longer constant. Notice that the result of Theorem 2 is not violated. In this respect, the maximum principle has a form quite different from the one for elliptic equations.

(ii) Theorem 2 can be combined with Lemma 2 to identify the entire region in which a solution which attains an interior maximum must be constant. Once we obtain a point Q at which $u = M$, the maximum, we know that $u \equiv M$ on the largest horizontal segment in E containing Q. Then Theorem 2 shows that all points in E below this segment must have $u \equiv M$. Lemma 2 shows that $u \equiv M$ on every horizontal segment containing such a point. If P is a point of E which can be connected with Q by a path in E consisting only of horizontal segments and "upward" pointing vertical segments, then $u(P) = M$ (Fig. 8). The portion of E in Figure 8 where u must have the value M if $u = M$ at the point Q is indicated by horizontal shading. The portion labelled A, B, and C in Fig. 8 are exterior to E.

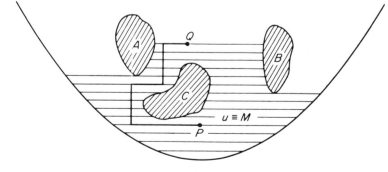

FIGURE 8

(iii) Since all the lemmas concerned only neighborhoods of interior points, it is sufficient to assume that a and b are bounded and L is uniformly parabolic in every closed subset of E.

We have just seen that a nonconstant solution u of a parabolic inequality $L[u] \geq 0$ can attain its maximum only at certain portions of the boundary. In the study of elliptic inequalities we found that the derivative normal to the boundary could never vanish at a point where the maximum is assumed (Theorem 7, Chapter 2). This important fact was used in several of the applications, especially in the proofs of the theorems on uniqueness of solutions of elliptic equations.

Under certain conditions, the solution u of a parabolic inequality has the property that its derivative normal to the boundary cannot vanish at a point where the maximum is attained. The precise statement of the result is given in the following theorem which, like its elliptic counterpart, has a number of applications.

THEOREM 3. Let E be a region and $E_{t_0} = \{(x,t) \in E : t \leq t_0\}$. Suppose u satisfies in E the uniformly parabolic inequality

$$a \frac{\partial^2 u}{\partial x^2} + b \frac{\partial u}{\partial x} - \frac{\partial u}{\partial t} \geq 0$$

where a and b are bounded. Suppose that u is continuously differentiable at the boundary point $P(x_0, t_0)$, that $u(P) = M$, that $u(x,t) < M$ for $(x,t) \in E_{t_0}$, that P lies on the boundary of a disk K tangent to ∂E, centered at (x_1, t_1) with $x_1 \neq x_0$, and that the portion of K below $t = t_0$, denoted K_{t_0}, lies in E_{t_0}. If $\partial / \partial \nu$ denotes any outward directional derivative from E_{t_0} at P, then

$$\frac{\partial u}{\partial \nu} > 0 \text{ at } P.$$

Proof. We construct a disk K_1 centered at P and of radius less than $|x_1 - x_0|$. See Fig. 9. We call C' the portion of ∂K_1 contained in K_{t_0} together

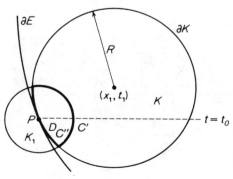

FIGURE 9

Sec. 2 *The One-Dimensional Parabolic Operator* 171

with its endpoints. Denoting by C'' the arc of ∂K which is in $K_1 \cap E_{t_0}$, we observe that the arcs C', C'' and the line segment $t = t_0$ form the boundary of a region D as shown in Fig. 9. By choosing a slightly smaller disk than K, if necessary, we can make $u < M$ on C'' except at P. Since $u < M$ on C', we can state the following three facts:

(i) $u < M$ on C'' except at P.
(ii) $u = M$ at P.
(iii) There exists a sufficiently small $\eta > 0$ such that

$$u \leq M - \eta \text{ on } C'.$$

We now introduce the auxiliary function

$$v(x, t) = e^{-\alpha[(x-x_1)^2 + (t-t_1)^2]} - e^{-\alpha R^2}$$

and note that

$$L[v] = 2\alpha e^{-\alpha[(x-x_1)^2 + (t-t_1)^2]}[2\alpha a(x - x_1)^2 - a - b(x - x_1) + (t - t_1)].$$

Thus for sufficiently large α, we have

$$L[v] > 0 \text{ for } (x, t) \text{ on } D \cup \partial D.$$

We construct the function

$$w = u + \epsilon v$$

and observe that for every positive ϵ, $L[w] = L[u] + \epsilon L[v] > 0$ in D. Because of fact (iii) above, we can choose ϵ so small that

$$w < M \text{ on } C'.$$

Since $v = 0$ on ∂K, we have, because of (i) above,

$$w < M \text{ on } C'' \text{ except at } P$$

and

$$w(P) = M.$$

Restricting our attention to the region D, we apply the maximum principle and conclude that the maximum of w on $D \cup \partial D$ occurs at the single point P. Therefore at P

$$\frac{\partial w}{\partial \nu} = \frac{\partial u}{\partial \nu} + \epsilon \frac{\partial v}{\partial \nu} \geq 0.$$

However, a computation shows that

$$\frac{\partial v}{\partial \nu} = \boldsymbol{\nu} \cdot \mathbf{n} \frac{\partial v}{\partial r} = -2\boldsymbol{\nu} \cdot \mathbf{n} \alpha R e^{-\alpha R^2} < 0.$$

We conclude that $\partial u / \partial \nu > 0$ at P, and the proof is complete.

Remark. The result of Theorem 3 may not hold if the normal to ∂E is parallel to the t-axis at a maximum point. Solutions u of parabolic inequalities may have regions in which u is constant. The boundary of such a region is perpendicular to the t-direction and, since u is continuously differentiable, it will have a vanishing normal derivative along the entire boundary of the domain in which $u \equiv$ constant. For example, if the maximum of u occurs at every point of the shaded region, as shown in Fig. 10, then $\partial u/\partial t$ will be zero along the entire length of the line l. Now the solution u, restricted to the region E^* consisting of points above the line l, will have its maximum along l and the normal derivative will vanish there.

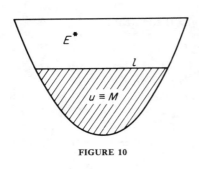

FIGURE 10

We now consider the maximum principle for inequalities of the form

$$(L + h)[u] \geq 0,$$

where L is uniformly parabolic and h is a prescribed function of x and t. The following result is an analogue of Theorems 6 and 8 of Chapter 2.

THEOREM 4. Suppose the hypotheses of Theorem 2 hold in a region E and that $h \leq 0$ in E. If the maximum M of u is attained at an interior point (x_1, t_1) and if $M \geq 0$, then $u \equiv M$ on all line segments $t =$ **constant** of E which lie directly below the horizontal segment of E containing (x_1, t_1). If a nonnegative maximum M occurs at a boundary point P, the conclusion of Theorem 3 holds at P.

Proof. We follow exactly the procedure used in the proof of Theorems 2 and 3. In the lemmas we choose the parameter α in the auxiliary function v so large that $(L + h)[v] > 0$. Because $h \leq 0$, we see that $(L + h)[w] \leq 0$ at a nonnegative maximum of w. The remainder of the argument is unchanged from that in Theorems 2 and 3.

For solutions of $(L + h)[u] \leq 0$ there is an associated minimum principle if the minimum m is nonpositive. The result follows by application of Theorem 4 to $(-u)$.

EXERCISES

1. Show that the conclusion of Lemma 1, page 164, is still valid if, in the hypotheses, the disk K is replaced by any convex region.

2. The function $u(x, t) = -(x^2 + 2xt)$ is a solution of the equation $x(\partial^2 u/\partial x^2) - (\partial u/\partial t) = 0$. Is the maximum principle valid in D: $-1 < x < 1, 0 < t \leq \frac{1}{2}$? Explain.

3. Show that any solution u of the following problem is unique:

$$\frac{\partial^2 u}{\partial x^2} - \frac{\partial u}{\partial t} = f(x, t) \text{ in } D: \{0 < x < a, 0 < t < T\},$$

$$u(x, 0) = g_1(x), \text{ for } 0 < x < a,$$

$$\frac{\partial u}{\partial x} = g_2(t), \text{ for } 0 \leq t \leq T, x = 0,$$

$$\frac{\partial u}{\partial x} = g_3(t), \text{ for } 0 \leq t \leq T, x = a,$$

where g_1, g_2, and g_3 are prescribed functions.

SECTION 3. THE GENERAL PARABOLIC OPERATOR

The operator

$$L \equiv \sum_{i,j=1}^{n} a_{ij}(\mathbf{x}, t) \frac{\partial^2}{\partial x_i \partial x_j} + \sum_{i=1}^{n} b_i(\mathbf{x}, t) \frac{\partial}{\partial x_i} - \frac{\partial}{\partial t}$$

is said to be **parabolic** at $(\mathbf{x}, t) \equiv (x_1, x_2, \ldots, x_n, t)$ if for fixed t the operator consisting of the first sum is elliptic at $(\mathbf{x}, t)$. That is, L is parabolic if there is a number $\mu > 0$ such that

$$\sum_{i,j=1}^{n} a_{ij}(\mathbf{x}, t) \xi_i \xi_j \geq \mu \sum_{i=1}^{n} \xi_i^2 \quad (1)$$

for all n-tuples of real numbers $(\xi_1, \xi_2, \ldots, \xi_n)$. The operator L is **uniformly parabolic** in a region E_T if (1) holds with the same number $\mu > 0$ for all $(\mathbf{x}, t)$ in E_T.

The results of Section 2 are extended to uniformly parabolic operators in a completely straightforward manner.

THEOREM 5. Let u satisfy the uniformly parabolic differential inequality

$$L[u] \equiv \sum_{i,j=1}^{n} a_{ij}(\mathbf{x}, t) \frac{\partial^2 u}{\partial x_i \partial x_j} + \sum_{i=1}^{n} b_i(\mathbf{x}, t) \frac{\partial u}{\partial x_i} - \frac{\partial u}{\partial t} \geq 0 \quad (2)$$

in a region $E_{\bar{t}} = \{(x_1, x_2, \ldots, x_n, t) \in E : t \leq \bar{t}\}$ where E is a domain, and suppose the coefficients of L are bounded. Suppose that the maximum of u in $E_{\bar{t}}$ is M and that it is attained at a point $P(x, t)$ of $E_{\bar{t}}$. Thus if Q is a point of E which can be connected to P by a path in E consisting only of horizontal segments and upward vertical segments, then $u(Q) = M$.

Proof. The result is derived in exactly the same way as was Theorem 2. Since the first term in the operator L is an elliptic operator in n-dimensional space, a transformation of coordinates reduces this portion of the operator to the Laplace operator at a single point (Theorem 4, Chapter 2). It follows at once that a solution of $L[u] > 0$ cannot have a maximum on $E_{\bar{t}}$. To extend this result to solutions of $L[u] \geq 0$, we establish the analogues of Lemmas 1, 2, and 3 of Section 2. We, replace the auxiliary function v in the proof of Lemma 1 by

$$v(\mathbf{x}, t) = e^{-\alpha[\sum_{i=1}^{n}(x_i - \bar{x}_i)^2 + (t-\bar{t})^2]} - e^{-\alpha R^2}.$$

The auxiliary function corresponding to that in Lemma 3 is

$$v(\mathbf{x}, t) = e^{-\sum_{i=1}^{n}(x_i - \bar{x}_i)^2 - \alpha(t - \bar{t})} - 1.$$

We replace disks by $(n + 1)$-dimensional balls, and the parabola (7) of Section 2 by the hyperparaboloid

$$\sum_{i=1}^{n}(x_i - \bar{x}_i)^2 + \alpha(t - \bar{t}) = 0.$$

The remaining details of the proof are left to the reader.

Remark. Theorem 5 is valid if the point $P(\bar{x}, \bar{t})$ is on a horizontal component $E(\bar{t})$ of the boundary ∂E of E, provided that u and the derivatives $\partial u/\partial x_i$, $\partial^2 u/\partial x_i \partial x_j$, and $\partial u/\partial t$ are all continuous on $E \cup E(\bar{t})$.

The next theorem is a direct extension to $(n + 1)$ dimensions of Theorem 3.

THEOREM 6. *Let u satisfy the uniformly parabolic inequality (2) with bounded coefficients in a domain E, and define $E_{\bar{t}} = \{(\mathbf{x}, t) \in E : t \leq \bar{t}\}$. Suppose the maximum M of u is attained at a point $P(\mathbf{x}, \bar{t})$ on the boundary ∂E. Assume that a sphere through P can be constructed which is tangent to ∂E at P and such that the part of its interior where $t \leq \bar{t}$ lies in $E_{\bar{t}}$, and that $u < M$ in $E_{\bar{t}}$. Also, suppose that the radial direction from the center of the sphere to P is not parallel to the t-axis. Then, if $\partial/\partial \nu$ denotes any directional derivative in an outward direction from $E_{\bar{t}}$, we have*

$$\frac{\partial u}{\partial \nu} > 0 \text{ at } P.$$

As in Theorem 4, we can apply the proofs of Theorems 5 and 6 to a solution of the differential inequality $(L + h)[u] \geq 0$ with $h \leq 0$ when $M \geq 0$. We then obtain the following theorem.

THEOREM 7. *The conclusions of Theorems 5 and 6 remain valid if u is a solution of $(L + h)[u] \geq 0$, provided $h \leq 0$ and $M \geq 0$.*

Remarks. (i) If $u(\mathbf{x})$ satisfies the elliptic inequality

$$(L + h)[u] \equiv \sum_{i,j=1}^{n} a_{ij}(\mathbf{x}) \frac{\partial^2 u}{\partial x_i \, \partial x_j} + \sum_{i=1}^{n} b_i(\mathbf{x}) \frac{\partial u}{\partial x_i} + h(\mathbf{x})u \geq 0, \qquad h \leq 0$$

in a domain D of x-space, it obviously also satisfies the parabolic inequality

$$L[u] - \frac{\partial u}{\partial t} + hu \geq 0$$

in the set $D \times (0, T]$ of $(\mathbf{x}, t)$-space. If u attains a maximum M at an interior point $\mathbf{x}_0$ in D or on ∂D, it attains this maximum at the point $(\mathbf{x}_0, \frac{1}{2}T)$ in $D \times (0, T]$ or on $\partial D \times (0, T]$. Hence Theorems 5, 6, and 7 imply the corresponding maximum principles for elliptic inequalities.

(ii) The change of variable $v = ue^{-\lambda t}$ replaces the inequality $(L + h)[u] \geq 0$ by $(L + h - \lambda)[v] \geq 0$. If h is bounded above, we can choose λ so large that $h - \lambda \leq 0$, so that a maximum principle applies to v. In particular, it follows that Theorem 7 with $M = 0$ holds without the restriction $h \leq 0$.

EXERCISES

1. Prove Theorem 5 in detail.
2. Prove Theorem 6 in detail.

SECTION 4. UNIQUENESS THEOREMS FOR BOUNDARY VALUE PROBLEMS

Let E be the rectangular domain in the x, t-plane determined by the inequalities

$$c < x < d, \qquad 0 < t < T.$$

(See Fig. 11.) We pose the problem of determining a function $v(x, t)$ which satisfies the uniformly parabolic equation

$$a \frac{\partial^2 v}{\partial x^2} + b \frac{\partial v}{\partial x} - \frac{\partial v}{\partial t} = f(x, t) \text{ in } E \tag{1}$$

and the boundary conditions

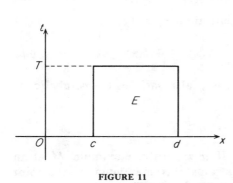

FIGURE 11

$$v(x, 0) = g_1(x) \text{ for } c \leq x \leq d,$$
$$v(c, t) = g_2(t) \text{ for } 0 \leq t < T,$$
$$v(d, t) = g_3(t) \text{ for } 0 \leq t < T.$$
(2)

The function f is prescribed throughout E with the functions g_i, $i = 1, 2, 3$ given on their respective domains of definition.

As in the case of elliptic equations, we shall not investigate conditions on the coefficients in equation (1) and on the boundary conditions (2) which guarantee the existence of a solution $v(x, t)$, but we shall show that it is possible to establish the uniqueness of a solution by means of the maximum principle alone. That is, we shall prove that **there can be at most one solution of equation (1) which satisfies the boundary conditions (2)**.

To establish this result, we assume that v_1 and v_2 are two functions which satisfy (1) and (2) with the same f and g_i, $i = 1, 2, 3$. We define

$$u = v_1 - v_2$$

and observe that

$$a \frac{\partial^2 u}{\partial x^2} + b \frac{\partial u}{\partial x} - \frac{\partial u}{\partial t} = 0 \text{ in } E$$

and

$$u(x, 0) = 0 \text{ for } c \leq x \leq d, \qquad u(c, t) = u(d, t) = 0 \text{ for } 0 \leq t < T.$$

According to the maximum principle as stated in Theorem 2, u cannot have a positive maximum in E, and so $u \leq 0$ everywhere. Applying the same reasoning to $(-u)$, we obtain $u \geq 0$ in E. Hence

$$u = v_1 - v_2 \equiv 0 \text{ in } E.$$

The result just established in the one-dimensional case will now be extended to solutions of general parabolic equations with less restrictive boundary conditions.

We consider a bounded domain D in n-dimensional Euclidean space and an interval $[0, T]$ of the t-axis. Let E_T denote the $(n+1)$-dimensional region $D \times (0, T]$. The portion of the boundary of E consisting of $\partial D \times (0, T)$ we denote by Γ.

THEOREM 8. *Let u be a solution of the uniformly parabolic equation*

$$(L + h)[u] \equiv \sum_{i,j=1}^{n} a_{ij}(\mathbf{x}, t) \frac{\partial^2 u}{\partial x_i \partial x_j} + \sum_{i=1}^{n} b_i(\mathbf{x}, t) \frac{\partial u}{\partial x_i} + h(\mathbf{x}, t)u - \frac{\partial u}{\partial t} = f(\mathbf{x}, t)$$

(3)

in E, and let the coefficients of L be bounded. Suppose $u(\mathbf{x}, t) \equiv u(x_1, x_2, \ldots, x_n, t)$ satisfies the boundary conditions

$$u(\mathbf{x}, 0) = g_1(\mathbf{x}) \text{ in } D \tag{4}$$

and

$$\alpha(\mathbf{x}, t) u(\mathbf{x}, t) + \beta(\mathbf{x}, t) \frac{\partial u}{\partial \nu} = g_2(\mathbf{x}, t) \tag{5}$$

for all $(\mathbf{x}, t)$ on Γ, where $\partial/\partial \nu$ is any directional derivative in an outward direction on Γ. Assume that $\alpha \geq 0$, $\beta \geq 0$ on Γ, that $\alpha^2 + \beta^2 > 0$ at each point, and that $h(\mathbf{x}, t)$ is bounded above. If v is another solution of (3) satisfying the same boundary conditions (4) and (5), then $v \equiv u$ in E.

Proof. The result follows as a simple application of the maximum principle. We define

$$w = u - v.$$

Then w satisfies

$$(L + h)[w] = 0,$$

and the initial and boundary conditions

$$w(\mathbf{x}, 0) = 0 \text{ in } D \text{ and } \alpha w + \beta \frac{\partial w}{\partial \nu} = 0 \text{ on } \Gamma. \tag{6}$$

By Remark (ii) at the end of Section 3, we may assume, without loss of generality, that $h(\mathbf{x}, t) \leq 0$. According to Theorem 7, the maximum of u must occur either at $t = 0$ or on Γ. If the maximum of w is positive, then it must occur on Γ. However, Theorem 7 states that at such a maximum point $\partial w/\partial \nu > 0$. Since α and β cannot vanish simultaneously, the condition

$$\alpha w + \beta \frac{\partial w}{\partial \nu} = 0$$

is violated at a positive maximum. Thus $w \leq 0$ throughout E. Applying the same reasoning to $-w$, we find $w \geq 0$. Therefore $w = u - v \equiv 0$ in E.

Remarks. (i) Note that if $\alpha \equiv 1$, $\beta \equiv 0$, then the problem (3), (4), (5) is a direct generalization of the one-dimensional boundary value problem for a rectangle.

(ii) Theorem 8 holds for more general domains E. In particular, the domain D may move with time provided its boundary points move with finite speed.

(iii) The fact that uniqueness of solutions holds regardless of the sign of h is in marked contrast to boundary value problems for elliptic equations, where eigenvalues may occur.

EXERCISES

1. Show that the problem

$$u_{xx} - tu_t + 2u = 0$$
$$u(x, 0) = 0, \quad 0 \leq x \leq \pi$$
$$u(0, t) = u(\pi, t) = 0, \quad t \geq 0$$

has the solutions $u = at \sin x$ for all values of the constant a. Why are the uniqueness theorems not applicable?

2. Show that the problem (polar coordinates in the plane)

$$u_{rr} + \frac{1}{r} u_r + \frac{1}{r^2} u_{\theta\theta} - u_t = 0, \quad 0 < r < 1, \quad t > 0$$
$$u(r, \theta, 0) = 0, \quad 0 \leq r \leq 1, \quad 0 \leq \theta \leq 2\pi$$
$$u_\theta(1, \theta, t) = 0, \quad 0 \leq \theta \leq 2\pi, \quad t > 0$$

has more than one solution. Why doesn't Theorem 8 apply?

SECTION 5. A THREE-CURVES THEOREM

The Hadamard three-circles theorem for subharmonic functions (Chapter 2, Section 12) does not have an exact analogue for functions which satisfy parabolic inequalities. However, by means of the maximum principle, it is possible to obtain a three-curves theorem which is similar to the Hadamard inequality. The result given here has a number of applications. As an example, we use it to establish the uniqueness of the solution of an initial value problem for the heat equation.

Let t_0 be a fixed positive constant, and consider the one-parameter family of parabolas

$$\frac{x^2}{t_0 - t} = \rho,$$

where the constant ρ takes on all positive values (see Fig. 12). Except for points on the t-axis, each point in the strip $\{0 < t < t_0, -\infty < x < \infty\}$ lies on exactly one member of this family.

We may consider ρ as a function of x and t. A computation yields

$$\frac{\partial^2 \rho}{\partial x^2} - \frac{\partial \rho}{\partial t} = \frac{2(t_0 - t) - x^2}{(t_0 - t)^2}.$$

We seek a function of ρ alone, say $\sigma(\rho)$, which is a solution of the heat equation. To determine σ, we write

$$\frac{\partial^2 \sigma}{\partial x^2} - \frac{\partial \sigma}{\partial t} \equiv \sigma''(\rho)\rho_x^2 + \sigma'(\rho)(\rho_{xx} - \rho_t).$$

A Three-Curves Theorem

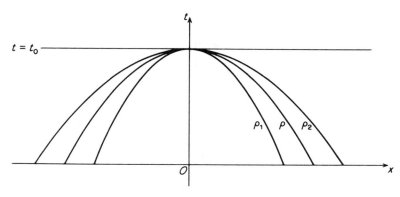

FIGURE 12

Thus σ must satisfy

$$\frac{\sigma''}{\sigma'} = -\frac{\rho_{xx} - \rho_t}{\rho_x^2} = \frac{1}{4} - \frac{1}{2\rho}$$

and, upon integration,

$$\log \sigma' = \tfrac{1}{4}\rho - \tfrac{1}{2}\log \rho.$$

Therefore,

$$\sigma' = \frac{1}{\sqrt{\rho}} e^{\rho/4},$$

and a second integration gives

$$\sigma(\rho) = \int_0^\rho \frac{1}{\sqrt{\rho_1}} e^{\rho_1/4} d\rho_1. \tag{1}$$

The function $\sigma(\rho)$ as given by (1) satisfies the heat equation for $t < t_0$ except for $x = 0$.

We consider a function $u(x, t)$ which satisfies

$$u_{xx} - u_t \geq 0$$

in a region D described as follows: D is bounded from below by the line $t = 0$ and from above by the line $t = \bar{t}$, where $\bar{t} < t_0$; D is bounded on its sides by the arcs of the parabolas $\rho = \rho_1$ and $\rho = \rho_2$ situated in the first quadrant (see Fig. 13). For $\rho_1 \leq \rho \leq \rho_2$, we define the functions

$$M_1(\rho) = \max_{\substack{x^2 = \rho(t_0 - t) \\ 0 \leq t \leq \bar{t}}} u(x, t), \qquad M_2 = \max_{\sqrt{\rho_1 t_0} \leq x \leq \sqrt{\rho_2 t_0}} u(x, 0),$$

and

$$M(\rho) = \max(M_1(\rho), M_2).$$

The function

$$\varphi(\rho) = a + b\sigma(\rho)$$

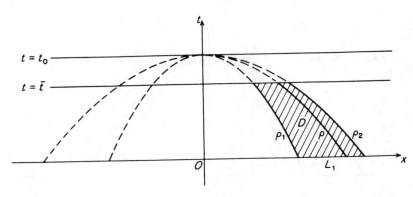

FIGURE 13

satisfies the heat equation. We now determine a and b by the relations

$$a + b\sigma(\rho_1) = M(\rho_1),$$
$$a + b\sigma(\rho_2) = M(\rho_2),$$

and we find

$$\varphi(\rho) = \frac{M(\rho_1)[\sigma(\rho_2) - \sigma(\rho)] + M(\rho_2)[\sigma(\rho) - \sigma(\rho_1)]}{\sigma(\rho_2) - \sigma(\rho_1)}.$$

The function

$$v = u - \varphi(\rho)$$

satisfies the inequality

$$\frac{\partial^2 v}{\partial x^2} - \frac{\partial v}{\partial t} \geq 0 \text{ in } D.$$

Furthermore, since

$$u(x, 0) \leq M_2$$

and

$$u(x, t) \leq M(\rho_1) \text{ for } x^2 = \rho_1(t_0 - t),$$
$$u(x, t) \leq M(\rho_2) \text{ for } x^2 = \rho_2(t_0 - t),$$

we find that $v \leq 0$ on the entire boundary of D below the line $t = \bar{t}$. Applying the maximum principle to v, we derive the formula

$$M(\rho) \leq \frac{M(\rho_1)[\sigma(\rho_2) - \sigma(\rho)] + M(\rho_2)[\sigma(\rho) - \sigma(\rho_1)]}{\sigma(\rho_2) - \sigma(\rho_1)}. \tag{2}$$

That is, $M(\rho)$ *is a convex function of* $\sigma(\rho)$. With the aid of (2) we can establish the following uniqueness result, which was first proved by A. N. Tikhonov [1].

THEOREM 9. Let $u(x, t)$ and $v(x, t)$ be solutions of $u_{xx} - u_t = f(x, t)$ in the strip $D: \{-\infty < x < \infty, 0 < t < T\}$, and suppose that u and v are continuous on $D \cup \partial D$. If $u(x, 0) = v(x, 0) = g(x)$, where g is a prescribed function, and if there are constants A and c such that

$$|u(x, t)|, |v(x, t)| \leq A e^{cx^2} \tag{3}$$

uniformly in t for $0 \leq t \leq T$, then $u(x, t) \equiv v(x, t)$ in D.

Proof. We employ the convexity inequality (2). We select $t_0 < 1/4c$ and consider the function $w(x, t) \equiv u(x, t) - v(x, t)$ in the domain D_1: $\{-\infty < x < \infty, 0 \leq t \leq t_0/2\}$. Then w satisfies the heat equation and $w(x, 0) \equiv 0$. We apply inequality (2) to w and let $\rho_2 \to \infty$. Since w is bounded by a multiple of e^{cx^2} and $\sigma(\rho)$ grows as rapidly as $e^{x^2/4t_0}$, we find that $M(\rho_2)/\sigma(\rho_2) \to 0$ as $\rho_2 \to \infty$. Therefore

$$M(\rho) \leq M(\rho_1).$$

Letting $\rho_1 \to 0$, we see that w takes on its maximum in the half-strip $\{x \geq 0, 0 \leq t \leq \frac{1}{2}t_0\}$ at $x = 0$. Applying the same reasoning for negative values of x, we conclude that w takes on its maximum in D_1 at $x = 0$. Then by Theorem 2 this maximum must be zero, and so $w \leq 0$ in the whole strip. Applying the same reasoning to $(-w)$, we find that $w \equiv 0$ in D_1. We may repeat the entire process using the line $t = \frac{1}{2}t_0$ as initial line and find that $w \equiv 0$ in $D_2: \{-\infty < x < \infty, t_0/2 \leq t \leq t_0\}$. After a finite number of steps, we conclude that $w \equiv 0$ in D.

Remarks. (i) In n dimensions, we can use the same method to derive an inequality such as (2) applicable to functions which satisfy $\Delta u - u_t \geq 0$ with the maximum taken over paraboloids of the form $(x_1^2 + \cdots + x_n^2)/(t_0 - t) = \rho$. Then we obtain a uniqueness theorem for solutions of the heat equation which grow no more rapidly than $Ae^{c(x_1^2 + x_2^2 + \cdots + x_n^2)}$.

(ii) The growth condition (3) is needed, since examples can be found of nonuniqueness of solutions if growth more rapid than (3) is allowed. See Tikhonov [1], Täcklind [1], Widder [1], Rosenbloom and Widder [1].

(iii) As in the case of elliptic equations, a general three-surfaces theorem can be obtained which is applicable to linear parabolic equations. A uniqueness theorem for parabolic equations with conditions as in Theorem 9 can also be obtained. However, it is also possible to derive a uniqueness theorem for unbounded domains by a direct application of the maximum principle. At the same time a Phragmèn-Lindelöf principle for parabolic equations may be established. We shall develop this method in Section 6.

EXERCISES

1. Establish a three-parabolas theorem for functions which satisfy $u_{xx} - u_t \geq 0$ with respect to the family of parabolas $x^2 + 2t = \rho$.

2. Establish a "three-paraboloid" theorem for functions which satisfy $\Delta u - u_t \geq 0$ with respect to the family of hypersurfaces $(x_1^2 + \cdots + x_n^2)/(t_0 - t) = \rho$.

3. Establish a three-parabolas theorem for functions which satisfy $au_{xx} + bu_x - u_t \geq 0$ (a, b constants) with respect to an appropriate class of curves of the form $(\alpha x^2 + \beta t)e^{\gamma x + \delta t} = \rho$.

4. Show by means of inequality (2) that the problem

$$u_{xx} - u_t = f(x, t), \quad \text{for } x > 0, \quad t > 0$$
$$u(x, 0) = g(x), \quad \text{for } x > 0$$
$$u(0, t) = h(t), \quad \text{for } t > 0$$

has at most one solution u such that $e^{-cx^2}u$ is bounded for some constant c.

5. Show by means of inequality (2) that the problem

$$u_{xx} - u_t = f(x, t), \quad \text{for } x > 0, \quad t > 0$$
$$u(x, 0) = g(x), \quad \text{for } x > 0$$
$$-u_x(0, t) + u(0, t) = h(t), \quad \text{for } t > 0$$

has at most one solution u such that $e^{-cx^2}u$ is bounded for some constant c.

SECTION 6. THE PHRAGMÈN-LINDELÖF PRINCIPLE

In Section 9 of Chapter 2, we discussed the solution of elliptic equations in unbounded domains. We found that boundary value problems are uniquely solvable only when the solutions are required to satisfy certain conditions at infinity. A similar situation prevails for solutions of parabolic equations. In Section 5, we saw that the uniqueness of the solution of the initial value problem for the heat equation can be established by means of a three-parabolas theorem. The result given there is valid for functions which satisfy a specific growth condition as $|x| \to \infty$. In this section we establish a maximum principle for functions which satisfy a parabolic inequality in an unbounded domain. This principle is then applied to extend to general parabolic equations the uniqueness result stated in Theorem 9 for the heat equation.

Let D be an unbounded domain in n-dimensional Euclidean space. We consider a function $u(\mathbf{x}, t) \equiv u(x_1, x_2, \cdots, x_n, t)$ defined in the region $E = D \times (0, T)$. Suppose that in E the function u satisfies the differential inequality

Sec. 6 The Phragmèn-Lindelöf Principle

$$(L + h)[u] \geq 0 \tag{1}$$

with $h = h(\mathbf{x}, t) \leq 0$ and L a uniformly parabolic operator given by

$$L \equiv \sum_{i,j=1}^{n} a_{ij}(\mathbf{x}, t) \frac{\partial^2}{\partial x_i \, \partial x_j} + \sum_{i=1}^{n} b_i \frac{\partial}{\partial x_i} - \frac{\partial}{\partial t}. \tag{2}$$

The maximum principle as given in Theorem 7 is applicable to the function u. However, since E is unbounded, we cannot always conclude that the maximum of u occurs either at $t = 0$ or on $\partial D \times (0, T)$, as is the case for bounded domains. We saw examples of such a situation in the study of elliptic equations in unbounded domains. (See the discussion on pages 93 and 94 of Chapter 2, Section 9.) Of course, if $u \to 0$ as $x_1^2 + x_2^2 + \cdots + x_n^2 \to \infty$ uniformly in t for $0 \leq t \leq T$, and if $u \leq 0$ for $t = 0$ and on $\partial D \times (0, T)$, then it is an easy consequence of Theorem 7 that $u \leq 0$ in E. Thus we can locate the maximum value for those functions which tend to a limit as $\mathbf{x} \to \infty$. As we show in the next theorem, this type of maximum principle is valid for a much broader class of functions.

THEOREM 10. Let D be an unbounded domain in n-dimensional space and let E be the domain $D \times (0, T)$. Suppose that u satisfies $(L + h)[u] \geq 0$ in E with L a uniformly parabolic operator of the form (2) with bounded coefficients and with $h(\mathbf{x}, t)$ bounded in E. Assume that u satisfies the growth condition

$$\liminf_{R \to \infty} e^{-cR^2} \left[\max_{\substack{x_1^2 + \cdots + x_n^2 = R^2 \\ 0 \leq t \leq T}} u(\mathbf{x}, t) \right] \leq 0 \tag{3}$$

for some positive constant c. If $u \leq 0$ for $t = 0$ and $u \leq 0$ on $\partial D \times (0, T)$, then $u \leq 0$ in E.

Proof. We let $r^2 = x_1^2 + x_2^2 + \cdots + x_n^2$ and define the function

$$v(\mathbf{x}, t) = u(\mathbf{x}, t) e^{-c\gamma r^2/(\gamma - ct) - \beta t},$$

where c is the constant in (3) and β, γ are constants to be determined. We find from a straightforward computation that

$$e^{-c\gamma r^2/(\gamma - ct) - \beta t}(L + h)[u] = L[v] + \frac{4c\gamma}{\gamma - ct} \sum_{i,j=1}^{n} a_{ij} x_i \frac{\partial v}{\partial x_j} + H(\mathbf{x}, t)v \geq 0,$$

where

$$H(\mathbf{x}, t) = \frac{4c^2 \gamma^2}{(\gamma - ct)^2} \sum_{i,j=1}^{n} a_{ij} x_i x_j + \frac{2c\gamma}{\gamma - ct} \sum_{i=1}^{n} (a_{ii} + b_i x_i) + h(\mathbf{x}, t)$$

$$- \beta - \frac{c^2 \gamma r^2}{(\gamma - ct)^2}.$$

Since the coefficients in the parabolic operator L are bounded, there is a constant M such that

$$\sum_{i,j=1}^{n} a_{ij} x_i x_j \leq Mr^2.$$

Therefore

$$H(\mathbf{x}, t) \leq -\frac{c^2 \gamma r^2}{(\gamma - ct)^2}(1 - 4\gamma M) + \left[\frac{2c\gamma}{\gamma - ct} \sum_{i=1}^{n}(a_{ii} + b_i x_i) + h(\mathbf{x}, t) - \beta\right].$$

We now define the constants

$$A = \sup_{0 \leq r \leq 1}\left|\sum_{i=1}^{n} b_i x_i\right|, \qquad B = \sup_{r \geq 1}\left|\sum_{i=1}^{n} b_i \frac{x_i}{r^2}\right|.$$

Then so long as $(\gamma - ct) > 0$, the function $H(\mathbf{x}, t)$ satisfies the inequality

$$H(\mathbf{x}, t) \leq -\frac{c^2 r^2 \gamma}{(\gamma - ct)^2}\left[1 - 4\gamma M - \frac{2(\gamma - ct)}{c} B\right]$$
$$- \left[\beta - \frac{2c\gamma}{\gamma - ct}\left(A + \sum_{i=1}^{n} a_{ii} - h\right)\right].$$

Now we select γ so small that the expression in the first bracket on the right above is always positive for t in the interval $[0, \gamma/2c]$. Next we select β so large that the expression in the second bracket is always positive. Hence for these choices of β and γ, we find $H(\mathbf{x}, t) \leq 0$ in $D \times [0, \gamma/2c]$.

We denote by D_R the portion of D inside the ball $x_1^2 + x_2^2 + \cdots + x_n^2 < R^2$. By Theorem 7, the function $v(\mathbf{x}, t)$ cannot have a positive maximum at an interior point of $D_R \times [0, \gamma/2c]$. For any $\epsilon > 0$, the condition (3) shows that $v < \epsilon$ on $\partial D_R \times [0, \gamma/2c]$ for some arbitrarily large R; also $v \leq 0$ for $t = 0$. Therefore $v < \epsilon$ in $D_R \times [0, \gamma/2c]$. Letting $R \to \infty$ and $\epsilon \to 0$, we find that $v \leq 0$ in $D \times [0, \gamma/2c]$. In particular, $v(\mathbf{x}, \gamma/2c) \leq 0$ for $\mathbf{x} \in D$. Now the entire argument above may be repeated with $t = \gamma/2c$ as the initial surface instead of $t = 0$. In this way we obtain $v \leq 0$ in $D \times [\gamma/2c, 2(\gamma/2c)]$. In a finite number of steps we obtain $v \leq 0$ in E, and hence also $u \leq 0$ in E.

Remarks. (i) From the details of the proof, we see that it is sufficient to assume that the quantities

$$r^{-2} \sum_{i,j=1}^{n} a_{ij} x_i x_j, \qquad (1 + r^2)^{-1} \sum_{i=1}^{n} b_i x_i$$

and h are bounded from above. Moreover, if h satisfies $h < K(1 + r^{2\delta})$, where K is a constant and $0 < \delta < 1$, it is easily seen that

$$h \leq K[1 + (\gamma/\delta)^{-\delta/(1-\delta)}(1 - \delta) + \gamma r^2],$$

and the proof can still be carried through.

(ii) If $h(\mathbf{x}, t) \equiv 0$, then we can subtract any constant from u and apply Theorem 10 to conclude that u must attain its maximum whether positive or negative either at $t = 0$ or on $\partial D \times [0, T]$. If $h(\mathbf{x}, t) \leq 0$, the same can be done for a nonnegative maximum.

(iii) For functions satisfying $(L + h)[u] \leq 0$, we have a corresponding minimum principle. In analogy with the result for elliptic operators, we call Theorem 10 a Phragmèn–Lindelöf Principle.

The following uniqueness result is an immediate consequence of Theorem 10.

THEOREM 11. Let $u(\mathbf{x}, t)$, $v(\mathbf{x}, t)$ be solutions of

$$(L + h)[u] = f(\mathbf{x}, t) \text{ in } D \times (0, T)$$

(*D* unbounded) and suppose that u and v are continuous on $(D \cup \partial D) \times [0, T]$. We assume that

$$u(\mathbf{x}, 0) = v(\mathbf{x}, 0) = g_1(\mathbf{x}) \text{ in } D$$
$$u(\mathbf{x}, t) = v(\mathbf{x}, t) = g_2(\mathbf{x}, t) \text{ on } \partial D \times [0, T],$$

and furthermore, that both $|u|$ and $|v|$ satisfy condition (3) for some constant $c > 0$. Then $u \equiv v$ in $D \times [0, T]$.

Proof. We define $w = u - v$ and observe that Theorem 10 is applicable to w. Thus $w \leq 0$ in $D \times [0, T]$. The same argument applies to $(-w)$, and so $w \equiv 0$.

Remarks. (i) We have assumed here and in Theorems 8 and 9 that the initial and boundary values are taken on continuously. It is not sufficient to assume, for example, that the function u is continuous in t for each $\mathbf{x}$. That is, the result is false if we only suppose that

$$\lim_{t \to 0^+} u(\mathbf{x}, t) = g_1(\mathbf{x}) \text{ for each fixed } \mathbf{x} \text{ in } D.$$

To see this we let $L = \partial^2/\partial x^2 - \partial/\partial t$ and choose

$$u = xt^{-3/2}\, e^{-x^2/4t}.$$

Then u satisfies $L[u] = 0$ for all $t > 0$ and $-\infty < x < \infty$. Also, we see that $\lim_{t \to 0^+} u(x, t) = 0$ for each fixed real number x. However, u is not identically zero. The above function u is actually unbounded in every neighborhood of $(0, 0)$.

(ii) The Phragmèn–Lindelöf principle (Theorem 10) and the uniqueness result (Theorem 11) can be extended to solution of parabolic inequalities and equations which satisfy the more general boundary condition

$$\alpha \frac{\partial u}{\partial \nu} + \beta u = g(\mathbf{x}, t) \text{ on } \partial D \times (0, T].$$

(iii) The above results and proofs are analogous to those used in establishing the Phragmèn–Lindelöf principle for elliptic operators. In fact, we can carry over the entire elliptic theory without change and find those exceptional sets where boundary data may, under appropriate conditions, be left unprescribed. In this connection it is interesting to note that if a function $u(\mathbf{x})$ satisfies an elliptic inequality $(L + h)[u] \geq 0$ then, since u is independent of t, it obviously satisfies the parabolic inequality $(L + h - \partial/\partial t)[u] \geq 0$. Suppose we have a Phragmèn-Lindelöf theorem which states that a nonconstant solution v of $(L + h - \partial/\partial t)[v] \geq 0$ in $D \times (0, T)$ is, in this set, less than its maximum either in D at $t = 0$ or on the subset $S \times (0, T)$ of the lateral boundary $\partial D \times (0, T)$. Then clearly $u(\mathbf{x})$ is less in D than its maximum on S. Thus, if $(\partial D - S) \times (0, T)$ is an exceptional set for the parabolic operator $L + h - \partial/\partial t$, then $\partial D - S$ is an exceptional set for the elliptic operator L.

On the other hand, if we have a sequence of positive functions w_k which can be used to establish a Phragmèn-Lindelöf principle (Theorem 19 of Chapter 2) for the elliptic operator $L + h$ with the exceptional set $\partial D - S$, the same functions establish a Phragmèn-Lindelöf principle for $L + h - \partial/\partial t$ with the exceptional set $(\partial D - S) \times (0, T)$.

(iv) A removable singularity theorem can be obtained for parabolic equations in the same way as it was for elliptic equations in Chapter 2, Section 9, provided a Phragmèn-Lindelöf theorem is known and a suitable boundary value problem can be solved.

(v) More general removable singularity theorems have been obtained by D. G. Aronson [2].

SECTION 7. NONLINEAR OPERATORS

In Chapters 1 and 2 we showed how the maximum principle could be employed to obtain results for nonlinear equations. The same methods may be used to yield results for nonlinear parabolic equations. We consider vectors $\mathbf{x} = (x_1, x_2, \ldots, x_n)$ and $\mathbf{p} = (p_1, p_2, \ldots, p_n)$ and the matrix $\mathbf{R} = (r_{ij})$, $i, j = 1, 2, \ldots, n$. Let $F(\mathbf{x}, t, u, \mathbf{p}, \mathbf{R})$ be a continuously differentiable function of its $n^2 + 2n + 2$ variables. We shall use the notation $F(\mathbf{x}, t, u, p_i, r_{ij})$ to denote the above function with p_i and r_{ij} denoting generic arguments of F. We say that F is **elliptic** with respect to a function $u(\mathbf{x}, t)$ at a given point $(\mathbf{x}, t)$ if, for all real vectors $\xi = (\xi_1, \xi_2, \ldots, \xi_n)$, we have

$$\sum_{i,j=1}^{n} \frac{\partial F}{\partial r_{ij}} \xi_i \xi_j > 0 \text{ for } \xi \neq \mathbf{0} \tag{1}$$

when the values $p_i = \partial u/\partial x_i$ and $r_{ij} = \partial^2 u/\partial x_i \partial x_j$ are substituted in the arguments of the partial derivatives of F appearing in (1). The function

F is **elliptic in a domain** E in $(\mathbf{x}, t)$-space if it is elliptic at each point of E. The nonlinear operator

$$L[u] \equiv F\left(\mathbf{x}, t, u, \frac{\partial u}{\partial x_i}, \frac{\partial^2 u}{\partial x_i \partial x_j}\right) - \frac{\partial u}{\partial t} \qquad (2)$$

is said to be **parabolic** whenever F is elliptic.

We can use the maximum principle, as given in Theorem 7, to compare solutions of nonlinear parabolic equations. Let u be a solution of

$$L[u] = f(\mathbf{x}, t),$$

with L given by (2) in a domain E in $(\mathbf{x}, t)$-space, and suppose that $w = w(\mathbf{x}, t)$ satisfies

$$L[w] \leq f \text{ in } E.$$

We form the function

$$v(\mathbf{x}, t) = u(\mathbf{x}, t) - w(\mathbf{x}, t)$$

and consider the inequality (using subscripts for partial derivatives)

$$F(\mathbf{x}, t, u, u_{x_i}, u_{x_i x_j}) - F(\mathbf{x}, t, w, w_{x_i}, w_{x_i x_j}) - \frac{\partial v}{\partial t} \geq 0.$$

We apply the mean-value theorem of multi-dimensional calculus. Letting θ be such that $0 < \theta < 1$ and evaluating the derivatives of F at the arguments $\theta u + (1-\theta)w$, $\theta u_{x_i} + (1-\theta)w_{x_i}$, $\theta u_{x_i x_j} + (1-\theta)w_{x_i x_j}$, we find

$$\sum_{i,j=1}^{n} \left(\frac{\partial F}{\partial r_{ij}}\right) \frac{\partial^2 v}{\partial x_i \partial x_j} + \sum_{i=1}^{n} \left(\frac{\partial F}{\partial p_i}\right) \frac{\partial v}{\partial x_i} + \left(\frac{\partial F}{\partial u}\right) v - \frac{\partial v}{\partial t} \geq 0. \qquad (3)$$

We assume that F is elliptic in E for all functions of the form $\theta u + (1-\theta)w$, $0 \leq \theta \leq 1$. Under this assumption, the left side of (3) is a linear parabolic operator for the function v. We may apply the maximum principle (Remark (ii) after Theorem 7) to conclude that if v is nonpositive initially and on the boundary, then v is nonpositive in E.

The above discussion establishes the following result on approximation.

THEOREM 12. Let D be a bounded domain in n-dimensional space and let $E = D \times (0, T]$. Suppose that u is a solution of $L[u] = f(\mathbf{x}, t)$ in E with L given by (2), and that u satisfies the initial and boundary conditions

$$u(\mathbf{x}, 0) = g_1(\mathbf{x}) \text{ in } D,$$

$$u(\mathbf{x}, t) = g_2(\mathbf{x}, t) \text{ on } \partial D \times (0, T).$$

We assume that z and Z satisfy the inequalities

$$L[Z] \leq f(\mathbf{x}, t) \leq L[z] \text{ in } E,$$

and that L is parabolic with respect to the functions $\theta u + (1-\theta)z$ and $\theta u + (1-\theta)Z$ for $0 \leq \theta \leq 1$. If

$$z(\mathbf{x}, 0) \le g_1(\mathbf{x}) \le Z(\mathbf{x}, 0) \text{ in } D,$$
$$z \le g_2 \le Z \text{ on } \partial D \times (0, T),$$

then

$$z(\mathbf{x}, t) \le u(\mathbf{x}, t) \le Z(\mathbf{x}, t) \text{ in } E.$$

Example. The one-dimensional flow of heat through a homogeneous medium is governed by the nonlinear equation

$$\frac{\partial u}{\partial t} = \frac{\partial}{\partial x}\left[k(u)\frac{\partial u}{\partial x}\right], \tag{4}$$

where k is a positive function with a bounded first derivative. We verify easily that (4) is parabolic for all functions u. Since any constant satisfies (4), we may apply Theorem 12 to conclude that for any solution u, the maximum and minimum values must occur either at the initial time or on the boundary.

SECTION 8. WEAKLY COUPLED PARABOLIC SYSTEMS

The maximum principle for functions satisfying a second-order parabolic inequality may be extended to certain systems of parabolic inequalities. We consider a set of k functions

$$u_1(\mathbf{x}, t), u_2(\mathbf{x}, t), \ldots, u_k(\mathbf{x}, t)$$

which will be treated as a k-vector $\mathbf{u}(\mathbf{x}, t)$. The vector $\mathbf{x}$ is an element in n-dimensional Euclidean space with components $x_1, x_2, \ldots, x_n$. In conjunction with the vector $\mathbf{u}$, we are given k uniformly parabolic operators

$$L_\nu \equiv \sum_{i,j=1}^n a_{ij}^{(\nu)}(\mathbf{x}, t)\frac{\partial^2}{\partial x_i \partial x_j} + \sum_{i=1}^n b_i^{(\nu)}(\mathbf{x}, t)\frac{\partial}{\partial x_i} - \frac{\partial}{\partial t}, \quad \nu = 1, 2, \ldots, k.$$

We also introduce a $k \times k$ matrix $H = H(\mathbf{x}, t)$ of functions with elements $\{h_{\mu\nu}(\mathbf{x}, t)\}$, $\mu, \nu = 1, 2, \ldots, k$.

The system of parabolic inequalities for which we shall establish a maximum principle has the form

$$\left.\begin{aligned} L_1[u_1] + \sum_{\nu=1}^k h_{1\nu} u_\nu &\ge 0, \\ L_2[u_2] + \sum_{\nu=1}^k h_{2\nu} u_\nu &\ge 0, \\ &\vdots \\ L_k[u_k] + \sum_{\nu=1}^k h_{k\nu} u_\nu &\ge 0. \end{aligned}\right\} \tag{1}$$

We observe that each inequality of (1) contains derivatives of just one component. The system is coupled only in the terms which are not differentiated; a system of this form is said to be **weakly coupled**. Such systems arise in the study of the simultaneous diffusion of several substances which decay spontaneously.

We make the additional hypothesis that the off-diagonal terms of the matrix H are nonnegative:

$$h_{\mu\nu} \geq 0 \text{ for } \mu \neq \nu, \quad \mu, \nu = 1, 2, \ldots, k. \tag{2}$$

We shall use the notation $\mathbf{u} < 0$ to mean that every component u_ν, $\nu = 1, 2, \ldots, k$ is negative. Similarly, $\mathbf{u} \leq 0$ means that every component is nonpositive.

A maximum principle is easy to prove for vector-valued functions $\mathbf{u}$ which satisfy the system of strict inequalities

$$L_\mu[u_\mu] + \sum_{\nu=1}^{k} h_{\mu\nu} u_\nu > 0, \quad \mu = 1, 2, \ldots, k \tag{3}$$

in a domain of the form $E = D \times (0, T)$ in $(n+1)$-dimensional space. We assume throughout that all the coefficients $a_{ij}^{(\mu)}$, $b_i^{(\mu)}$, and $h_{\mu\nu}$ are bounded in E. We now show that if $\mathbf{u} < 0$ at $t = 0$, and on $\partial D \times (0, T)$, then $\mathbf{u} < 0$ in E. To do this, we suppose that the inequality $\mathbf{u} < 0$ is violated at a point $(\mathbf{x}, t)$ in E. Let $\bar{t}$ be the least upper bound of values of t at which $\mathbf{u}(\mathbf{x}, t) < 0$ in E. Then by continuity we see that $\mathbf{u}(\mathbf{x}, \bar{t}) \leq 0$ and, at some point $(\bar{\mathbf{x}}, \bar{t})$ in E, one of the components of $\mathbf{u}$ must vanish. That is, for at least one component, u_τ, we have $u_\tau(\bar{\mathbf{x}}, \bar{t}) = 0$. Since $\mathbf{u} \leq 0$ for $t \leq \bar{t}$,

$$L_\tau[u_\tau] \leq 0 \text{ at } (\bar{\mathbf{x}}, \bar{t}) \tag{4}$$

and, on account of hypothesis (2),

$$\sum_{\nu=1}^{k} h_{\tau\nu} u_\nu \leq 0 \text{ at } (\bar{\mathbf{x}}, \bar{t}). \tag{5}$$

Inequalities (4) and (5) together contradict the τ-th inequality in (3), and we conclude that $\mathbf{u} < 0$ in E.

In order to obtain a maximum principle for nonpositive functions $\mathbf{u}$ which satisfy (1) rather than (3), we first choose a constant β so that

$$\beta - \sum_{\nu=1}^{k} h_{\mu\nu}(\mathbf{x}, t) > 0 \text{ for } \mu = 1, 2, \ldots, k$$

throughout E; this selection is always possible since the elements of H are assumed bounded in E. Then if $\mathbf{u}$ satisfies inequalities (1) in E, we see that for any $\epsilon > 0$,

$$L_\mu[u_\mu - \epsilon e^{\beta t}] + \sum_{\nu=1}^{k} h_{\mu\nu}[u_\nu - \epsilon e^{\beta t}] > 0 \text{ in } E, \quad \mu = 1, 2, \ldots, k.$$

We introduce the vector function $\mathbf{v}(\mathbf{x}, t)$ with components $v_\mu = u_\mu - \epsilon e^{\beta t}$, $\mu = 1, 2, \ldots, k$, and note that the inequality $\mathbf{u} \leq 0$ at $t = 0$ and on

$\partial D \times (0, T)$ implies that $\mathbf{v} < 0$ on the same set of points. The maximum principle established above for functions satisfying the system of strict inequalities is valid for $\mathbf{v}$ for each $\epsilon > 0$. Letting $\epsilon \to 0$ we conclude that $\mathbf{u} \leq 0$ in E. It now follows from the inequalities $h_{\mu\nu} \geq 0$ for $\mu \neq \nu$ that each component u_μ satisfies the inequality

$$L_\mu[u_\mu] + h_{\mu\mu} u_\mu \geq 0 \text{ in } E.$$

For any constant β we can rewrite this inequality as

$$L_\mu[e^{\beta t} u_\mu] + (h_{\mu\mu} + \beta) e^{\beta t} u_\mu \geq 0 \text{ in } E.$$

We choose β so large that $h_{\mu\mu} + \beta \geq 0$ in E. Since $u_\mu \leq 0$ in E, we see that

$$L_\mu[e^{\beta t} u_\mu] \geq 0 \text{ in } E,$$
$$e^{\beta t} u_\mu \leq 0 \text{ at } t = 0 \text{ and on } \partial D \times (0, T). \tag{6}$$

We conclude from Theorem 5 that if $u_\mu = 0$ at some interior point $(\mathbf{x}_0, t_0)$, then $u_\mu \equiv 0$ for $t \leq t_0$. In this way we obtain the following strong maximum principle:

THEOREM 13. Suppose that $\mathbf{u}$ satisfies the uniformly parabolic system of inequalities (1) in a bounded domain $E = D \times (0, T)$. If $\mathbf{u} \leq 0$ at $t = 0$ and on $\partial D \times (0, T)$ and if H satisfies the conditions (2), then $\mathbf{u} \leq 0$ in E. Moreover, if $u_\mu = 0$ at an interior point $(\mathbf{x}_0, t_0)$, then $u_\mu \equiv 0$ for $t \leq t_0$.

By applying Theorem 6 to the inequalities (6), we find the next result.

THEOREM 14. Suppose that in a domain $E = D \times (0, T)$, $\mathbf{u}$ satisfies inequalities (1), that H satisfies (2), and that $\mathbf{u} \leq 0$. If some component u_τ vanishes at a boundary point P of $\partial D \times (0, T)$, and if there is a ball K in E having P on its boundary and such that $u_\tau < 0$ in K, then

$$\frac{\partial u_\tau}{\partial \nu} > 0 \text{ at } P$$

for any outward directional derivative $\partial/\partial \nu$.

Theorems 13 and 14 show that zero cannot be a maximum of a nonpositive solution $\mathbf{u}$ of (1) except under special circumstances. We shall now extend the above theorem to functions $\mathbf{u}$ having a nonnegative maximum M. We simply apply the same methods to the vector function having $u_\nu - M$, $\nu = 1, 2, \ldots, k$ as its components. To do so, we require the hypothesis that the matrix H have the property

$$\sum_{\nu=1}^{k} h_{\mu\nu} \leq 0, \quad \mu = 1, 2, \ldots, k. \tag{7}$$

For M a constant, we define the vector $\mathbf{M}$ as one with all k of its components equal to M.

THEOREM 15. Suppose that u satisfies inequalities (1), that H satisfies (2) and (7), and that $\mathbf{u} - \mathbf{M} \leq 0$ in D at $t = 0$ and on $\partial D \times [0, T]$. Then if $M \geq 0$, $\mathbf{u} - \mathbf{M} \leq 0$ in $E = D \times (0, T)$. If $u_\tau = M$ at an interior point $(\mathbf{x}, t_0)$ of E, then $u_\tau \equiv M$ for $t \leq t_0$. If $u_\tau = M$ at a boundary point P with the properties described in Theorem 14, and $u_\tau < M$ in the appropriate ball K, then $\partial u_\tau/\partial v > 0$ at P where $\partial/\partial v$ is any outward directional derivative.

Remarks. (i) If we introduce the change of variables

$$\bar{u}_\mu = u_\mu e^{-ct}, \qquad \mu = 1, 2, \ldots, k, \tag{8}$$

we find that

$$e^{-ct} L_\mu[u_\mu] = L_\mu[\bar{u}] - c\bar{u}_\mu, \qquad \mu = 1, 2, \ldots, k.$$

Thus whenever u satisfies a system (1), the vector $\bar{\mathbf{u}}$ satisfies a system of the same form but with $h_{\mu\mu}$ replaced by $h_{\mu\mu} - c$ in each of the inequalities. By choosing c sufficiently large it is always possible to fulfill conditions (7) (assuming, as always, that the elements of H are bounded).

(ii) Suppose $u_\tau = M$ at an interior point $(\mathbf{x}_0, t_0)$. Then $u_\tau \equiv M$ for $t \leq t_0$ and the τth inequality (1) becomes

$$\sum_{\nu=1}^{k} h_{\tau\nu} u_\nu \geq 0.$$

By (2), (7), and the fact that $\mathbf{u} \leq \mathbf{M}$ we have

$$\sum_{\nu=1}^{k} h_{\tau\nu} u_\nu \leq M \sum_{\nu=1}^{k} h_{\tau\nu} \leq 0.$$

Thus all these inequalities must be equalities. In particular, $\sum_{\nu=1}^{k} h_{\tau\nu} \equiv 0$ for $t \leq t_0$. Moreover, $u_\nu \equiv M$ for $t \leq t_0$ provided $h_{\tau\nu} \neq 0$ at some point $(\mathbf{x}, t_0)$. Even if $h_{\tau\nu}(\mathbf{x}, t_0) \equiv 0$, if there is an index μ such that $h_{\tau\mu}(\mathbf{x}, t_0) \neq 0$ and $h_{\mu\nu}(\mathbf{x}, t_0) \neq 0$ we conclude that $u_\mu \equiv 0$ and hence that $u_\nu \equiv 0$. Thus except under some very special circumstances we can conclude that if $u_\tau(\mathbf{x}_0, t_0) = M$, then $\mathbf{u} \equiv \mathbf{M}$ for $t \leq t_0$, and all the row sums of H vanish.

Theorem 15 yields uniqueness theorems and bounds for solutions of the problem

$$\left. \begin{array}{r} L_\mu[u_\mu] + \sum_{\nu=1}^{k} h_{\mu\nu} u_\nu = f_\mu \text{ in } D \times (0, T) \\ u_\mu(\mathbf{x}, 0) = g_\mu \text{ in } D \\ u_\mu = \bar{h}_\mu \text{ on } \Gamma_1 \times (0, T) \\ \dfrac{\partial u_\mu}{\partial v_\mu} + \sum_{\nu=1}^{k} s_{\mu\nu} u_\nu = \bar{h}_\mu \text{ on } \Gamma_2 \times (0, T) \end{array} \right\} \quad \mu = 1, 2, \ldots, k,$$

where D is a domain in n-dimensional Euclidean space and $\Gamma_1 \cup \Gamma_2 = \partial D$. Each $\partial/\partial v_\mu$ is an outward directional derivative. The quantities $f_\mu, g_\mu, \bar{h}_\mu$,

$\bar{h}_\mu$ are prescribed functions, H satisfies condition (2), and the matrix S with elements $\{s_{\mu\nu}\}$ has the properties that

$$s_{\mu\nu} \leq 0 \text{ for } \mu \neq \nu \text{ and } \sum_{\nu=1}^{k} s_{\mu\nu} \geq 0, \mu = 1, \ldots, k.$$

If the time-independent vector function $\mathbf{u}(x)$ satisfies the **weakly coupled elliptic system** of inequalities

$$\sum_{i,j=1}^{n} a_{ij}^{(\mu)} \frac{\partial^2 u_\mu}{\partial x_i \partial x_j} + \sum_{i=1}^{n} b_i^{(\mu)} \frac{\partial u_\mu}{\partial x_i} + \sum_{\nu=1}^{k} h_{\mu\nu} u_\nu \geq 0, \qquad \mu = 1, 2, \ldots, k$$

in a domain D, then $\mathbf{u}$ also satisfies (1) in $D \times [0, T]$. If H satisfies conditions (2) and (7), we can apply Theorem 15 to establish the maximum principle: *If $\mathbf{u} \leq \mathbf{M}$ with $\mathbf{M} \geq 0$ on ∂D, then $\mathbf{u} \leq \mathbf{M}$ in D.* In this way we get a maximum principle for elliptic systems. We observe that the change of variables (8) is no longer appropriate; hence conditions (7) are a genuine restriction for elliptic systems.

Remarks. (i) The above results can be generalized to give comparison theorems for weakly coupled nonlinear parabolic systems of the form

$$F_\mu\left(\mathbf{x}, t, \mathbf{u}, \frac{\partial u_\mu}{\partial x_1}, \ldots, \frac{\partial u_\mu}{\partial x_n}, \frac{\partial^2 u_\mu}{\partial x_1^2}, \frac{\partial^2 u_\mu}{\partial x_1 \partial x_2}, \ldots, \frac{\partial^2 u_\mu}{\partial x_n^2}\right) - \frac{\partial u_\mu}{\partial t} = f_\mu,$$

$$\mu = 1, 2, \ldots, k.$$

(ii) The maximum principle fails if the coupling is stronger and occurs, say, in the first derivative terms. For example, the system

$$u_{1xx} - u_{1t} \geq 0,$$
$$u_{2xx} - 9u_{1x} - u_{2t} \geq 0,$$

is satisfied in $E: \{0 \leq x \leq 1, 0 \leq t \leq 1\}$ by the functions

$$u_1 = -e^{x+t}, \qquad u_2 = t - 4(x - \tfrac{1}{2})^2.$$

However, u_1 and u_2 are nonpositive on the boundary of E below $t = 1$, while u_2 is positive on the line $x = \tfrac{1}{2}$. Thus weak coupling cannot be strengthened appreciably if the maximum principle is to hold.

(iii) A different maximum principle for another class of weakly coupled parabolic systems has been given by Szepticky [1] and Stys [1].

(iv) An application of a weakly coupled parabolic system and some of its properties are discussed by Habetler and Martino [1].

EXERCISES

1. Show that there is at most one solution of the problem

$$\frac{\partial^2 u_1}{\partial x^2} - \frac{\partial u_1}{\partial t} + u_1 + u_2 = 0,$$

$$\frac{\partial^2 u_2}{\partial x^2} - 2\frac{\partial u_2}{\partial t} + 2u_1 = 0 \text{ for } 0 < x < 1, \quad t > 0,$$

$$u_1(0, t) = u_1(1, t) = u_2(0, t) = \frac{\partial u_2}{\partial x}(1, t) = 0,$$

$$u_1(x, 0) = \sin \pi x,$$

$$u_2(x, 0) = 0.$$

2. Find upper and lower bounds for $u_2(1, 1)$ in Problem 1.
3. Show that if there is a vector field $\mathbf{w}(\mathbf{x}) > 0$ on $D \cup \partial D$ such that

$$L_\mu[w_\mu] + \sum_{\nu=1}^{k} h_{\mu\nu} w_\nu \leq 0 \text{ in } D, \mu = 1, 2, \ldots, k,$$

and if $\mathbf{u}(\mathbf{x})$ satisfies the elliptic system of inequalities

$$L_\mu[u_\mu] + \sum_{\nu=1}^{k} h_{\mu\nu} u_\nu \geq 0, \quad h_{\mu\nu} \geq 0 \text{ for } \mu \neq \nu,$$

where

$$L_\mu = \sum_{i,j=1}^{n} a_{ij}^{(\mu)}(\mathbf{x}) \frac{\partial^2}{\partial x_i \partial x_j} + \sum_{i=1}^{n} b_i^{(\mu)}(\mathbf{x}) \frac{\partial}{\partial x_i},$$

then $\mathbf{u} \leq M\mathbf{w}$ on ∂D implies $\mathbf{u} \leq M\mathbf{w}$ in D when $M \geq 0$. *Hint:* Obtain a system of inequalities for u_ν/w_ν.

BIBLIOGRAPHICAL NOTES

The weak maximum principle for the heat equation was discovered by E. E. Levi [1, 2]. A weak maximum principle for general parabolic equations was given by M. Picone [2, 3]. The strong maximum principle is due to L. Nirenberg [1]. The positivity of the outward directional derivative was found by R. Vyborny [1] and independently by A. Friedman [2] and C. Pucci [5]. A. Friedman extended the results to weak solutions of parabolic inequalities.

A three-cylinders theorem for parabolic equations which is analogous to the three-spheres theorem of Landis [2, 3] has been given by A. Ya. Glagelova [1].

The uniqueness theorem for unbounded domains which we have stated is due to A. N. Tikhonov [1]. A somewhat stronger result was given by S. Täcklind [1].

Extensions of the maximum principle to unbounded domains and corresponding uniqueness theorems under various assumptions about the coefficients have been derived by M. Picone [3], M. Krzyzański [1, 2, 3], W. Mlak [1], R. Výborný [1], P. Besala [1], I. Łojckzyk-Królikiewicz [1], W. Bodanko [1], D. G. Aronson [3], and D. G. Aronson and P. Besala [1].

A different uniqueness theorem for positive solutions was given for the heat equation by D. V. Widder [1] and extended to general parabolic equations by D. G. Aronson and P. Besala [1].

A theorem of Phragmèn-Lindelöf type has been obtained by A. Friedman [1]. Removable singularity thoerems have been given by D. G. Aronson [2] and B. Pini [1].

Extensions of maximum principles to nonlinear parabolic equations were made by M. Nagumo and S. Simoda [1], H. Westphal [1], O. A. Oleĭnik and T. D. Wentzel [1, 2], P. Besala [4], J. Kadlec and R. Výborný [1], S. Kaplan [1], J. Szarski [5], R. M. Redheffer [4], and I. I. Kolodner and R. N. Pederson [1].

Results on the approximation of solutions and error bounds by means of the maximum principle have been given by L. Collatz [1,2,3], C. Pucci [2], H. Westphal [1], K. Nickel [2], J. Schröder [1, 2] R. M. Redheffer [4], and W. Walter [1, 2, 3].

Maximum principles for weakly coupled systems of nonlinear parabolic equations have been given by J. Szarski [3, 4, 5], W. Mlak [1], J. Schröder [1, 2], P. Besala [2, 3] and A. McNabb [1]. A different kind of maximum principle in which the Euclidean length of the solution vector rather than the maximum of its components is bounded has been found by P. Szeptycki [1] and T. Stys [1]. R. K. Juberg [1] has shown that this Euclidean length can be bounded by a constant times the maximum of its initial and boundary values for a solution of a system of nonlinear parabolic equations which are coupled in the first derivatives of the unknown function.

J. Hadamard [1] and B. Pini [2] proved a Harnack inequality for the heat equation. A Harnack inequality for a very general class of second order linear parabolic equations was given by J. Moser [2]. Extensions to nonlinear equations were obtained by D. G. Aronson and J. B. Serrin [1] and by N. S. Trudinger [2]. Harnack inequalities and Liouville theorems for parabolic systems were given by S. D. Eidelman [1].

K. Nickel [1], O. A. Oleĭnik [2], and W. Velte [1] have applied the maximum principle to the study of the boundary layer in the flow of a viscous fluid. An application to the study of the asymptotic behavior of the solution of a singular perturbation problem was found by D. G. Aronson [1].

M. Picone [1], G. Fichera [2, 3], P. Hartman and R. Sacksteder [1], and O. A. Oleĭnik [3, 4] have obtained and used maximum principles for solutions of elliptic-parabolic equations of the form

$$\sum_{i,j=1}^{n} a_{ij}(\mathbf{x}) \frac{\partial^2 u}{\partial x_i \partial x_j} + \sum_{i=1}^{n} b_i(\mathbf{x}) \frac{\partial u}{\partial x_i} + h(\mathbf{x})u = 0$$

where the matrix a_{ij} is only known to be positive semidefinite, so that

$$\sum_{i,j=1}^{n} a_{ij} \xi_i \xi_j \geq 0$$

for every vector ξ.

Other presentations of the maximum principle may be found in the books of L. Bers, F. John, and M. Schechter [1], P. Garabedian [1], G. Hellwig [1], I. G. Petrovsky [1] and H. F. Weinberger [4]. Extensive surveys of results on parabolic partial differential equations are to be found in the books of A. Friedman [1], W. Walter [3], and J. Szarski [5], and in the survey articles of E. M. Landis [1, 3], A. M. Il'in, A. S. Kalashnikov, and O. A. Oleĭnik [1], and O. A. Oleĭnik and S. N. Kružkov [1].

CHAPTER 4

HYPERBOLIC EQUATIONS

SECTION 1. THE WAVE EQUATION

The solutions of hyperbolic equations and inequalities do not exhibit the type of maximum principle that was studied in the preceding chapters. Even in the simplest case of the wave equation in two independent variables*

$$u_{xx} - u_{tt} = 0, \tag{1}$$

it is easily seen that the maximum of a nonconstant solution u in a domain D may occur at an interior point. For example, we observe that the function

$$u = \sin x \sin t$$

satisfies the above equation, and that it attains its maximum in the square $0 < x < \pi$, $0 < t < \pi$, at the center $(\pi/2, \pi/2)$.

In order to find a possible maximum principle we investigate the nature of well-posed boundary and initial value problems for hyperbolic equations.

The wave equation (1) describes the transverse motion of a homogeneous string under tension. The most elementary problem for such a mechanical system is the initial value problem. In this problem we prescribe u and $\partial u/\partial t$ at $t = 0$ on some interval $2\alpha \leq x \leq 2\beta$. We shall show that the motion is then uniquely determined within the so-called characteristic triangle; that is, the triangle whose sides are composed of the interval $2\alpha \leq x \leq 2\beta$, $t = 0$ and the lines

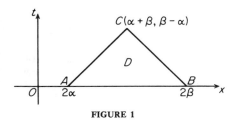

FIGURE 1

---

*In this chapter we shall frequently indicate partial derivatives by subscripts, so that, for instance, u_{xx} means $\partial^2 u/\partial x^2$.

making angles $\pm \pi/4$ with the x-axis and passing through $(2\alpha, 0)$ and $(2\beta, 0)$ (see Fig. 1). Our demonstration consists of obtaining the explicit solution of this problem by Riemann's method.

Let u be a twice continuously differentiable function, and let

$$L[u] \equiv u_{xx} - u_{tt}$$

be given in the triangle D with vertices at $A(2\alpha, 0)$, $B(2\beta, 0)$ and $C(\alpha + \beta, \beta - \alpha)$. We proceed in a purely formal way and consider the expression

$$\iint_D L[u]\, dx\, dt = \iint_D (u_{xx} - u_{tt})\, dx\, dt.$$

A simple application of Stokes' theorem* gives

$$\iint_D L[u]\, dx\, dt = \int_A^B u_t\, dx + \int_B^C (u_x\, dt + u_t\, dx) + \int_C^A (u_x\, dt + u_t\, dx).$$

Since $dx = -dt$ along segment BC and $dx = dt$ along CA, we have

$$\iint_D L[u]\, dx\, dt = \int_A^B u_t\, dx - \int_B^C (u_x\, dx + u_t\, dt) + \int_C^A (u_x\, dx + u_t\, dt).$$

The second two line integrals may now be integrated, and we get

$$\iint_D L[u]\, dx\, dt = \int_A^B u_t\, dx + u(A) + u(B) - 2u(C)$$

or

$$u(C) = \tfrac{1}{2}[u(A) + u(B)] + \tfrac{1}{2}\int_A^B u_t\, dx - \tfrac{1}{2}\iint_D L[u]\, dx\, dt. \tag{2}$$

Thus the value of u at C is uniquely determined by $u(2\alpha, 0)$, $u(2\beta, 0)$, $\partial u/\partial t$ for $2\alpha < x < 2\beta$, $t = 0$, and by $L[u]$ in D.

In particular, we see that if

$$L[u] \geq 0 \text{ in } D \tag{3}$$

and

$$\frac{\partial u}{\partial t}(x, 0) \leq 0, \qquad 2\alpha \leq x \leq 2\beta, \tag{4}$$

then

$$u(C) \leq \tfrac{1}{2}[u(A) + u(B)].$$

---

*The two dimensional **Stokes' theorem** states that for any bounded domain D with smooth boundary ∂D and any continuously differentiable functions $p(x, t)$ and $q(x, t)$ we have

$$\iint_D \left(\frac{\partial p}{\partial x} + \frac{\partial q}{\partial t}\right) dx\, dt = \oint_{\partial D} (p\, dt - q\, dx),$$

where the integral around ∂D is taken in a counter-clockwise sense. This identity follows from the divergence theorem and the fact that the outward unit normal vector has the components $(dt/ds, -dx/ds)$.

If we take any point C' within the characteristic triangle ABC, we can construct the right isosceles triangle $A'B'C'$ with A' and B' on the x-axis and the right angle at C'. Then we find in the same way that

$$u(C') \leq \tfrac{1}{2}[u(A') + u(B')].$$

It is clear from this inequality that the values of u in the triangle ABC cannot exceed the maximum of the values of u on the initial line segment AB. Thus, **if u satisfies (3) and (4), its maximum on $D \cup \partial D$ must occur on the initial line AB.**

This result is a weak maximum principle, since it does not give any information about whether or not u may achieve its maximum at an interior point. As a matter of fact, the function

$$u(x, t) = \cos x \cos t$$

satisfies $L[u] = 0$; also,

$$\left.\frac{\partial u}{\partial t}\right|_{t=0} = 0.$$

The maximum of u occurs on the initial line at $(0, 0)$ and $(2\pi, 0)$, but it is also achieved at (π, π).

Relation (2) shows that if the function $u_t(x, 0)$ is strictly negative on AB or if $L[u] > 0$ in D, then the value of u at C is strictly less than the average of the values at A and B. In this situation we see that if M denotes the maximum of u on AB, then $u < M$ in D.

EXERCISES

1. Show that if $L[u] \equiv u_{xx} - u_{tt} + u \geq 0$ in a domain D as shown in Fig. 1, and if $u(x, 0) \leq M < 0$, $u_t(x, 0) \leq 0$, then $u < 0$ in D.

2. Show that if $|u_{xx} - u_{tt}| \leq A$ in a domain D as shown in Fig. 1, and if $|u(x, 0)| \leq B$, $|u_t(x, 0)| \leq C$, then

$$|u(x, t)| \leq B + tC + \tfrac{1}{2}At^2.$$

SECTION 2. THE WAVE OPERATOR WITH LOWER ORDER TERMS

We now consider the operator

$$(L + h)[u] \equiv u_{xx} - u_{tt} + d(x, t)u_x + e(x, t)u_t + h(x, t)u$$

and apply the same procedure which we used for the wave equation.

If D again denotes the same triangular domain ABC (Fig. 1) and u satisfies $(L + h)[u] \geq 0$ in D, we have from Stokes' theorem:

$$0 \leq \iint_D [u_{xx} - u_{tt} + du_x + eu_t + hu]\, dx\, dt$$

$$= \iint_D (h - d_x - e_t)u\, dx\, dt + \int_A^B [u_t - eu]\, dx$$

$$+ \int_B^C [(u_x + du)\, dt + (u_t - eu)\, dx]$$

$$+ \int_C^A [(u_x + du)\, dt + (u_t - eu)\, dx].$$

Again noting that $dx = -dt$ along BC and $dx = dt$ along AC, we get the relation

$$\iint_D (h - d_x - e_t)u\, dx\, dt + \int_A^B (u_t - eu)\, dx - \int_B^C (u_x\, dx + u_t\, dt)$$

$$+ \int_C^A (u_x\, dx + u_t\, dt) + \int_B^C u(d + e)\, dt + \int_C^A u(d - e)\, dt \geq 0.$$

Performing the integrations in the third and fourth integrals, we see that

$$u(C) \leq \frac{1}{2}[u(A) + u(B)] + \frac{1}{2}\iint_D (h - d_x - e_t)u\, dx\, dt$$

$$+ \frac{1}{2}\int_A^B (u_t - eu)\, dx + \frac{1}{2}\int_B^C u(d + e)\, dt + \frac{1}{2}\int_A^C u(e - d)\, dt.$$

If, somehow, all the integrals on the right are nonpositive, then the weak maximum principle must hold for the function u. We suppose that

$$\left.\begin{aligned}
u_t - eu &\leq 0 \text{ on } AB, \\
u &< 0 \text{ on } AB, \\
h - d_x - e_t &\geq 0 \text{ in } D, \\
d + e &\geq 0 \text{ in } D, \\
e - d &\geq 0 \text{ in } D.
\end{aligned}\right\} \quad (1)$$

Suppose that u is not negative somewhere in D. Then there is a point $P(x_0, t_0)$ where u vanishes and such that u is negative for all points (x, t) with $t < t_0$. In this case $u \leq 0$ in the triangle $A'B'P$ similar to ABC with $A'B'$ on the x-axis. We have

$$0 = u(P) \leq \frac{1}{2}[u(A') + u(B')] + \frac{1}{2}\iint_{A'B'P} (h - d_x - e_t)u\, dx\, dt$$

$$+ \frac{1}{2}\int_{A'}^{B'} (u_t - eu)\, dx + \frac{1}{2}\int_{B'}^P u(d + e)\, dt + \frac{1}{2}\int_{A'}^P u(e - d)\, dt. \quad (2)$$

Under these circumstances a contradiction is reached since, because of (1), the right side of (2) is negative. Therefore $u < 0$ in D. The inequality in formula (2) applies at any point P in D, and the inequality is strengthened if we drop those terms containing integrals. Thus we see that $u(P)$ cannot be greater than the average of the values of u at two points on AB.

We wish to relax the strict inequality $u < 0$ on AB in (1) and replace it by $u \leq 0$ on AB. For this purpose we suppose that e is bounded and that h is bounded from above. Then a computation shows that for α sufficiently large, the inequalities*

$$(L + h)[\mathfrak{e}^{\alpha t}] \equiv [-\alpha^2 + e\alpha + h]\mathfrak{e}^{\alpha t} \leq 0$$

$$\frac{\partial}{\partial t}(\mathfrak{e}^{\alpha t}) - e\mathfrak{e}^{\alpha t} \geq 0$$

hold. Therefore, if conditions (1) are replaced by the conditions

$$\left. \begin{array}{r} u_t - eu \leq 0 \text{ on } AB, \\ u \leq 0 \text{ on } AB, \\ h - d_x - e_t \geq 0 \text{ in } D, \\ d + e \geq 0 \text{ in } D, \\ e - d \geq 0 \text{ in } D, \end{array} \right\} \quad (3)$$

we find that for any $\epsilon > 0$ the quantity $u - \epsilon\mathfrak{e}^{\alpha t}$ is bounded by its maximum on AB. Letting $\epsilon \to 0$, we conclude that u assumes its maximum on AB. We have established the principle: **If $(L + h)[u] \geq 0$ in D and conditions (3) hold, then $u \leq 0$ in D.**

EXERCISES

1. Let D be the domain given by:
$$\{x - t > 0, x + t - 1 < 0, t > 0\}.$$

Suppose u satisfies
$$L[u] = u_{xx} - u_{tt} + tu_x + xu_t \geq 0$$
in D. If $u(x, 0) \equiv 0$ and $u_t(x, 0) \leq 0$, show that $u \leq 0$ in D. Is the same result true if L is replaced by the operator
$$\bar{L}[u] \equiv u_{xx} - u_{tt} + tu_t ?$$

---

*In this chapter, we shall use the German $\mathfrak{e}$ for the natural base of logarithms and italic e for a coefficient in the operator L.

2. Suppose u is a solution of
$$u_{xx} - u_{tt} + u_t = 0$$
in the domain D described in Exercise 1. If $|u(x, 0)| \le B$, $|u_t(x, 0)| \le C$, show that
$$|u(x, t)| \le B + C(e^t - 1)$$
in D.

SECTION 3. THE TWO-DIMENSIONAL HYPERBOLIC OPERATOR

We shall now extend the result of the preceding section to the general second-order operator
$$L[u] \equiv au_{xx} + 2bu_{xt} + cu_{tt} + du_x + eu_t, \tag{1}$$
where $a, b,$ and c are twice continuously differentiable and d and e are continuously differentiable functions of x and t. L is said to be **hyperbolic** at a point (x, t) if
$$b^2 - ac > 0$$
there. It is hyperbolic in a domain D if it is hyperbolic at each point of D, and **uniformly hyperbolic** in D if there is a constant μ such that $b^2 - ac \ge \mu > 0$ in D.

The **characteristic curves** or **characteristics** of L are the solutions of the ordinary differential equation
$$c\left(\frac{dx}{dt}\right)^2 - 2b\frac{dx}{dt} + a = 0.$$
Solving for dx/dt, we obtain the two differential equations
$$\frac{dx}{dt} = \frac{-b \pm \sqrt{b^2 - ac}}{-c}.$$
Because L is hyperbolic, the quantity under the square root is positive. Hence through each point (x, t) of D there are two characteristics C_+ and C_- corresponding to the two signs in front of the square root (provided $c \ne 0$).

We assume that L is hyperbolic in the half-plane $t \ge 0$, and that there is a constant c_0 such that
$$c \le c_0 < 0$$
there. Let C be any point in $t > 0$ and construct the two characteristic curves from C to the x-axis. We denote by A the point where the C_+ curve hits the x-axis, and by B the point where the C_- curve hits it (see Fig. 2).

(We assume that a, b, and c are sufficiently smooth to guarantee the existence of the characteristic curves AC and BC.)

The segment AB and the two characteristics AC and BC form a (curvilinear) **characteristic triangle**.

We associate with L the **adjoint operator**

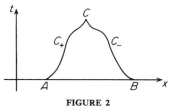

FIGURE 2

$$L^*[v] \equiv (av)_{xx} + 2(bv)_{xt} + (cv)_{tt} - (dv)_x - (ev)_t$$
$$= av_{xx} + 2bv_{xt} + cv_{tt} + (2a_x + 2b_t - d)v_x + (2b_x + 2c_t - e)v_t$$
$$+ (a_{xx} + 2b_{xt} + c_{tt} - d_x - e_t)v.$$

This operator has the same characteristic curves as L. Moreover, it has the property that any functions u and v which have continuous second derivatives satisfy the identity

$$vL[u] - uL^*[v] = \frac{\partial}{\partial x}[a(vu_x - uv_x) + b(vu_t - uv_t) - (a_x + b_t - d)uv]$$
$$+ \frac{\partial}{\partial t}[b(vu_x - uv_x) + c(vu_t - uv_t) - (b_x + c_t - e)uv]$$

Putting $v \equiv 1$ and applying Stokes' theorem, we find that

$$\iint_{ABC} (L[u] - uL^*[1])\, dx\, dt = \oint \{[au_x + bu_t - u(a_x + b_t - d)]\, dt$$
$$- [bu_x + cu_t - u(b_x + c_t - e)]\, dx\}, \quad (2)$$

where the integral on the right is over the boundary $ABCA$ of the curvilinear triangle ABC.

On the segment AB we have $dt = 0$. On the C_- characteristic BC we have

$$dx = -\frac{1}{c}(-b - \sqrt{b^2 - ac})\, dt,$$

and therefore along C_-

$$a\, dt - b\, dx = -\sqrt{b^2 - ac}\, dx, \quad b\, dt - c\, dx = -\sqrt{b^2 - ac}\, dt.$$

Consequently,

$$\int_B^C \{[au_x + bu_t]\, dt - [bu_x + cu_t]\, dx\} = -\int_B^C \sqrt{b^2 - ac}\, (u_x\, dx + u_t\, dt)$$
$$= -\sqrt{b^2 - ac}\, u(C) + \sqrt{b^2 - ac}\, u(B) + \int_B^C u\frac{d}{dt}\sqrt{b^2 - ac}\, dt.$$

Similarly,

$$\int_C^A \{[au_x + bu_t]\, dt - [bu_x + cu_t]\, dx\} = \int_C^A \sqrt{b^2 - ac}\,(u_x\, dx + u_t\, dt)$$

$$= \sqrt{b^2 - ac}\, u(A) - \sqrt{b^2 - ac}\, u(C) + \int_A^C u\frac{d}{dt}\sqrt{b^2 - ac}\, dt.$$

Combining these results, we find that

$$2\sqrt{b^2 - ac}\, u(C) = \sqrt{b^2 - ac}\, u(A) + \sqrt{b^2 - ac}\, u(B) + \int_A^C uK_+ \, dt$$

$$+ \int_B^C uK_- \, dt + \iint_{ABC} (uL^*[1] - L[u])\, dx\, dt$$

$$+ \int_A^B [u(b_x + c_t - e) - bu_x - cu_t]\, dx, \qquad (3)$$

where

$$K_\pm \equiv K_\pm(x, t) \equiv (\sqrt{b^2 - ac})_t + \frac{b}{c}(\sqrt{b^2 - ac})_x$$

$$+ \frac{1}{c}(b_x + c_t - e)\sqrt{b^2 - ac}$$

$$\pm \left[-\frac{1}{2c}(b^2 - ac)_x + a_x + b_t - d - \frac{b}{c}(b_x + c_t - e) \right]. \qquad (4)$$

In addition to the hyperbolic operator L we are given a function $h(x, t)$, and we wish to establish a maximum principle for functions u which satisfy the differential inequality $(L + h)[u] \geq 0$ in a domain D which has a portion of the x-axis as part of its boundary.

We suppose that u and the **conormal derivative**

$$\frac{\partial u}{\partial \nu} \equiv -b\frac{\partial u}{\partial x} - c\frac{\partial u}{\partial t}$$

are given at $t = 0$. We observe that since $c \leq c_0 < 0$, the coefficient of $\partial u/\partial t$ in the conormal derivative is positive. In particular, when $b = 0$ and $c = -1$, as in the preceding section, then $\partial u/\partial \nu \equiv \partial u/\partial t$.

We call a domain D in the half-plane $t > 0$ an **admissible domain** if it has the property that for each point C of D the corresponding characteristic triangle ABC with AB on the x-axis is also in D. For example, one can let D be the characteristic triangle corresponding to a particular point in $t > 0$. One can also let D be the infinite strip $0 \leq t < t_0$. More generally, D is admissible if it is the finite or infinite union of characteristic triangles.

At each point C of D we have the identity (3), which may be written in the form

$$2\sqrt{b^2 - ac}\, u(C) = \sqrt{b^2 - ac}\, u(A) + \sqrt{b^2 - ac}\, u(B) + \int_A^C uK_+\, dt$$
$$+ \int_B^C uK_-\, dt + \iint_{ABC} (u\{L^*[1] + h\} - (L + h)[u])\, dx\, dt$$
$$+ \int_A^B \left[\frac{\partial u}{\partial \nu} + (b_x + c_t - e)u \right] dx. \tag{5}$$

We now suppose that h and the coefficients of L have the properties
$$K_+ \geq 0, \quad K_- \geq 0, \quad L^*[1] + h \geq 0 \text{ in } D. \tag{6}$$
where K_+ and K_- are defined in Eq. (4). Let u be any function which satisfies the inequalities
$$\left. \begin{array}{c} (L + h)[u] > 0 \text{ in } D, \\ \dfrac{\partial u}{\partial \nu} + (b_x + c_t - e)u \leq 0 \text{ on } \Gamma_0, \\ u < 0 \text{ on } \Gamma_0, \end{array} \right\} \tag{7}$$
where Γ_0 denotes the portion of the boundary of D situated on the x-axis.

We wish to show that conditions (6) and (7) imply that $u < 0$ throughout D. To do this, we suppose, on the contrary, that $u \geq 0$ somewhere in D. Let C be one of the points with smallest t-coordinate of the closed subset of D where $u \geq 0$. Then $u(C) = 0$ and $u \leq 0$ in the characteristic triangle of D having C as its vertex (Fig. 2). Applying identity (5), we find

$$\left. \begin{array}{l} 0 = 2\sqrt{b^2 - ac}\, u(C) = \sqrt{b^2 - ac}\, u(A) + \sqrt{b^2 - ac}\, u(B) \\ + \int_A^C uK_+\, dt + \int_B^C uK_-\, dt \\ + \iint_{ABC} \{u(L^*[1] + h) - (L + h)[u]\}\, dx\, dt \\ + \int_A^B \left\{ \dfrac{\partial u}{\partial \nu} + (b_x + c_t - e)u \right\} dx. \end{array} \right\} \tag{8}$$

By hypothesis, each term on the right in (8) is nonpositive. Moreover, $(L + h)[u] > 0$ in D. Hence the right side of (8) is negative, a contradiction. We conclude that $u < 0$ throughout D.

To obtain a more useful result, we must relax the inequalities in (7) and replace them by the conditions
$$\left. \begin{array}{c} (L + h)[u] \geq 0 \text{ in } D, \\ \dfrac{\partial u}{\partial \nu} + (b_x + c_t - e)u \leq 0 \text{ on } \Gamma_0, \\ u \leq 0 \text{ on } \Gamma_0. \end{array} \right\} \tag{9}$$

In analogy with the method used in Section 2, we compute

$$(L + h)[e^{\alpha t}] \equiv (c\alpha^2 + e\alpha + h)e^{\alpha t},$$

$$\frac{\partial}{\partial \nu}(e^{\alpha t}) = -c\alpha e^{\alpha t}.$$

Since $c \leq c_0 < 0$, we can choose α so large that

$$(L + h)[e^{\alpha t}] < 0$$

throughout any bounded admissible subdomain D' of D. Also, for α sufficiently large, we have

$$\frac{\partial}{\partial \nu}(e^{\alpha t}) + (b_x + c_t - e)\alpha t \geq 0$$

on Γ_0', the portion of the boundary of D' situated on the x-axis. Then for any $\epsilon > 0$, we find

$$(L + h)[u - \epsilon e^{\alpha t}] > 0 \text{ in } D',$$

$$\frac{\partial}{\partial \nu}[u - \epsilon e^{\alpha t}] + (b_x + c_t - e)(u - \epsilon e^{\alpha t}) \leq 0 \text{ on } \Gamma_0',$$

$$u - \epsilon e^{\alpha t} < 0 \text{ on } \Gamma_0'.$$

These three conditions imply that $u - \epsilon e^{\alpha t} < 0$ in D' for any $\epsilon > 0$. Letting $\epsilon \to 0$, we conclude that $u \leq 0$ in D'. Since D' is an arbitrary bounded admissible subdomain of D, we obtain $u \leq 0$ in D. This argument establishes the following maximum principle.

THEOREM 1. Let L be a hyperbolic operator of the form (1) with bounded coefficients and with $c \leq c_0 < 0$. Let D be an admissible domain whose boundary on the x-axis is denoted by Γ_0. Suppose that h and the coefficients of L satisfy the inequalities (6). If u is twice continuously differentiable in D, once continuously differentiable in $D \cup \Gamma_0$, and if u satisfies the three inequalities

$$(L + h)[u] \geq 0 \text{ in } D,$$

$$\frac{\partial u}{\partial \nu} + (b_x + c_t - e)u \leq 0 \text{ on } \Gamma_0,$$

$$u \leq 0 \text{ on } \Gamma_0,$$

then $u \leq 0$ in D.

Remarks. (i) We see easily from the proof of Theorem 1 that the condition $(L + h)[u] \geq 0$ can be replaced by the weaker condition

$$\iint_{ABC} (L + h)[u] \, dx \, dt \geq 0$$

for every characteristic triangle ABC having its base on Γ_0.

(ii) Having established that $u \leq 0$, we obtain from (5) the nonnegative upper bound

$$2\sqrt{b^2 - ac}\, u(C) \leq \sqrt{b^2 - ac}\, u(A) + \sqrt{b^2 - ac}\, u(B)$$
$$- \iint (L + h)[u]\, dx\, dt + \int_A^B \left\{\frac{\partial u}{\partial \nu} + (b_x + c_t - e)u\right\} dx. \quad (10)$$

Suppose now that $u \leq M$ on Γ_0. We can obtain a maximum principle for u if $u - M$ satisfies the hypotheses of Theorem 1. We observe that

$$(L + h)[u - M] = (L + h)[u] - hM.$$

Thus if we wish $(L + h)[u - M]$ to be nonnegative in D whenever $[L + h][u] \geq 0$ in D, we must have

$$hM \leq 0 \text{ in } D.$$

If M is a positive constant, then h must be nonpositive in D, while if M is negative, we must have $h \geq 0$ in D. Similarly, the condition

$$\frac{\partial}{\partial \nu}(u - M) + (b_x + c_t - e)(u - M) \leq 0 \text{ on } \Gamma_0$$

follows from the two conditions

$$\frac{\partial u}{\partial \nu} + (b_x + c_t - e)u \leq 0 \text{ on } \Gamma_0,$$
$$-M(b_x + c_t - e) \leq 0 \text{ on } \Gamma_0.$$

In this way we obtain the following modification of Theorem 1.

THEOREM 2. Suppose the hypotheses of Theorem 1 hold with the condition $u \leq 0$ on Γ_0 replaced by $u \leq M$ on Γ_0. If, in addition, the inequalities $hM \leq 0$ in D and $M(b_x + c_t - e) \geq 0$ on Γ_0 hold, then $u \leq M$ in D.

Remarks. (i) Theorem 2 with $M \geq 0$ states that any nonnegative maximum of u must occur on the initial line. This result is closely related to the weak maximum principles for elliptic and parabolic operators. The analogue of the above result for M negative can be proved in the same way for parabolic operators, but it is false for elliptic operators.

(ii) As in Remark (i) following Theorem 1, we can replace the conditions $(L+h)[u] \geq 0$ and $Mh \leq 0$ by the integral condition $M \iint_{ABC} h\, dx\, dt \leq \iint_{ABC} (L + h)[u]\, dx\, dt$ for every characteristic triangle ABC in D with base on Γ_0.

(iii) Under the hypotheses of Theorem 2 we can also apply the inequality (10) to $u - M$.

Theorem 1 is rather special in that it applies only to operators $L + h$ whose coefficients happen to satisfy the three inequalities (6). We now show how to extend the maximum principle to a more general class of operators. First, we observe that the hyperbolicity of an operator $L + h$ is preserved if it is multiplied by any positive function $v(x, t)$. However, such multiplication alters the quantities K_+, K_-, and $L^*[1] + h$. Furthermore, the operator $\partial/\partial \nu + (b_x + c_t - e)$ is changed. The three inequalities (6) for the operator $v(L + h)$ are

$$2\sqrt{b^2 - ac}\left[v_t - \frac{1}{c}(\sqrt{b^2 - ac} - b)v_x\right] + vK_+ \geq 0,$$

$$2\sqrt{b^2 - ac}\left[v_t + \frac{1}{c}(\sqrt{b^2 - ac} + b)v_x\right] + vK_- \geq 0, \qquad (11)$$

$$(L^* + h)[v] \geq 0.$$

The quantity $(b_x + c_t - e)$, which occurs in the second condition of (9) concerning the conormal derivative, is replaced by

$$(b_x + c_t - e)v - \frac{\partial v}{\partial \nu}.$$

A restatement of Theorem 1 applied to an operator $L + h$ which has been multiplied by a positive function $v(x, t)$ yields the following result.

THEOREM 3. **Suppose there is a positive function $v(x, t)$ defined in an admissible domain D such that the coefficients of the hyperbolic operator L given by (1) satisfy the inequalities (11). If u satisfies**

$$(L + h)[u] \geq 0 \text{ in } D,$$

$$v\frac{\partial u}{\partial \nu} + u\left[(b_x + c_t - e)v - \frac{\partial v}{\partial \nu}\right] \leq 0 \text{ on } \Gamma_0,$$

$$u \leq 0 \text{ on } \Gamma_0,$$

then $u \leq 0$ in D.

Remark. The inequality (10) with $L + h$ replaced by $v(L + h)$ gives the nonnegative upper bound

$$u(C) \leq \frac{1}{2\sqrt{b^2 - ac}\, v(C)}\left[\sqrt{b^2 - ac}\, vu(A) + \sqrt{b^2 - ac}\, vu(B)\right.$$

$$- \iint_{ABC} v(L + h)[u]dx\, dt$$

$$\left. + \int_A^B \left\{v\frac{\partial u}{\partial \nu} + \left[(b_x + c_t - e)v - \frac{\partial v}{\partial \nu}\right]u\right\}dx\right]. \qquad (12)$$

Theorem 3 is useful only if we are able to find a positive function v with the desired properties. We shall now show that for any hyperbolic operator L there is a function v which satisfies conditions (11) in a sufficiently small strip $0 \leq t \leq t_0$. We let

$$v(x, t) = 1 + \alpha t - \beta t^2. \tag{13}$$

A computation shows that conditions (11) are

$$\left. \begin{array}{l} 2\sqrt{b^2 - ac}\,(\alpha - 2\beta t) + (1 + \alpha t - \beta t^2)K_+ \geq 0, \\ 2\sqrt{b^2 - ac}\,(\alpha - 2\beta t) + (1 + \alpha t - \beta t^2)K_- \geq 0, \\ -2c\beta + (2b_x + 2c_t - e)(\alpha - 2\beta t) \\ + (a_{xx} + 2b_{xt} + c_{tt} - d_x - e_t + h)(1 + \alpha t - \beta t^2) \geq 0. \end{array} \right\} \tag{14}$$

Since all the coefficients and their derivatives which appear in the above expressions are supposed bounded and since $-c$ and $\sqrt{b^2 - ac}$ have positive lower bounds, the first two expressions above are positive at $t = 0$ if α is chosen sufficiently large. The third expression is positive at $t = 0$ if β is chosen sufficiently large. With these values of α and β there is a number $t_0 > 0$ such that $v(x, t) > 0$ and all the inequalities (11) hold for $0 \leq t \leq t_0$. Theorem 3 is valid in this strip.

With v given by (13), the condition on the conormal derivative becomes

$$\frac{\partial u}{\partial \nu} + (b_x + c_t - e + c\alpha)u \leq 0 \text{ at } t = 0.$$

If we select a constant k so large that

$$k \geq -[b_x + c_t - e + c\alpha] \text{ on } \Gamma_0, \tag{15}$$

then the conormal derivative condition of Theorem 3 is satisfied whenever $u \leq 0$ and

$$\frac{\partial u}{\partial \nu} - ku \leq 0.$$

In this way we obtain the following maximum principle for a strip adjacent to the x-axis.

THEOREM 4. Suppose that the coefficients of the operator L given by (1) are bounded and have bounded first and second derivatives. Let D be an admissible domain. If t_0 and k are selected in accordance with (14) and (15), then any function u which satisfies

$$(L + h)[u] \geq 0 \text{ in } D$$

$$\frac{\partial u}{\partial \nu} - ku \leq 0 \text{ on } \Gamma_0$$

$$u \leq 0 \text{ on } \Gamma_0$$

also satisfies $u \leq 0$ in the part of D which lies in the strip $0 \leq t \leq t_0$. The constants t_0 and k depend only on lower bounds for $-c$ and $\sqrt{b^2 - ac}$ and on bounds for the coefficients of L and their derivatives.

EXERCISES

1. Determine explicitly the C_+ and C_- families of characteristics for the "generalized Tricomi operator"

$$L \equiv t^n \frac{\partial^2}{\partial x^2} - \frac{\partial^2}{\partial t^2} + d\frac{\partial}{\partial x} + e\frac{\partial}{\partial t}, \qquad t > 0.$$

2. Compute the quantities $K_\pm$ for the operator

$$L \equiv 2\frac{\partial^2}{\partial x \, \partial t} - (1 + x^2)\frac{\partial^2}{\partial t^2} - x\frac{\partial}{\partial x} + (1 - t)\frac{\partial}{\partial t}.$$

Find the largest admissible domain D in the half-plane $t \geq 0$ in which the inequalities (6) are satisfied.

3. Show that if $h \geq 0$, the inequalities (6) are satisfied everywhere for the operator

$$(L + h) \equiv \frac{\partial^2}{\partial x^2} - \frac{\partial^2}{\partial t^2} + x^2\frac{\partial}{\partial t} + h.$$

4. Construct a function v satisfying inequalities (11) with respect to the operator

$$(L + h) \equiv \frac{\partial^2}{\partial x^2} - \frac{\partial^2}{\partial t^2} - 1$$

for $t_0 < \pi/2$.

SECTION 4. BOUNDS AND UNIQUENESS IN THE INITIAL VALUE PROBLEM

We consider the problem of determining a function $w(x, t)$ which satisfies

$$\left.\begin{aligned}(L + h)[w] &\equiv aw_{xx} + 2bw_{xt} + cw_{tt} + dw_x + ew_t + hw = f(x) \text{ in } D,\\ w(x, 0) &= g(x) \text{ on } \Gamma_0,\\ \frac{\partial w}{\partial \nu} &\equiv -bw_x - cw_t = \gamma(x) \text{ on } \Gamma_0.\end{aligned}\right\} \quad (1)$$

The problem of finding a function fulfilling conditions (1) is called the **initial value problem** or the **Cauchy problem** for w. Suppose that we can find a function $w_1(x, t)$ which satisfies

$$\left.\begin{aligned}(L + h)[w_1] &\leq f(x) \text{ in } D,\\ w_1(x, 0) &\geq g(x) \text{ on } \Gamma_0,\\ \frac{\partial w_1}{\partial \nu} - kw_1 &\geq \gamma(x) - kg(x) \text{ on } \Gamma_0,\end{aligned}\right\} \quad (2)$$

where k is the constant given by inequality (15) of Section 3. If w satisfies

(1) and if w_1 satisfies (2), we may apply Theorem 4 to $w - w_1$ and conclude that
$$w(x, t) \leq w_1(x, t) \text{ in } D', \tag{3}$$
with D' an admissible subdomain of D situated in the strip $0 \leq t \leq t_0$. Similarly, if $w_2(x, t)$ satisfies
$$(L + h)[w_2] \geq f \text{ in } D,$$
$$w_2 \leq g \text{ on } \Gamma_0,$$
$$\frac{\partial w_2}{\partial \nu} - k w_2 \leq \gamma(x) - kg(x) \text{ on } \Gamma_0,$$
then
$$w_2(x, t) \leq w(x, t) \text{ in } D'. \tag{4}$$

Inequalities (3) and (4) show that problem (1) has at most one solution. For, if there were two solutions w and $\bar{w}$, then we could apply (2) to both solutions, getting $\bar{w} \leq w \leq \bar{w}$, so that $w \equiv \bar{w}$ for $0 \leq t \leq t_0$. We now set up a new initial value problem at $t = t_0$ with $w = \bar{w}$ and $(\partial w/\partial \nu) = (\partial \bar{w}/\partial \nu)$ at $t = t_0$. We apply (2) again and find that $w \equiv \bar{w}$ for $t_0 \leq t \leq t_1$. Continuing in this fashion, we obtain $w \equiv \bar{w}$ throughout D. We conclude that **the initial value problem (1) has at most one solution.**

By means of the maximum principle we can get bounds for solutions of (1) in terms of bounds for the coefficients of L and the given functions f, g, h, and γ. We first choose a positive constant ρ so large that
$$\left.\begin{array}{r}2(-c)\rho^2 e^{2\rho t} \geq 1 \\ (-c)\rho - e \geq 0 \\ (-c)\rho^2 - e\rho - h \geq 0\end{array}\right\} \text{ for } 0 \leq t \leq t_0,$$
$$(-c)\rho \geq k \text{ for } t = 0,$$
$$(-c)\rho \geq 1 \text{ for } t = 0,$$
where t_0 and k are the constants occuring in Theorem 4. A computation then shows that
$$(L + h)[e^{\rho t}] \leq 0 \quad \text{ for } 0 \leq t \leq t_0,$$
$$(L + h)[e^{\rho t} - 1] \leq 0 \quad \text{ for } 0 \leq t \leq t_0,$$
$$(L + h)[(e^{\rho t} - 1)^2]$$
$$= (e^{\rho t} - 1)^2(c\rho^2 + e\rho + h) + (e^{2\rho t} - 1)(c\rho^2 + e\rho) + 2c\rho^2 e^{\rho t}$$
$$\leq -1 \quad \text{ for } 0 \leq t \leq t_0,$$
$$\frac{\partial}{\partial \nu}(e^{\rho t}) - k(e^{\rho t}) \geq 0 \text{ for } t = 0,$$
$$\frac{\partial}{\partial \nu}(e^{\rho t} - 1) - k(e^{\rho t} - 1) \geq 1 \text{ for } t = 0.$$

We select constants M_1, M_2, and M_3 such that
$$|f| \leq M_1, \qquad |g| \leq M_2, \qquad |\gamma(x) - kg(x)| \leq M_3,$$
and we apply Theorem 4 to the functions $\pm w - M_1(e^{\rho t} - 1)^2 - M_2 e^{\rho t} - M_3(e^{\rho t} - 1)$. We get the estimate
$$|w(x, t)| \leq M_1(e^{\rho t} - 1)^2 + M_2 e^{\rho t} + M_3(e^{\rho t} - 1) \text{ in } D',$$
where D' is any admissible domain contained in the strip $0 \leq t \leq t_0$.

Better bounds can be obtained by using functions more complicated than $e^{\rho t}$ and by employing the inequalities (10) and (12) of Section 3.

EXERCISES

1. Show that the problem (1) has the property that if f, g, and γ are only altered slightly, then the solution is only altered slightly.

2. Let D be the domain
$$\{x - t > 0,\ x + t - 2 < 0,\ t > 0\}.$$
Let u be a solution of the problem
$$u_{xx} - u_{tt} + xu_x - tu_t = 0,$$
$$u(x, 0) = 1 + x^2, \qquad 0 \leq x \leq 2,$$
$$u_t(x, 0) = 0, \qquad 0 \leq x \leq 2.$$
Find a bound for the solution u in D of the form $Me^{\rho t}$.

3. Show how to obtain a constant K such that the inequalities
$$|(L + h)[u]| \leq A \text{ in } D$$
$$|u(x, 0)| \leq B, \qquad \left|\frac{\partial u(x, 0)}{\partial v}\right| \leq C$$
imply that
$$|u| \leq K(A + B + C) \text{ in } D,$$
where D is a bounded admissible domain in the strip $0 \leq t \leq t_0$.

SECTION 5. RIEMANN'S FUNCTION

Suppose that, corresponding to each point C of an admissible domain, we can find a function $R(C; x, t)$ defined on the closure of the characteristic triangle ABC which satisfies the conditions

$$\left.\begin{array}{r}(L^* + h)[R] = 0 \text{ in } ABC,\\ 2\sqrt{b^2 - ac}\left[R_t - \frac{1}{c}(-b + \sqrt{b^2 - ac})R_x\right] + K_+R = 0 \text{ on } AC,\\ 2\sqrt{b^2 - ac}\left[R_t - \frac{1}{c}(-b - \sqrt{b^2 - ac})R_x\right] + K_-R = 0 \text{ on } BC,\\ 2\sqrt{b^2 - ac}\,R(C) = 1.\end{array}\right\} \quad (1)$$

Formula (5) of Section 3 with $L + h$ replaced by $R(L + h)$ gives the representation,

$$u(C) = \sqrt{b^2 - ac}\,Ru(A) + \sqrt{b^2 - ac}\,Ru(B) - \iint_{ABC} R(L + h)[u]\,dx\,dt$$

$$+ \int_A^B \{u[bR_x + cR_t + (b_x + c_t - e)R] - R[bu_x + cu_t]\}\,dx. \quad (2)$$

Since the right-hand side of (2) involves precisely the data given in the initial value problem (1) of Section 4, this formula for $u(C)$ is an explicit solution of the problem. The function $R(C; x, t)$ is called the **Riemann function** (or radiation kernel) of the initial value problem for the operator $L + h$.

It is clear from (2) that the maximum principle enunciated in Theorem 1 will hold if and only if

$$R > 0$$

in each characteristic triangle ABC of D and

$$\frac{\partial R}{\partial \nu} \equiv -bR_x - cR_t \leq 0 \text{ on } \Gamma_0.$$

More generally, we see that the conclusion of Theorem 3 for a given function v holds if and only if

$$R > 0 \quad (3)$$

in each characteristic triangle, and

$$v\frac{\partial R}{\partial \nu} \leq R\frac{\partial v}{\partial \nu} \text{ on } \Gamma_0. \quad (4)$$

In particular, a positive function v which satisfies the inequalities (11) of Section 3 can exist only if $R > 0$ in each triangle ABC. On the other hand, if $R > 0$ in each triangle ABC, the function R itself satisfies the requirements for v, so that Theorem 3 holds for at least one function v.

Suppose now that the Riemann function $R(C; x, t)$ exists and that v

is any function which satisfies inequalities (11) of Section 3. Putting $u(x, 0) = \partial u(x, 0)/\partial v = 0$ and comparing the representation (2) above with inequality (12) of of Section 3, which holds whenever $(L + h)[u] \geq 0$, we see that

$$\frac{1}{2\sqrt{[b^2 - ac](C)}} v(x, t) \leq R(C; x, t). \tag{5}$$

Thus **the Riemann function maximizes at each point of the curvilinear triangle ABC the values of all positive functions v which satisfy inequalities (11) of Section 3 and which are normalized so that $2\sqrt{b^2 - ac}\, v(C) = 1$.**

We return now to the inequality (4). We observe that the conditions (1) for R and the conditions (11) of Section 3 for v do not involve the initial line $t = 0$. We could use any other initial time below the point C. It follows that the inequality

$$v(-bR_x - cR_t) \leq R(-bv_x - cv_t)$$

holds throughout the characteristic triangle ABC.

More generally, we can prescribe initial values on any arc Γ extending from the characteristic AC to the characteristic BC, if Γ has the property that it crosses any characteristic curve no more than once. This initial value problem can be solved in terms of R by the analogue of the representation (2) in which the integral along AB is replaced by an integral along Γ. A maximum principle can be established by introducing a new coordinate system (ξ, τ) such that Γ has the equation $\tau = 0$. The inequality (4) applied to such curves Γ shows that **the ratio R/v is a nonincreasing function of t along any characteristic in the triangle ABC.**

We now observe that the second and third equations for R in (1) involve only directional derivatives along AC and BC, respectively. They may be written in the form

$$\frac{d}{dt} \log R = -K_+/2\sqrt{b^2 - ac} \text{ on } AC,$$

$$\frac{d}{dt} \log R = -K_-/2\sqrt{b^2 - ac} \text{ on } BC,$$

where d/dt denotes the total derivative in each case. Once the characteristics AC and BC are known, we may integrate these equations, using the initial condition $2\sqrt{b^2 - ac}\, R(C) = 1$ to determine R along AC and BC.

Then the fact that R/v is a nonincreasing function of t along each characteristic gives a lower bound for R in terms of v at each point of the triangle ABC. Generally this lower bound is an improvement over (5). Substituting this improved bound and the inequality (4) in the representation formula (2), we get a better estimate for $u(C)$ than the one given by (12) in Section 3.

EXERCISES

1. Show that the Riemann function for the wave operator
$$L \equiv \frac{\partial^2}{\partial x^2} - \frac{\partial^2}{\partial t^2}$$
at any point C is the function $R \equiv 1/2$.

2. Show that the function
$$R(\xi, \tau; x, t) = \tfrac{1}{2} J_0(\sqrt{(t-\tau)^2 - (x-\xi)^2}),$$
where J_0 is the Bessel function of order zero, is the Riemann function for the operator
$$L + h \equiv \frac{\partial^2}{\partial x^2} - \frac{\partial^2}{\partial t^2} - 1.$$

SECTION 6. INITIAL-BOUNDARY VALUE PROBLEMS

So far we have been concerned with initial value problems. It often occurs in practice that in addition to the initial value and derivative of a function u on a finite segment of the x-axis, we prescribe either u or a derivative of u at the ends of this interval. For example, in the case of a vibrating string, we prescribe either that the ends are fixed or that they move according to a given law. A problem in which such data are given is called an initial-boundary value problem.

We consider the hyperbolic differential equation
$$(L + h)[u] \equiv au_{xx} + 2bu_{xt} + cu_{tt} + du_x + eu_t + hu = f \qquad (1)$$
in the domain
$$D: x_1 < x < x_2, \qquad 0 < t < T.$$
We are given the initial values
$$\left. \begin{array}{l} u(x, 0) = g_1(x) \text{ for } x_1 < x < x_2, \\ \dfrac{\partial u}{\partial \nu}(x, 0) \equiv -bu_x - cu_t = g_2(x) \text{ for } x_1 < x < x_2, \end{array} \right\} \qquad (2)$$
and the boundary conditions
$$\left. \begin{array}{l} u(x_1, t) = 0 \text{ for } 0 < t < T, \\ u(x_2, t) = 0 \text{ for } 0 < t < T. \end{array} \right\} \qquad (3)$$
(See Fig. 3.) We could prescribe nonzero values for $u(x_1, t)$ and $u(x_2, t)$. In such a case, we simply define a new function

$$\bar{u}(x, t) = u(x, t) - u(x_1, t) - \frac{x - x_1}{x_2 - x_1}[u(x_2, t) - u(x_1, t)],$$

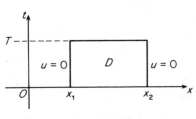

FIGURE 3

and the problem with nonzero data is reduced to one of the form (2), (3) for the function $\bar{u}$.

We shall suppose that

$$a(x_1, 0) > 0, \quad a(x_2, 0) > 0$$

and that $c(x, t) \leq c_0 < 0$ in D. Then t increases along C_+ characteristics and decreases along C_- characteristics going into D from points on the boundary $x = x_1$. The opposite applies when going into D from points on $x = x_2$. Since there are points in D whose C_+ characteristics end on the line $x = x_1$, D is not admissible in the sense of Section 3 if we take for Γ_0 the line $t = 0$. On the other hand, on the line $x = x_1$ we know u but not its derivatives. Therefore we cannot apply the maximum principle if we include $x = x_1$ as part of Γ_0.

Instead we employ the following artifice. We define u outside the interval $x_1 < x < x_2$ by reflection as an odd function about $x = x_1$ and about $x = x_2$. That is, we let

$$u(x, t) = -u(2x_1 - x, t) \text{ for } 2x_1 - x_2 < x < x_1,$$
$$u(x, t) = -u(2x_2 - x, t) \text{ for } x_2 < x < 2x_2 - x_1,$$

and then more generally,

$$u(x, t) = u\!\left(x - 2n(x_2 - x_1), t\right)$$

for

$$2x_1 - x_2 + 2n(x_2 - x_1) \leq x \leq 2x_1 - x_2 + 2(n + 1)(x_2 - x_1)$$
$$n = 0, \pm 1, \pm 2, \ldots.$$

In order that u still satisfy the differential equation (1), we continue $a, c, e,$ and h by reflection as even functions; that is,

$$a(x, t) = a(2x_1 - x, t) \text{ for } 2x_1 - x_2 < x < x_1,$$
$$a(x, t) = a\!\left(x - 2n(x_2 - x_1), t\right)$$

for

$$2x_1 - x_2 + 2n(x_2 - x_1) < x < 2x_1 - x_2 + 2(n + 1)(x_2 - x_1),$$

and so forth, and we continue $b, d, f, g_1,$ and g_2 as odd functions:

$$b(x, t) = -b(2x_1 - x, t) \text{ for } 2x_1 - x_2 < x < x_1,$$
$$b(x, t) = b\!\left(x - 2n(x_2 - x_1), t\right)$$

for
$$2x_1 - x_2 + 2n(x_2 - x_1) < x < 2x_1 - x_2 + 2(n+1)(x_2 - x_1),$$
and so forth.

In order that L^*, the adjoint of L, should have bounded coefficients, we impose the requirement:
$$a_x = c_x = b = d = 0 \text{ at } x = x_1 \text{ and } x = x_2.$$
Moreover, we require that
$$g_1(x_1) = g_2(x_1) = g_1(x_2) = g_2(x_2) = 0.$$
Then the extended function u satisfies the differential equation $L[u] = f$ in the admissible domain $-\infty < x < \infty$, $0 < t < T$. While the conditions $(L + h)[u] \geq 0$, $u(x, 0) \leq 0$, and $\partial u/\partial v + (b_x + c_t - e)u \leq 0$ are not preserved when f, g_1, and g_2 are continued as odd functions, the maximum principles of Section 3 can be applied as in Section 4 to obtain a uniqueness theorem and estimates for the solution of an initial-boundary value problem. (Sather [4] has given a more direct generalization to the initial-boundary value problem of Theorem 1, which holds for a restricted t-interval.)

If at one or both of the boundaries $x = x_1$, $x = x_2$, the condition $\partial u/\partial x = 0$ is imposed instead of $u = 0$, we simply continue u as an even function rather than an odd function. The coefficients are continued as before, and the uniqueness and estimates of solutions again follow as in Section 4.

EXERCISE

Suppose the operator $L + h$ as given by (1) has constant coefficients. If f, g_1, and g_2 are also constant, are the methods of this section always applicable? Explain.

SECTION 7. ESTIMATES FOR SERIES SOLUTIONS

The hyperbolic differential equation
$$L[u] \equiv u_{xx} - u_{tt} - 2tu_t = 0 \tag{1}$$
clearly has the product solutions
$$u(x, t) = \cos\alpha x \, \phi_{\alpha^2}(t),$$
where α is any constant and $\phi_{\alpha^2}(t)$ is any solution of the ordinary differential equation
$$\phi''_{\alpha^2} + 2t\phi'_{\alpha^2} + \alpha^2 \phi_{\alpha^2} = 0. \tag{2}$$

If $\alpha^2 = 4k + 2$ with k an integer, this differential equation is easily seen to possess the solution $(d/dt)^{2k}[e^{-t^2}]$, which is an even function of t. If we impose the initial conditions

$$\phi_{4k+2}(0) = 1, \qquad \phi'_{4k+2}(0) = 0,$$

we can write the solution in the form

$$\phi_{4k+2}(t) = \frac{(-1)^k k!}{(2k)!} H_{2k}(t) e^{-t^2},$$

where $H_{2k}(t)$ is the Hermite polynomial

$$H_{2k}(t) = e^{t^2}\left(\frac{d}{dt}\right)^{2k}[e^{-t^2}].$$

We see that the function

$$u(x, t) = \frac{(-1)^k k!}{(2k)!} \cos \sqrt{4k+2}\, x \, H_{2k}(t) e^{-t^2} \tag{3}$$

is the solution of the problem

$$L[u] \equiv u_{xx} - u_{tt} - 2tu_t = 0$$
$$u(x, 0) = \cos \sqrt{4k+2}\, x \tag{4}$$
$$u_t(x, 0) = 0.$$

An easy computation shows that for the above operator L, we have $K_+ = K_- = -2t$, so that the conditions $K_\pm \geq 0$ of Theorem 1 are violated. However, if we let $v = e^{t^2/2}$ we see that

$$2\sqrt{b^2 - ac}\,(v_t \pm v_x) + K_\pm v = 0,$$

and

$$L^*[v] = v_{xx} - v_{tt} + 2tv_t + 2v = (t^2 + 1)e^{t^2/2} > 0,$$

so that the conditions (11) of Theorem 3 are satisfied. Moreover,

$$(b_x + c_t - e)v - \frac{\partial v}{\partial v} = 0 \text{ at } t = 0,$$

so that the conditions on u in Theorem 3 become

$$L[u] \geq 0 \text{ in } D,$$
$$u \leq 0 \text{ at } t = 0,$$
$$\frac{\partial u}{\partial t} \leq 0 \text{ at } t = 0.$$

We observe that these conditions are satisfied by the functions $u - 1$ and $-u - 1$, with u given by (3). It follows from Theorem 3 that $|u| \leq 1$ everywhere. In particular, putting $x = 0$, we obtain the bound

$$|H_{2k}(t)| \leq \frac{(2k)!}{k!} e^{t^2}$$

Sec. 7 Estimates for Series Solutions

for Hermite polynomials. If we restrict our attention to the characteristic triangle $t + |x| < t_0$, $t > 0$, with $\sqrt{4k+2}\, t_0 < \pi$, we observe that $\cos \sqrt{4k+2}\, t_0 - u \leq 0$ at $t = 0$. It follows that $u \geq \cos \sqrt{4k+2}\, t_0$ in this triangle. Putting $x = 0$ and letting t tend to t_0, we obtain the lower bound

$$\frac{(-1)^k k!}{(2k)!} H_{2k}(t_0) \geq e^{t_0^2} \cos \sqrt{4k+2}\, t_0 \text{ for } 0 \leq t_0 \leq \frac{\pi}{\sqrt{4k+2}}.$$

We now consider the initial-boundary value problem

$$u_{xx} - u_{tt} - 2t u_t = 0 \text{ for } 0 < x < \frac{\pi}{2\sqrt{2}},\ t > 0$$

$$u_x(0, t) = 0$$

$$u\!\left(\frac{\pi}{2\sqrt{2}}, t\right) = 0$$

$$u(x, 0) = f(x)$$

$$u_t(x, 0) = 0.$$

The method of separation of variables leads to the solution

$$u(x, t) = \sum_{n=0}^{\infty} a_n \frac{[2n(n+1)]!}{[4n(n+1)]!} H_{4n(n+1)}(t) e^{-t^2} \cos \sqrt{2}\,(2n+1)x$$

where a_n is the Fourier coefficient

$$a_n = \frac{4\sqrt{2}}{\pi} \int_0^{\pi/(2\sqrt{2})} f(x) \cos \sqrt{2}\,(2n+1)x\, dx.$$

In order to investigate the convergence of this series solution, we observe that we may apply Theorem 3 to any finite sum

$$\sum_{n=M+1}^{N} a_n \frac{[2n(n+1)]!}{[4n(n+1)]!} H_{4n(n+1)}(t) e^{-t^2} \cos \sqrt{2}\,(2n+1)x,$$

which is a solution of the differential equation everywhere, and whose t-derivative vanishes at $t = 0$. We see that

$$\left| \sum_{n=M+1}^{N} a_n \frac{[2n(n+1)]!}{[4n(n+1)]!} H_{4n(n+1)}(t) e^{-t^2} \cos \sqrt{2}\,(2n+1)x \right|$$

$$\leq \max_{0 < x < \pi/2\sqrt{2}} \left| \sum_{M+1}^{N} a_n \cos \sqrt{2}\,(2n+1)x \right|. \tag{5}$$

Suppose now that the Fourier series $\sum a_n \cos \sqrt{2}\,(2n+1)x$ converges to $f(x)$ uniformly. Then for any given $\epsilon > 0$ there is an M_ϵ such that whenever $M \geq M_\epsilon$,

$$\left| f(x) - \sum_0^M a_n \cos \sqrt{2}\,(2n+1)x \right| < \epsilon.$$

It follows that
$$\left|\sum_{M+1}^{N} a_n \cos \sqrt{2}(2n+1)x\right| < 2\epsilon \text{ for } N > M \geq M_\epsilon.$$
Therefore by inequality (5) above,
$$\left|\sum_{M+1}^{N} a_n \frac{[2n(n+1)]!}{[4n(n+1)]!} H_{4n(n+1)}(t)e^{-t^2} \cos \sqrt{2}(2n+1)x\right|$$
$$< 2\epsilon \text{ for } N > M \geq M_\epsilon.$$

In other words, the series solution for $u(x, t)$ converges uniformly whenever the series for $f(x)$ does. Letting $N \to \infty$, we obtain the bound
$$\left|u(x, t) - \sum_0^M a_n \frac{[2n(n+1)]!}{[4n(n+1)]!} H_{4n(n+1)}(t)e^{-t^2} \cos \sqrt{2}(2n+1)x\right|$$
$$\leq \max_{0 \leq x \leq \pi/(2\sqrt{2})} \left|f(x) - \sum_0^M a_n \cos \sqrt{2}(2n+1)x\right|$$

for the error made in replacing $u(x, t)$ by a finite partial sum.

The above methods can, of course, be applied to many partial differential equations which have product solutions. For a related result on the convergence of certain double series, see Weinberger [2].

EXERCISES

1. Show that the solution $\phi(t)$ of the initial value problem
$$\phi'' - e(t)\phi' + \phi = 0$$
$$\phi(0) = 1$$
$$\phi'(0) = 0$$
satisfies the inequality $|\phi(t)| \leq 1$ whenever $e(0) = 0$ and the function $(1/e) + \frac{1}{2}t$ is nondecreasing for $t > 0$.

2. Show that the condition $e(0) = 0$ in Exercise 1 can be replaced by $e(0) \geq 0$.

SECTION 8. THE TWO-CHARACTERISTIC PROBLEM

In Sections 3 and 6 we treated the initial value problem and the mixed initial-boundary value problem. In both cases, the function u and its first derivatives are prescribed on a segment Γ_0 of the x-axis. We can extend these results to the case where Γ_0 is a more general curve. To do so, it is only necessary to define an admissible domain properly.

Sec. 8   The Two-Characteristic Problem

A curve Γ written in parametric form
$$\Gamma : x = x(\sigma), \quad t = t(\sigma)$$
is a characteristic of the hyperbolic operator
$$L[u] \equiv a u_{xx} + 2b u_{xt} + c u_{tt} + d u_x + e u_t \tag{1}$$
if at each point of Γ the equation
$$a\left(\frac{dt}{d\sigma}\right)^2 - 2b \frac{dt}{d\sigma}\frac{dx}{d\sigma} + c\left(\frac{dx}{d\sigma}\right)^2 = 0$$
holds. By the quadratic formula, this differential equation may be written in either of the two forms:
$$\left. \begin{array}{l} a\dfrac{dt}{d\sigma} - (b \pm \sqrt{b^2 - ac})\dfrac{dx}{d\sigma} = 0, \\[6pt] c\dfrac{dx}{d\sigma} + (-b \pm \sqrt{b^2 - ac})\dfrac{dt}{d\sigma} = 0. \end{array} \right\} \tag{2}$$

The forms (2) are equivalent except at those points of Γ where the coefficients of $dt/d\sigma$ and $dx/d\sigma$ in one of the equations vanish simultaneously. Equations (2) give two families of characteristics, one corresponding to each choice of sign. Those characteristics obtained by using the plus sign are called C_+ characteristics, and those obtained by using the minus sign are called C_- characteristics.

Suppose we are given a curve Γ_0. A domain D is said to be **admissible** with respect to the operator (1) and Γ_0 if to each point C of D corresponds a unique (curvilinear) characteristic triangle bounded by the C_+ characteristic and the C_- characteristic which intersect at C and by a segment AB of the curve Γ_0. (See Fig. 4.). In particular, no characteristic in D can intersect Γ_0 twice.

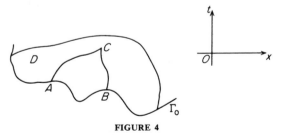

FIGURE 4

Each characteristic going from Γ_0 into D has a parametric representation with a parameter σ, say. We choose σ so that it is increasing as we go from Γ_0 into D. We now make a geometric hypothesis relating to the characteristics passing through each point C of D. We suppose that the characteristic triangle ABC corresponding to any point C of D lies to the right of the C_+ characteristic and to the left of the C_- characteristic on its boundary when these characteristics are traversed from Γ_0 toward C. Thus in Fig. 4, AC is a C_+ characteristic and BC is a C_- characteristic. This geometric hypothesis replaces the assumption made in Section 3 that $c < 0$. Since replacing L by $-L$ also interchanges the C_+ and C_- characteristics, this hypothesis is easily fulfilled by replacing the operator $(L + h)$ by $-(L + h)$, if necessary.

We may repeat the procedures of Section 3 and establish maximum principles under hypotheses on the behavior of u and its first derivatives along Γ_0 which are analogous to those given in Theorems 1 through 4.

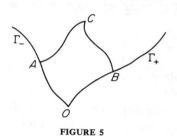

FIGURE 5

In this section we shall be concerned only with the special case where Γ_0 is composed of two characteristics Γ_- and Γ_+ emanating from a point O. (See Fig. 5.) According to the geometric hypothesis concerning the two families of characteristics given above, we see that Γ_+ is a C_+ characteristic and Γ_- is a C_- characteristic. We consider the hyperbolic operator L given by (1) together with its adjoint L^*, and we shall employ formula (2) of Section 3.

Let C be a point of D, arc AC a C_+ characteristic, arc BC a C_- characteristic, and suppose the curvilinear quadrilateral $AOBC$ is entirely in D (Fig. 5). We describe each of the four arcs OA, AC, OB, and BC in parametric form

$$x = x(\sigma), \quad t = t(\sigma)$$

with σ increasing in the direction from O to C. Following the methods of Section 3, we obtain from (2) of that section and the hypotheses that AC is a C_+ characteristic and BC a C_- characteristic the formula:

$$2\sqrt{b^2 - ac}\, u(C) = 2\sqrt{b^2 - ac}\, u(O) + \int_A^C u\tilde{K}_+ \, d\sigma_+ + \int_B^C u\tilde{K}_- \, d\sigma_-$$
$$+ \iint_{AOBC} \{u(L^* + h)[1] - (L + h)[u]\} \, dx \, dt$$
$$+ \int_0^A \left\{2\sqrt{b^2 - ac}\, \frac{du}{d\sigma_-} + K_-^* u\right\} d\sigma_- \qquad (3)$$
$$+ \int_0^B \left\{2\sqrt{b^2 - ac}\, \frac{du}{d\sigma_+} + K_+^* u\right\} d\sigma_+.$$

We have used the abbreviations:

$$\left.\begin{aligned}\tilde{K}_\pm &= \frac{d}{d\sigma_\pm}(\sqrt{b^2 - ac}) \pm \left[(a_x + b_t - d)\frac{dt}{d\sigma_\pm} - (b_x + c_t - e)\frac{dx}{d\sigma_\pm}\right], \\ K_\pm^* &= \frac{d}{d\sigma_\pm}(\sqrt{b^2 - ac}) \mp \left[(a_x + b_t - d)\frac{dt}{d\sigma_\pm} - (b_x + c_t - e)\frac{dx}{d\sigma_\pm}\right],\end{aligned}\right\} \quad (4)$$

and we have denoted the parameter on the C_+ characteristics by σ_+ and that on the C_- characteristics by σ_-. The quantities $\tilde{K}_\pm$ are related to $K_\pm$ as defined in Equation (4) of Section 3 by the formula $\tilde{K}_\pm = K_\pm dt/d\sigma_\pm$. The line integrals in (3) are independent of the particular parameters σ_+, σ_- used. Each parameter increases upon going from O toward C.

We shall now establish a maximum principle for a function u in D which satisfies the inequalities

$$\left.\begin{aligned} (L+h)[u] &> 0 \text{ in } D, \\ u &< 0 \text{ on } \Gamma_+ \text{ and } \Gamma_-, \\ 2\sqrt{b^2-ac}\,\frac{du}{d\sigma_+} + K_+^* u &\leq 0 \text{ on } \Gamma_+, \\ 2\sqrt{b^2-ac}\,\frac{du}{d\sigma_-} + K_-^* u &\leq 0 \text{ on } \Gamma_-. \end{aligned}\right\} \quad (5)$$

We assume that the coefficients of the operator L satisfy in D the following inequalities:

$$\left.\begin{aligned} (L^* + h)[1] &\geq 0, \\ \tilde{K}_+ &\geq 0, \\ \tilde{K}_- &\geq 0, \end{aligned}\right\} \quad (6)$$

with $\tilde{K}_\pm$ defined in (4).

The argument to establish the maximum principle proceeds as before. We have $u < 0$ on Γ_+ and Γ_-, and we suppose that $u \geq 0$ somewhere in D. We shall reach a contradiction. Let C be a point of D such that $u(C) = 0$ and $u \leq 0$ in the curvilinear quadrilateral $AOBC$ bounded by characteristics (Fig. 5). The inequalities (5) and (6) inserted in (3) show that we must have $u(C) < 0$, and a contradiction is reached. Hence $u < 0$, throughout D. As in Section 3, we can relax the inequalities $(L+h)[u] > 0$ and $u < 0$ to $(L+h)[u] \geq 0$ and $u \leq 0$. We observe that if $u(O) \leq 0$, the last two inequalities in (5) imply that $u \leq 0$ on Γ_+ and Γ_-. In this way we get the following theorem.

THEOREM 5. Let L be a hyperbolic operator in a domain D which is admissible with respect to two characteristic arcs Γ_+ and Γ_-. If the coefficients of L satisfy the inequalities (6) and if u satisfies the inequalities

$$(L+h)[u] \geq 0 \text{ in } D,$$

$$2\sqrt{b^2-ac}\,\frac{du}{d\sigma_+} + K_+^* u \leq 0 \text{ on } \Gamma_+,$$

$$2\sqrt{b^2-ac}\,\frac{du}{d\sigma_-} + K_-^* u \leq 0 \text{ on } \Gamma_-,$$

$$u(O) \leq 0,$$

where $\tilde{K}_{\pm}$ and $K_{\pm}^{*}$ are defined in Equation (4) and O is the intersection of Γ_{+} and Γ_{-}, then

$$u \leq 0 \text{ in } D.$$

Theorem 5 is valid for functions which are nonpositive on Γ_{+} and Γ_{-}. If u is bounded on Γ_{+} and Γ_{-} by some positive constant M, we may apply Theorem 5 to the function $u - M$ and obtain the following result.

THEOREM 6. If the coefficients of the hyperbolic operator L satisfy the inequalities (6) in an admissible domain D, if $h \leq 0$ in D, $K_{+}^{*} \geq 0$ on Γ_{+}, $K_{-}^{*} \geq 0$ on Γ_{-}, and if u satisfies the inequalities

$$(L + h)[u] \geq 0 \text{ in } D,$$

$$2\sqrt{b^2 - ac}\,\frac{du}{d\sigma_{+}} + K_{+}^{*}u \leq 0 \text{ on } \Gamma_{+},$$

$$2\sqrt{b^2 - ac}\,\frac{du}{d\sigma_{-}} + K_{-}^{*}u \leq 0 \text{ on } \Gamma_{-},$$

$$u(O) \leq M$$

with $M \geq 0$, then

$$u \leq M \text{ in } D.$$

We observe that if $K_{+}^{*} \geq 0$ on Γ_{+} and $u(O) \leq 0$, then the inequality $du/d\sigma_{+} \leq 0$ on Γ_{+} implies the condition on Γ_{+} of Theorem 5. A similar result is true on Γ_{-} if $K_{-}^{*} \geq 0$. Applying these facts to $u - M$ when $h \leq 0$, we obtain another theorem under the hypotheses about the coefficients which were made in Theorem 6.

THEOREM 6'. Suppose that the coefficients of L satisfy inequalities (6) in a domain D which is admissible with respect to two characteristics Γ_{+} and Γ_{-}. Suppose further that

$$h \leq 0 \text{ in } D, K_{+}^{*} \geq 0 \text{ on } \Gamma_{+}, K_{-}^{*} \geq 0 \text{ on } \Gamma_{-}.$$

If u satisfies the inequalities

$$(L + h)[u] \geq 0 \text{ in } D$$

$$\frac{du}{d\sigma_{+}} \leq 0 \text{ on } \Gamma_{+}, \qquad \frac{du}{d\sigma_{-}} \leq 0 \text{ on } \Gamma_{-},$$

then

$$u(x, t) \leq \max(0, u(O)) \text{ in } D$$

where O is the point of intersection of Γ_{+} and Γ_{-}. If, in addition, $h \equiv 0$ in D, then $u(x, t) \leq u(O)$.

To fix the ideas involved in Theorems 5, 6 and 6', we present a special case. Consider the operator

Sec. 8 The Two-Characteristic Problem

$$L_1[u] \equiv -u_{xt} + du_x + eu_t \tag{7}$$

with the positive x-axis as Γ_+ and the positive t-axis as Γ_-. The sign of the first coefficient in L_1 is selected so that Γ_+ is a C_+ characteristic and Γ_- is a C_- characteristic. (See Fig. 6.) We choose the parameter $\sigma_+ = x$ on the C_+ characteristics $t = \text{constant}$, and $\sigma_- = t$ on the C_- characteristics $x = \text{constant}$. In the particular case of the operator (7), the inequalities (6) become

$$\left. \begin{array}{l} L_1^*[1] \equiv -d_x - e_t \geq 0, \\ \tilde{K}_+ \equiv e \geq 0, \quad \tilde{K}_- \equiv d \geq 0. \end{array} \right\} \tag{8}$$

Correspondingly, we find that $K_+^* \equiv -e$, $K_-^* \equiv -d$. Theorem 5 applied to the operator (7) yields the following maximum principle: If d and e satisfy the inequalities (8) for $x \geq 0$, $t \geq 0$ and if u satisfies

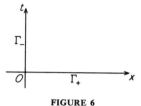

FIGURE 6

$$\left. \begin{array}{l} L_1[u] \geq 0 \text{ in } D, \\ u_x(x, 0) - e(x, 0)u(x, 0) \leq 0, \\ u_t(0, t) - d(0, t)u(0, t) \leq 0, \\ u(0, 0) \leq 0, \end{array} \right\} \tag{9}$$

then

$$u(x, t) \leq 0 \text{ for } x \geq 0,\ t \geq 0.$$

Since $d \geq 0$ and $e \geq 0$, Theorems 6 and 6' apply if and only if $d(0, t)$ and $e(x, 0)$ vanish identically. In this case, Theorems 6 and 6' are identical.

In Section 3 we obtained maximum principles for the initial value problem provided the coefficients of the operator satisfy certain inequalities. These restrictions were relaxed by multiplying the given operator $(L + h)$ by a positive function v, and then by selecting v so that the coefficients of L satisfy the appropriate inequalities. A completely analogous process is applicable to the two-characteristic problem.

If D is an admissible domain with respect to two characteristics Γ_+ and Γ_-, we multiply $(L + h)$ by a positive function v and note that the inequalities (6) now become

$$\left. \begin{array}{l} (L^* + h)[v] \geq 0 \text{ in } D, \\ 2\sqrt{b^2 - ac}\,\dfrac{dv}{d\sigma_+} + \tilde{K}_+ v \geq 0 \text{ in } D, \\ 2\sqrt{b^2 - ac}\,\dfrac{dv}{d\sigma_-} + \tilde{K}_- v \geq 0 \text{ in } D, \end{array} \right\} \tag{10}$$

with $\tilde{K}_\pm$ defined in (4). The quantities $K_\pm^*$ remain unchanged. The next two theorems are appropriate generalizations of Theorems 5 and 6'.

THEOREM 7. **Suppose there is a positive function v defined in an admissible domain D which satisfies inequalities (10). If u satisfies**

$$(L + h)[u] \geq 0 \text{ in } D,$$

$$2\sqrt{b^2 - ac}\,\frac{du}{d\sigma_+} + K_+^* u \leq 0 \text{ on } \Gamma_+,$$

$$2\sqrt{b^2 - ac}\,\frac{du}{d\sigma_-} + K_-^* u \leq 0 \text{ on } \Gamma_-,$$

$$u(O) \leq 0,$$

then

$$u \leq 0 \text{ in } D.$$

THEOREM 8. **Suppose there is a positive function v defined in an admissible domain D which satisfies inequalities (10). In addition, suppose that $K_+^* \geq 0$ on Γ_+, $K_-^* \geq 0$ on Γ_-, and $h \leq 0$. If $u(x, t)$ satisfies the inequalities**

$$(L + h)[u] \geq 0 \text{ in } D,$$

$$\frac{du}{d\sigma_+} \leq 0 \text{ on } \Gamma_+, \qquad \frac{du}{d\sigma_-} \leq 0 \text{ on } \Gamma_-,$$

then

$$u(x, t) \leq \max\,(0, u(O)) \text{ in } D,$$

where O is the point of intersection of Γ_+ and Γ_-. If $h \equiv 0$ in D, then $u \leq u(O)$ in D.

Theorems 7 and 8 are useful only when a positive function v with the desired properties can be found. For this purpose we first show how to construct such a function for the special operator

$$L_1 + h = -\frac{\partial^2}{\partial x\,\partial t} + d\frac{\partial}{\partial x} + e\frac{\partial}{\partial t} + h$$

in a neighborhood of the characteristics $\Gamma_+\colon t = 0$, $\Gamma_-\colon x = 0$. For the operator $L_1 + h$, conditions (10) become

$$-v_{xt} - dv_x - ev_t - (d_x + e_t - h)v \geq 0,$$
$$v_x + ev \geq 0,$$
$$v_t + dv \geq 0.$$

We set

$$v = e^{\alpha(x+t)} + \gamma x e^{-\beta t}$$

Sec. 8   *The Two-Characteristic Problem*   225

with the positive constants α and β chosen so that

$$\alpha + d > 0, \quad \alpha + e > 0, \quad \beta - d > 0,$$

and with γ positive and so large that

$$-[(\alpha + d)(\alpha + e) + d_x + e_t - de - h]e^{\alpha t} + \gamma(\beta - d)e^{-\beta t} > 0$$

on the line segment s_1: $\{x = 0, \ 0 \leq t \leq t_0\}$. Then the inequalities (10) are strictly satisfied on s_1. Hence by continuity there is a domain D_1: $0 < x < \delta, 0 < t < t_0$, where the inequalities (10) hold. We note that α, β, γ and δ, and hence also D_1, depend only on bounds for d, e, h, d_x and e_t. By interchanging x and t, we can construct a function v for a rectangle D_2 with any finite segment s_2: $\{t = 0, \ 0 \leq x \leq x_0\}$ on its boundary. Combining D_1 and D_2 we obtain an admissible domain D whose boundary contains s_1 and s_2 and in which Theorem 7 holds.

To obtain a function v for the general operator $(L + h)$ given by (1), we proceed by transforming coordinates. We introduce new coordinates (ξ, η) which are constant along characteristics; that is, $\xi = \sigma_+$ on Γ_+ and is constant along C_- characteristics, while $\eta = \sigma_-$ on Γ_- and is constant along C_+ characteristics. The quantities (ξ, η) are called *characteristic coordinates*. We choose $\sigma_+ = \sigma_- = \xi = \eta = 0$ at the intersection O of Γ_+ and Γ_-. An application of the chain rule to $L + h$ shows that in the (ξ, η) system the coefficients of $\partial^2/\partial \xi^2$ and $\partial^2/\partial \eta^2$ in the operator $(L + h)$ are zero and the coefficient of $\partial^2/\partial \xi \, \partial \eta$ is negative. Therefore, choosing the functions

$$v = e^{\alpha(\xi + \eta)} + \gamma \xi e^{-\beta \eta}$$
and
$$v = e^{\alpha(\xi + \eta)} + \gamma \eta e^{-\beta \xi}$$

with α, β, and γ sufficiently large, we can satisfy inequalities (10) in an admissible domain D. The size of D will depend on bounds for the coefficients of $L + h$ and their first two derivatives and on a positive lower bound for $\sqrt{b^2 - ac}$. The existence of v is thereby established, a fact which we use in the next theorem.

THEOREM 9. For any finite characteristic segments Γ_+ and Γ_- there exists an admissible domain D, its size depending on bounds for the coefficients of $L + h$ and their first two derivatives and on a positive lower bound for $b^2 - ac$, such that Γ_+ and Γ_- lie on the boundary of D and Theorem 7 is valid in D. If, in addition, $K_+^* \geq 0$ on Γ_+, $K_-^* \geq 0$ on Γ_-, and $h \leq 0$, then Theorem 8 is valid in D.

All the preceding maximum principles require certain hypotheses on the coefficients of the operator $L + h$ as well as hypotheses on the function u. We now show that a variety of theorems can be obtained by strengthening some hypotheses and weakening others.

THEOREM 10. Suppose the coefficients of $L + h$ satisfy the inequalities

$$(L^* + h)[1] \geq 0 \text{ in } D,$$

$$\tilde{K}_+ \geq 0, \quad \tilde{K}_- \geq 0 \text{ in } D,$$

$$\tilde{K}_- = 0 \text{ on } \Gamma_-.$$

If u satisfies the inequalities

$$(L + h)[u] \geq 0 \text{ in } D,$$

$$2\sqrt{b^2 - ac}\,\frac{du}{d\sigma_+} + K_+^* u \leq 0 \text{ on } \Gamma_+,$$

$$u \leq 0 \text{ on } \Gamma_-,$$

where $\tilde{K}_\pm$ and $K_\pm^*$ are defined in (4), then

$$u \leq 0 \text{ in } D.$$

Proof. An integration by parts along Γ_- yields the identity

$$\int_O^A \left(2\sqrt{b^2 - ac}\,\frac{du}{d\sigma_-} + K_-^* u\right) d\sigma = 2\sqrt{b^2 - ac}\, u(A)$$
$$- 2\sqrt{b^2 - ac}\, u(O) - \int_O^A \tilde{K}_- u\, d\sigma_-.$$

Inserting this relation in formula (3), we obtain

$$u(C) = \frac{1}{2\sqrt{[b^2 - ac]_C}} \Bigg[2\sqrt{b^2 - ac}\, u(A) + \int_A^C u\tilde{K}_+\, d\sigma_+ + \int_B^C u\tilde{K}_-\, d\sigma_-$$
$$+ \iint_{AOBC} \{u(L^* + h)[1] - (L + h)[u]\}\, dx\, dt - \int_O^A u\tilde{K}_-\, d\sigma_-$$
$$+ \int_O^B \left\{2\sqrt{b^2 - ac}\,\frac{du}{d\sigma_+} + K_+^* u\right\} d\sigma_+ \Bigg].$$

The remainder of the argument proceeds as in the proof of Theorem 5.

Remarks. (i) We note that Theorem 10 differs from Theorem 5 in that no hypotheses are made about $du/d\sigma_-$ on Γ_-. On the other hand, the coefficients of L are required to satisfy $\tilde{K}_- = 0$ on Γ_-.

(ii) Clearly, an analogous result holds if we assume that $\tilde{K}_+ = 0$ on Γ_+ instead of $\tilde{K}_- = 0$ on Γ_-.

(iii) If both $\tilde{K}_+ = 0$ on Γ_+ and $\tilde{K}_- = 0$ on Γ_-, we can replace the conditions on u, $du/d\sigma_+$, and $du/d\sigma_-$ by the condition

$$\sqrt{b^2 - ac}\, u(A) + \sqrt{b^2 - ac}\, u(B) - \sqrt{b^2 - ac}\, u(O) \leq 0$$

for all A on Γ_- and all B on Γ_+.

(iv) Note that the condition

$$2\sqrt{b^2 - ac}\,\frac{du}{d\sigma_+} + K_+^* u \leq 0 \text{ on } \Gamma_+$$

together with $u(O) \leq 0$ implies $u \leq 0$ on Γ_+.

If we multiply the operator $(L + h)$ by a positive function v, the following extension of Theorem 10 is easily established.

THEOREM 11. Suppose there is a positive function v which satisfies inequalities (10) in D. In addition, suppose that v satisfies

$$2\sqrt{b^2 - ac}\,\frac{dv}{d\sigma_-} + \tilde{K}_- v = 0 \text{ on } \Gamma_-.$$

If u satisfies

$$(L + h)[u] \geq 0 \text{ in } D,$$

$$2\sqrt{b^2 - ac}\,\frac{du}{d\sigma_+} + K_+^* u \leq 0 \text{ on } \Gamma_+,$$

$$u \leq 0 \text{ on } \Gamma_-,$$

then

$$u \leq 0 \text{ in } D.$$

The analogue of Theorem 6' may be derived by adjusting the hypotheses. We state the result.

THEOREM 12. Suppose there is a positive function v which satisfies inequalities (10) in D. In addition, suppose that

$$2\sqrt{b^2 - ac}\,\frac{dv}{d\sigma_-} + \tilde{K}_- v = 0 \text{ on } \Gamma_-,$$

$$K_+^* \geq 0 \text{ on } \Gamma_+,$$

$$h \leq 0 \text{ in } D,$$

and that M is a nonnegative constant. If u satisfies

$$(L + h)[u] \geq 0 \text{ in } D,$$

$$\frac{du}{d\sigma_+} \leq 0 \text{ on } \Gamma_+,$$

$$u \leq M \text{ on } \Gamma_-,$$

then

$$u \leq M \text{ in } D.$$

We now investigate the particular forms which Theorems 11 and 12 take for the special operator

$$(L_1 + h)[u] = -u_{xt} + du_x + eu_t + hu \tag{11}$$

228   Hyperbolic Equations   Chap. 4

with the positive x-axis selected for Γ_+ and the positive t-axis for Γ_-. (See Fig. 7.) We easily verify that

$$2\sqrt{b^2-ac}\,\frac{dv}{d\sigma_+} + \tilde{K}_+ v = v_x + ev,$$

$$2\sqrt{b^2-ac}\,\frac{dv}{d\sigma_-} + \tilde{K}_- v = v_t + dv.$$

We choose

$$v = e^{\beta x - \int_0^t d(x,\tau)d\tau} \tag{12}$$

where β is yet to be selected. A simple computation shows that

$$v_t + dv \equiv 0.$$

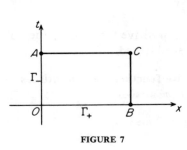

FIGURE 7

We now choose β so large that

$$v_x + ev \equiv \left[\beta - \int_0^t d_x(x,\tau)d\tau + e\right] v$$

is nonnegative in D. Another computation yields

$$(L_1^* + h)[v] = (de - e_t + h)v.$$

We now apply Theorem 11 to the operator $L_1 + h$. We find that if u satisfies

$$(L_1 + h)[u] \geq 0 \text{ in } D$$
$$u_x - eu \leq 0 \text{ on } \Gamma_+$$
$$u(0, t) \leq 0$$

and if

$$de - e_t + h \geq 0 \text{ in } D, \tag{13}$$

then

$$u \leq 0 \text{ in } D.$$

We may also apply Theorem 12 to the operator $L_1 + h$, using the function v given by (12). We note that $K_+^* = -e$. Therefore, if the condition $-e(x, 0) \geq 0$ is satisfied on Γ_+, and if the inequalities $h \leq 0$ and (13) hold in D, then we may apply Theorem 12. The result states that if u satisfies

$$(L_1 + h)[u] \geq 0 \text{ in } D,$$
$$u_x \leq 0 \text{ on } \Gamma_+,$$
$$u \leq M \text{ on } \Gamma_-,$$

where $M \geq 0$, then

$$u \leq M \text{ in } D.$$

This is a result of Agmon, Nirenberg, and Protter [1]. Another result comes from Theorem 10 ($v \equiv 1$). It states that u satisfies the same maximum principle if instead of (13)

$$\left.\begin{array}{r} d \geq 0 \\ e \geq 0 \\ -d_x - e_t + h \geq 0 \end{array}\right\} \text{ in } D,$$

$$d(0, t) = 0.$$

Similar considerations apply to the more general hyperbolic operator

$$(L + h)[u] = au_{xx} + 2bu_{xt} + cu_{tt} + du_x + eu_t + hu.$$

We define v by the condition

$$2\sqrt{b^2 - ac}\,\frac{dv}{d\sigma_-} + \tilde{K}_- v = 0$$

on each C_- characteristic. We see easily that the initial values of v on Γ_+ may be assigned so that

$$2\sqrt{b^2 - ac}\,\frac{dv}{d\sigma_+} + \tilde{K}_+ v \geq 0$$

in any bounded admissible domain D. The condition

$$(L^* + h)[v] \geq 0$$

becomes

$$(-b - \sqrt{b^2 - ac})P_x - cP_t + aQ_x - (-b + \sqrt{b^2 - ac})Q_t$$
$$- \tfrac{1}{2}(aQ^2 - 2bPQ + cP^2) + Qd - Pe - 2h \leq 0, \quad (14)$$

where we have put

$$P \equiv \frac{\partial}{\partial x}\left(\frac{a}{\sqrt{b^2 - ac}}\right) + \frac{\partial}{\partial t}\left(\frac{b}{\sqrt{b^2 - ac}}\right) - \frac{d}{\sqrt{b^2 - ac}}$$

$$Q \equiv \frac{\partial}{\partial x}\left(\frac{b}{\sqrt{b^2 - ac}}\right) + \frac{\partial}{\partial t}\left(\frac{c}{\sqrt{b^2 - bc}}\right) - \frac{e}{\sqrt{b^2 - ac}}.$$

In the special case of the operator (11), inequality (14) reduces to (13). If (14) holds, then Theorem 11 is applicable. If, in addition, $K_+^* \geq 0$ on Γ_+, then Theorem 12 is also valid.

Remarks. (i) If we wish to interchange the roles of Γ_+ and Γ_-, it is only necessary to replace e_t by d_x in (13) or to interchange the coefficients $-b - \sqrt{b^2 - ac}$ and $-b + \sqrt{b^2 - ac}$ in (14). Of course, K_+^* is to be replaced by K_-^*.

(ii) If, contrary to our convention, Γ_+ is to the left of Γ_-, it is only necessary to replace h by $-h$ in (13) or (14). The quantity K_+^* remains unchanged but the inequality $(L + h)[u] \geq 0$ has to be replaced by $(L + h)[u] \leq 0$ and the condition $h \leq 0$ in Theorem 12 becomes $h \geq 0$.

(iii) The following example shows that condition (13) or (14) is needed. (See also Agmon, Nirenberg, and Protter [1].) Suppose that $de - e_t + h < 0$ on an interval $(0, t_0]$ of the t-axis. Let

$$u(x, 0) \equiv 0,$$

$$u(0, t) = -1 + \cos(2\pi t/t_0).$$

We write the equation $(L_1 + h)[u] = 0$ in the form

$$\left(\frac{\partial}{\partial t} - d\right)[u_x - eu] = (de - e_t + h)u.$$

At $x = 0$ the right-hand side is positive for $0 < t < t_0$. Since $u_x - eu = 0$ at $(0, 0)$, it follows that

$$u_x(0, t_0) = u_x(0, t_0) - e(0, t_0) u(0, t_0) > 0.$$

Therefore u becomes positive in D, contrary to Theorem 11 and, if $e(x, 0) \leq 0$, to Theorem 12.

EXERCISES

1. Calculate the quantities $\tilde{K}_\pm$ and $K_\pm^*$ for the operator

$$L[u] \equiv u_{xx} - u_{tt} + (1 + x^2)u_x - 2tu_t.$$

2. Verify that the coefficients of the operator

$$L[u] \equiv tu_{xx} - u_{tt}$$

satisfy the inequalities (6) for $t > 0$.

3. Show that solutions of $u_{xt} = 0$ satisfy a weak but not a strong maximum principle in any admissible domain D bounded by the positive x- and t-axes.

4. Determine a function v so that inequalities (10) are satisfied for the operator

$$L[u] \equiv -u_{xt} - tu_x - xu_t$$

in the domain D: $\{0 < x < 1, 0 < t < 1\}$.

5. Determine a function v so that the hypotheses of Theorem 11 are valid for the operator

$$L[u] \equiv -u_{xt} + (1 + t)u_x - (1 - x^2)u_t$$

in the domain D: $\{1 < x < 2, 0 < t < 1\}$.

6. (a) Determine the values of α, if any, for which inequality (14) is satisfied with respect to the operator

$$L[u] \equiv t^{2\alpha} u_{xx} - u_{tt} \text{ for } t > 0.$$

(b) Determine inequalities for d and e such that inequality (14) is satisfied for the operator

$$L[u] \equiv tu_{xx} - xu_{tt} + du_x + eu_t, \qquad x > 0, \qquad t > 0.$$

SECTION 9. THE GOURSAT PROBLEM

We again consider the differential inequality $(L + h)[u] \geq 0$ in a domain D, where L is given by Equation (1) in Section 8. We suppose that one part of the boundary of D is a C_+ characteristic Γ_+, and another portion consists of a curve Γ_1: $x = x_1(\sigma)$, $t = t_1(\sigma)$ whose tangent vector $(dx_1/d\sigma, dt_1/d\sigma)$ satisfies the inequality

$$a \left(\frac{dt_1}{d\sigma}\right)^2 - 2b \frac{dx_1}{d\sigma} \frac{dt_1}{d\sigma} + c \left(\frac{dx_1}{d\sigma}\right)^2 \geq 0.$$

The domain D is admissible if it has the property that through each point C of D there is a C_+ characteristic AC and a C_- characteristic BC lying in D, with A on Γ_1 and B on Γ_+. (See Fig. 8.) The problem of determining a solution w of $(L + h)[w] = f$ whose values are prescribed on Γ_+ and Γ_1 is called the **Goursat problem**.

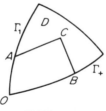

FIGURE 8

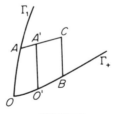

FIGURE 9

Now suppose that for each point C of D there is a characteristic quadrangle $A'O'BC$ to which we can apply Theorem 12. (See Fig. 9) That is, we suppose there is a positive function v which satisfies the inequalities (10) and the relation

$$2\sqrt{b^2 - ac} \frac{dv}{d\sigma_-} + \tilde{K}_- v = 0$$

on $O'A'$. We also assume that $h \leq 0$ and $(L + h)[u] \geq 0$ in D. If $K_*^* \geq 0$ and $du/d\sigma_+ \leq 0$ on Γ_+, then u cannot attain a nonnegative maximum at C, an arbitrary point of the domain D. Consequently, under these conditions a nonnegative maximum of u must occur either on Γ_1 or Γ_+. Since

$du/d\sigma_+ \leq 0$ on Γ_+, this maximum must occur on Γ_1. The next theorem states this principle for the operator $L_1 \equiv -(\partial^2/\partial x\, \partial t) + d(\partial/\partial x) + e(\partial/\partial t)$.

THEOREM 13. Suppose that the following conditions hold:

(i) $\quad (L_1 + h)[u] \equiv -u_{xt} + du_x + eu_t + hu \geq 0$

in a domain D in the first quadrant lying between the x-axis and a curve Γ_1 of nonnegative slope.

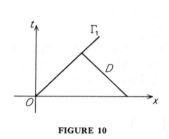

FIGURE 10

(ii) $\quad de - e_t + h \geq 0,$

(iii) $\quad e(x, 0) \leq 0,$

(iv) $\quad u_x(x, 0) \leq 0,$

(v) $\quad u \leq M$ on Γ_1, where $M \geq 0$,

and

(vi) $\quad h \leq 0.$

Then

$$u \leq M \text{ in } D.$$

This is the original maximum principle of Agmon, Nirenberg, and Protter [1].

More generally, we have the same result if u satisfies $(L + h)[u] \geq 0$ and the coefficients of L satisfy inequality (14) of Section 8 and $K_+^* \geq 0$ on Γ_+.

EXERCISES

1. Determine explicitly the C_+ and C_- characteristic families for the operator $L[u] \equiv t^\alpha u_{xx} - x^\beta u_{tt}$, $x > 0$, $t > 0$. Describe an admissible domain for the Goursat problem.

2. Determine conditions on the function $K(t) > 0$ such that the maximum principle of Theorem 13 is valid for the operator $K(t)u_{xx} - u_{tt}$ in a domain D bounded by one characteristic and one noncharacteristic.

SECTION 10. COMPARISON THEOREMS

As in the case of elliptic equations, we can apply the various theorems on hyperbolic equations to the ratio u/φ where u satisfies a differential

Sec. 10   Comparison Theorems   233

inequality and φ is any positive function. Putting

$$\bar{u} = \frac{u}{\varphi},$$

we find that

$$(\bar{L} + \bar{h})[\bar{u}] \equiv \frac{1}{\varphi}(L + h)[u] = L[\bar{u}] + \frac{2}{\varphi}(a\varphi_x + b\varphi_t)\bar{u}_x$$
$$+ \frac{2}{\varphi}(b\varphi_x + c\varphi_t)\bar{u}_t + \frac{(L + h)[\varphi]}{\varphi}\bar{u}.$$

Thus, the inequality $(L + h)[u] \geq 0$ becomes

$$(\bar{L} + \bar{h})[\bar{u}] \geq 0,$$

where $\bar{h} = (L + h)[\varphi]/\varphi$. The inequalities for v are simply multiplied by φ. Hence the conditions on v remain unchanged. However, K_+^* is replaced by $2\sqrt{b^2 - ac}\, d\varphi/d\sigma_+ + K_+^*\varphi$ and, as seen above, h goes into $(L + h)[\varphi]/\varphi$. Thus if we have a function φ satisfying

$$\left.\begin{array}{c} \varphi > 0 \text{ in } D \cup \Gamma_+ \cup \Gamma_-, \\ 2\sqrt{b^2 - ac}\, \dfrac{d\varphi}{d\sigma_+} + K_+^*\varphi \geq 0 \text{ on } \Gamma_+, \\ (L + h)[\varphi] \leq 0 \text{ in } D, \end{array}\right\} \quad (1)$$

then Theorem 12 applies to $\bar{u} = u/\varphi$.

It is not difficult to construct a function φ with the properties (1). For example, in case

$$(L_1 + h)[u] \equiv -u_{xt} + du_x + eu_t + hu,$$

we need only put

$$\varphi = e^{\beta(x+t)}$$

and then choose β so large thst

$$(L + h)[\varphi] = (-\beta^2 + \beta d + \beta e + h)\varphi \leq 0 \text{ in } D.$$
$$2\sqrt{b^2 - ac}\, \frac{d\varphi}{dx} + K_+^*\varphi = (\beta - e)\varphi \geq 0 \text{ for } t = 0.$$

Thus, in the case of this equation, the only condition needed to obtain a maximum principle for $e^{-\beta(x+t)}u$ is (13) of Section 8. Similarly we can show that in the general case the inequality (14) of Section 8 is sufficient to obtain a maximum principle for u/ϕ with some $\phi > 0$. The example in Remark (iii) at the end of Section 8 shows that condition (13) or (14) is needed.

EXERCISE

Given the operator

$$(L + h)[u] \equiv -u_{xt} + xu_x + xtu_t + xu,$$

verify that inequality (13) of Section 8 is satisfied for $t \geq 0$. Then find a value of β such that $e^{\beta(x-t)}u$ satisfies a maximum principle in the domain D: $\{0 < x < 1, 0 < t < 1\}$.

SECTION 11. THE WAVE EQUATION IN HIGHER DIMENSIONS

So far we have been concerned with hyperbolic equations in only two dimensions. In many physical problems, however, one is concerned with the initial value problem for a hyperbolic differential equation in three or more dimensions. The simplest example is the wave equation

$$\square u \equiv \frac{\partial^2 u}{\partial x^2} + \frac{\partial^2 u}{\partial y^2} - \frac{\partial^2 u}{\partial t^2} = f(x, y, t) \tag{1}$$

in three dimensions. This is the equation obeyed by small vertical displacements of a stretched membrane with normal force f per unit area acting on it.

The initial value problem consists of prescribing u and $\partial u/\partial t$ as functions of x and y at $t = 0$ and seeking $u(x, y, t)$ for $t > 0$. This problem can be solved explicitly (see, for example, Courant and Hilbert [1, p. 205] or Garabedian [1, p. 205]).

$$u(x, y, t) = \frac{\partial}{\partial t} \frac{1}{2\pi} \iint_{\xi^2 + \eta^2 < t^2} \frac{u(x + \xi, y + \eta, 0)}{\sqrt{t^2 - \xi^2 - \eta^2}} d\xi d\eta$$

$$+ \frac{1}{2\pi} \iint_{\xi^2 + \eta^2 < t^2} \frac{u_t(x + \xi, y + \eta, 0)}{\sqrt{t^2 - \xi^2 - \eta^2}} d\xi d\eta$$

$$- \frac{1}{2\pi} \int_0^t d\tau \iint_{\xi^2 + \eta^2 < (t-\tau)^2} \frac{\square u(x + \xi, y + \eta, \tau)}{\sqrt{(t - \tau)^2 - \xi^2 - \eta^2}} d\xi d\eta. \tag{2}$$

It follows that if

$$u(x, y, 0) = 0,$$
$$u_t(x, y, 0) \leq 0,$$
$$\square u \geq 0,$$

then
$$u \leq 0 \quad \text{for} \quad t > 0.$$

This result is a very simple maximum principle. However, such a maximum principle cannot hold for general initial values $u(x, y, 0)$. If, for example, we set $r^2 = x^2 + y^2$ and define u to be the solution of the equation $\Box u = 0$ with the initial conditions

$$u(x, y, 0) = \begin{cases} e^{-\alpha(r-2)^2/(r-1)(3-r)} & \text{for } 1 \leq r \leq 3, \\ 0 & \text{for } r \leq 1, \\ 0 & \text{for } r \geq 3, \end{cases}$$

$$u_t(x, y, 0) = 0,$$

then we see that $0 \leq u(x, y, 0) \leq 1$ for any $\alpha > 0$ and that $u(x, y, 0)$ is infinitely differentiable. On the other hand, for $t > 3$, we have from (2) that

$$u(0, 0, t) = -t \int_1^3 (t^2 - r^2)^{-3/2} e^{-\alpha(r-2)^2/(r-1)(3-r)} r \, dr, \quad t > 3.$$

Thus,
$$\lim_{\alpha \to 0} u(0, 0, t) = -t[(t^2 - 9)^{-1/2} - (t^2 - 1)^{-1/2}].$$

It follows that for α sufficiently near 0 and t sufficiently near 3, $-u(0, 0, t)$ will be arbitrarily large. Hence the wave equation cannot satisfy a maximum principle with the above data.

To obtain conditions for a maximum principle, we differentiate the differential equation (1) with respect to t and then apply the formula (2) to u_t. We find

$$u_t(x, y, t) = \frac{\partial}{\partial t} \frac{1}{2\pi} \iint_{\xi^2 + \eta^2 < t^2} \frac{u_t(x + \xi, y + \eta, 0)}{\sqrt{t^2 - \xi^2 - \eta^2}} d\xi \, d\eta$$

$$+ \frac{1}{2\pi} \iint_{\xi^2 + \eta^2 < t^2} \frac{u_{tt}(x + \xi, y + \eta, 0)}{\sqrt{t^2 - \xi^2 - \eta^2}} d\xi \, d\eta \quad (3)$$

$$- \frac{1}{2\pi} \int_0^t d\tau \iint_{\xi^2 + \eta^2 < (t-\tau)^2} \frac{\frac{\partial}{\partial t} \Box u(x + \xi, y + \eta, \tau)}{\sqrt{(t - \tau)^2 - \xi^2 - \eta^2}} d\xi \, d\eta.$$

We note that for $t = 0$ the differential equation may be written

$$u_{tt}(x, y, 0) = \Delta u(x, y, 0) - f(x, y, 0), \quad (4)$$

so that $u_{tt}(x, y, 0)$ is determined by the given data.

We now integrate (3) from $t = 0$ to an arbitrary value of t to obtain the formula

$$u(x, y, t) = u(x, y, 0) + \frac{1}{2\pi} \iint_{\xi^2+\eta^2<t^2} \frac{u_t(x+\xi, y+\eta, 0)}{\sqrt{t^2 - \xi^2 - \eta^2}} d\xi\, d\eta$$

$$+ \frac{1}{2\pi} \int_0^t d\tau \iint_{\xi^2+\eta^2<\tau^2} \frac{u_{tt}(x+\xi, y+\eta, 0)}{\sqrt{\tau^2 - \xi^2 - \eta^2}} d\xi\, d\eta$$

$$- \frac{1}{2\pi} \int_0^t dt_1 \int_0^{t_0} d\tau \iint_{\xi^2+\eta^2<(t-\tau)^2} \frac{\frac{\partial}{\partial t}\Box u(x+\xi, y+\eta, \tau)}{\sqrt{(t-\tau)^2 - \xi^2 - \eta^2}} d\xi\, d\eta. \tag{5}$$

This formula is an alternative solution formula equivalent to (2). We have, of course, assumed that f is differentiable with respect to t. The formula (5) involves only integrals of given quantities and $u_{tt}(x, y, 0)$, which can be computed by (4).

It is now clear from (5) that if

$$\left.\begin{array}{l} \dfrac{\partial}{\partial t} \Box u \geq 0, \\[4pt] u_t(x, y, 0) \leq 0, \\[4pt] u_{tt}(x, y, 0) \leq 0, \end{array}\right\} \tag{6}$$

then

$$u(x, y, t) \leq u(x, y, 0).$$

It is, in fact, sufficient that these inequalities hold for $(\bar{x}, \bar{y}, \bar{t})$ in the characteristic cone

$$(\bar{x} - x)^2 + (\bar{y} - y)^2 < (\bar{t} - t)^2, \quad 0 \leq \bar{t} < t,$$

corresponding to the point (x, y, t).

We thus have a maximum principle (see Weinstein [2, 3]):

THEOREM 14. *If u satisfies the conditions (6) in a domain D in the half-space $t \geq 0$ having the property that it contains the characteristic cone of each of its points, then the maximum of u in D is attained on the initial plane $t = 0$.*

Similar results hold in higher dimensions. We define

$$\Box u(x_1, x_2, \ldots, x_n, t) \equiv \sum_{i=1}^n \frac{\partial^2 u}{\partial x_i^2} - \frac{\partial^2 u}{\partial t^2},$$

and we find that if

$$\frac{\partial^{n-1}}{\partial t^{n-1}} \Box u \geq 0 \quad \text{in} \quad D,$$

$$\frac{\partial^i u}{\partial t^i}(x_1, \ldots, x_n, 0) \leq 0, \quad i = 1, 2, \ldots, n,$$

where D is a domain of $t \geq 0$ containing the characteristic cone of each of its points, then the maximum of u in D must occur at $t = 0$.

The quantities $\partial^2 u/\partial t^2, \ldots, \partial^n u/\partial t^n$ at $t = 0$ may again be found from the differential equation and the initial values of u and $\partial u/\partial t$. Thus

$$\frac{\partial^2 u}{\partial t^2}(x_1, \ldots, x_n, 0) = \Delta u(x_1, \ldots, x_n, 0) - \Box u(x_1, \ldots, x_n, 0)$$

$$\frac{\partial^3 u}{\partial t^3}(x_1, \ldots, x_n, 0) = \Delta \frac{\partial u}{\partial t}(x_1, \ldots, x_n, 0) - \frac{\partial}{\partial t}\Box u(x_1, \ldots, x_n, 0),$$

and so forth (see Weinberger [3]).

The above results may be used to estimate the error made in an approximation. Consider, for example, the problem

$$\Box u \equiv u_{xx} + u_{yy} - u_{tt} = 0,$$
$$u(x, y, 0) = g(x, y),$$
$$u_t(x, y, 0) = 0.$$

We are interested in u at a particular point, say $P(0, 0, 1)$. Then we may alter $g(x, y)$ in any way we like outside the disk $x^2 + y^2 \leq 1$ without affecting the value $u(0, 0, 1)$ given by (5). In particular, we can continue g as a periodic function of period 2π outside the square $|x| < \pi$, $|y| < \pi$.

If the function g, as altered and extended, is sufficiently smooth, it can be uniformly approximated by a partial sum of its Fourier series

$$g \cong \sum_{m,n=-N}^{N} a_{mn} e^{i(mx+ny)}.$$

We can then expect to approximate u by the function

$$v = \sum_{m,n=-N}^{N} a_{mn} e^{i(mx+ny)} \cos\sqrt{m^2 + n^2}\, t.$$

Suppose

$$\left|g - \sum_{-N}^{N} a_{mn} e^{i(mx+ny)}\right| \leq \epsilon \text{ for } |x| \leq \pi, \qquad |y| \leq \pi.$$

Can we say that $|u - v|$ is small at $P(0, 0, 1)$? We know that

$$\Box(u - v) = 0,$$

that

$$\frac{\partial}{\partial t}(u - v) = 0 \text{ at } t = 0,$$

and that

$$|u - v| \leq \epsilon \text{ at } t = 0.$$

This is not enough information to apply the maximum principle. We need to know that $(\partial^2/\partial t^2)(u - v) \leq 0$ at $t = 0$ to conclude that $u - v \leq \epsilon$ at P. Now by the differential equation, we have

$$\frac{\partial^2}{\partial t^2}(u - v)\Big|_{t=0} = \Delta g + \sum_{-N}^{N}(m^2 + n^2)a_{mn}\, e^{i(mx+ny)}.$$

By the theorem on differentiation of Fourier series, the sum on the right is a partial sum of the Fourier series for $-1\,\Delta g$. Hence, if Δg is smooth (for example, continuously differentiable), we can choose N so that

$$\left|\left[\frac{\partial^2(u - v)}{\partial t^2}\right]_{t=0}\right| \leq \epsilon_1,$$

where ϵ_1 is another small constant.

We now consider the function

$$w = u - v - \tfrac{1}{2}\epsilon_1 t^2.$$

Clearly this function satisfies

$$\frac{\partial}{\partial t}\Box w = 0,$$

$$\frac{\partial w}{\partial t}(x, y, 0) = 0,$$

$$\frac{\partial^2 w}{\partial t^2}(x, y, 0) \leq 0,$$

$$w(x, y, 0) \leq \epsilon.$$

Hence by the maximum principle.

$$w(x, y; t) \leq \epsilon.$$

This means that

$$u - v \leq \epsilon + \tfrac{1}{2}\epsilon_1 t^2,$$

The same bound holds for $v - u$. Thus we obtain the error bound

$$|u - v| < \epsilon + \tfrac{1}{2}\epsilon_1 t^2$$

at the point $P(0, 0, 1)$, and, in fact, at all points of the cone $t^2 \leq x^2 + y^2$. Approximations and error bounds at other points can be obtained by altering u outside a larger disk and continuing it as a function of larger period.

The maximum principle of Theorem 14 has been extended by Weinberger [3] to a class of three-dimensional equations which includes the wave equation, and to a more general class of hyperbolic equations in any number of dimensions by Sather [1, 2, 3]. Sather's results in n dimensions show, in particular, that the conditions

$$\frac{\partial^{n-1}\Box u}{\partial t^{n-1}} \geq 0, \qquad \frac{\partial^i u(\mathbf{x}, 0)}{\partial t^i} \leq 0, \qquad 1 \leq i \leq n$$

of Theorem 14 may be replaced by

$$\frac{\partial^N \Box u}{\partial t^N} \geq 0, \quad \frac{\partial^i u(\mathbf{x}, 0)}{\partial t^i} = 0 \text{ for } 0 \leq i \leq N, \frac{\partial^{N+1} u(\mathbf{x}, 0)}{\partial t^{N+1}} \leq 0,$$

where N is any integer such that $N \geq \frac{1}{2}(n-2)$. (For a similar result see Carroll [2].)

EXERCISES

1. Verify that (2) is a solution of $\Box u = f$.
2. State a maximum principle like that of Theorem 14 for solutions of $L[u] \equiv u_{xx} + u_{yt} = 0$ with respect to the initial value problem.

BIBLIOGRAPHICAL NOTES

The first maximum principle for a hyperbolic equation was obtained by P. Germain and R. Bader [1] for the Tricomi equation $yu_{xx} + u_{yy} = 0$. This result was extended to more general equations by S. Agmon, L. Nirenberg, and M. H. Protter [1]. These theorems were used to obtain uniqueness theorems for problems concerning equations of mixed type, which are elliptic in one part of the domain and hyperbolic in another. Another maximum principle for the equation of mixed type $K(y)u_{xx} + u_{yy} = 0$ was discovered by C. S. Morawetz [1], who used it to establish a different uniqeness theorem. Extensive discussions of equations of mixed type are found in the books of F. Tricomi [1] and A. V. Bitsadze [1].

A maximum principle for the initial value problem in two dimensions was given by H. F. Weinberger [2] and extended by M. H. Protter [1].

A different maximum principle for the two-characteristic problem was found by J. Schröder [1].

More general maximum principles for initial-boundary value problems in two dimensions have been obtained by D. Sather [4] and L. E. Payne and D. Sather [1].

Comparison theorems and error bounds for nonlinear hyperbolic equations in two dimensions have been given by W. Walter [2, 3] and H. H. Gloistehn [2]. References to other literature may be found in the book of W. Walter [3].

A maximum principle for the class of equations $u_{tt} + (k/t)u_t - \Delta u = 0$ (the Euler-Poisson-Darboux equation) in n dimensions was discovered by A. Weinstein [2, 3]. R. Carroll [1, 2] improved and extended this result to the equation $u_{tt} + (k/t)u_t - A[u] = 0$, where A is a differential operator in the variables $x_1, x_2, \ldots, x_n$ with constant coefficients which satisfies a hypothesis somewhat weaker than ellipticity.

A maximum principle for a class of equations with variable coefficients in three dimensions was given by H. F. Weinberger [3] and improved and extended to a larger class of equations in n dimensions by D. Sather [1, 2, 3].

BIBLIOGRAPHY

Agmon, S.
1. Maximum properties for solutions of higher order elliptic equations. *Bull. Amer. Math. Soc.*, **66** (1960), pp. 77-80.
2. Unicité et convexité dans les problèmes différentielles. *Sem. d'Analyse Sup., Univ. de Montreal*, 1965.

Agmon, S., L. Nirenberg, and M. H. Protter
1. A maximum principle for a class of hyperbolic equations and applications to equations of mixed elliptic-hyperbolic type. *Comm. on Pure and Appl. Math.*, **6** (1953), pp. 455-470.

Aleksandrov, A. D.
1. Certain estimates for the Dirichlet problem. *Dokl. Akad. Nauk S.S.S.R.*, **134** (1960), pp. 1001-1004. Translated in *Soviet Math.*, **1** (1960), pp. 1151-1154.
2. The method of normal map in uniqueness problems and estimations for elliptic equations. *Seminari 1962-1963 di Analisi, Algebra, Geometria, e Topologia, Istituto Nazionale di Alta Matematica*, Edizioni Cremonese, Rome, **2** (1965), pp. 744-796.
3. Estimates for the solution of linear second order equations. *Vestnik, Leningrad University, Math., Mech, and Astron. series*, (1966), No. 1.

Antohin, Ju. T.
1. The Dirichlet problem for a second order elliptic equation in an unbounded region. *Trudy Mat. Inst. Steklov*, **60** (1961), pp. 22-41.

Aronson, D. G.
1. Linear parabolic differential equations containing a small parameter. *J. of Rat. Mech. Anal.*, **5** (1956), pp. 1003-1014.
2. Removable singularities for linear parabolic equations. *Arch. for Rat. Mech. and Anal.*, **17** (1964), pp. 79-84.
3. Uniqueness of positive weak solutions of second order parabolic equations. *Ann. Polon. Math.*, **16** (1964-65), pp. 285-303.

Aronson, D. G., and P. Besala
1. Uniqueness of positive solutions of parabolic equations with unbounded coefficients. *Colloquium Math.*, (in print).

Aronson, D. G., and J. B. Serrin
1. Local behavior of solutions of quasilinear parabolic equations. *University of Minnesota Report*, 1966.

Barta, J.
1. Sur la vibration fondamentale d'une membrane. *Comptes Rendus de l'Acad. des Sci., Paris*, **204** (1937), pp. 472-473.

2. Über die elastische Grundschwingung eingespannter Stäbe. *Ingenieur Archiv*, **8** (1937), pp. 35-37.

Bernstein, S. N.
1. Sur la généralisation du problème de Dirichlet. *Math Ann.*, **69** (1910), pp. 82-136.
2. Über ein geometrisches Theorem und seine Anwendung auf die partiellen Differential gleichungen vom elliptischen Typus. *Math Zeit.*, **26** (1927), pp. 551-558.
3. On the equations of the calculus of variations. *Uspekhi Mat. Nauk*, **8** (1941), pp. 32-74.

Bers, L.
1. On mildly nonlinear partial differential equations of elliptic type, *J. of Research of Nat. Bureau of Standards*, **51** (1953), pp. 229-236.
2. Existence and uniqueness of a subsonic flow past a given profile. *Comm. on Pure and Appl. Math.*, **7** (1954), pp. 441-504.

Bers, L., F. John, and M. Schechter
1. Partial Differential Equations. New York: Interscience Publishers, Inc., 1964.

Bers, L., and L. Nirenberg
1. On linear and nonlinear elliptic boundary value problems in the plane. *Convegne Internazionalle sulle Equazioni à Derivati Parziali*, (Trieste, 1954), Edizioni Cremonese (1955), pp. 141-167.

Besala, P.
1. A remark on a problem of M. Krzyżański concerning second order parabolic equations. *Colloq. Math.*, **10** (1963), pp. 161-164.
2. On solutions of Fourier's first problem for a system of non-linear parabolic equations in an unbounded domain. *Ann. Polon. Math.*, **13** (1963), pp. 247-265.
3. Concerning solutions of an exterior boundary-value problem for a system of non-linear parabolic equations. *Ann. Polon. Math.*, **14** (1964), pp. 289-301.
4. On solutions of non-linear second order elliptic equations defined in unbounded domains. *Atti. Sem. Mat. Fis. Università di Modena*, **13** (1964), pp. 74-86.
5. On weak differential inequalities. *Ann. Polon. Math.*, **16** (1965), pp. 185-194.
6. Limitations of solutions of non-linear parabolic equations in unbounded domains. *Ann. Polon. Math.*, **17** (1965), pp. 25-47.

Bitsadze, A. V.
1. Equations of the Mixed Type. New York: Pergamon Press, 1964.

Blohina, G. N.
1. Theorems of Phragmèn-Lindelöf type for a second order linear elliptic equation. *Dokl. Akad. Nauk. S.S.S.R.*, **162** (1965), pp. 727-730. Translated in *Soviet Math.*, **6** (1965), pp. 720-723.

Bodanko, W.
1. Sur le problème de Cauchy et les problèmes de Fourier pour les équations

paraboliques dans un domaine non borné. *Ann. Polon. Math.*, **18** (1966), pp. 79-94.

Bohn, E., and L. K. Jackson
1. The Liouville theorem for a quasilinear elliptic partial differential equation. *Trans. Amer. Math. Soc.*, **104** (1962), 392-397.

Calabi, E.
1. An extension of E. Hopf's maximum principle with an application to Riemannian geometry. *Duke Math. J.*, **25** (1958), pp. 45-56.

Carroll, R.
1. L'équation d'Euler-Poisson-Darboux et les distributions sousharmoniques. *Comptes Rendus de l'Acad. des Sci.*, Paris, **246** (1958), pp. 2560-2562.
2. Some singular Cauchy problems. *Annali di Mat. Pura ed Appl.*, ser. 4, **56** (1961), pp. 1-31.

Clark, C., and C. A. Swanson
1. Comparison theorems for elliptic differential equations. *Proc. Amer. Math. Soc.*, **16** (1965), pp. 886-890.

Coddington, E. A., and N. Levinson
1. Theory of Ordinary Differential Equations. New York: McGraw-Hill Book Company, 1955.

Collatz, L.
1. Fehlerabschätzung für Näherungslösungen parabolischer Differentialgleichungen. *Anais Acad. Brasileira de Ciéncias*, **28** (1956), pp. 1-9.
2. Fehlerabschätzungen bei Randwertaufgaben partieller Differentialgleichungen mit unendlichem Grundgebiet. *Zeitschr. für Ang. Math. und Phys.*, **9a** (1958), pp. 118-128.
3. Approximation in partial differential equations. *On Numerical Approximation*, Madison, Wis.: University of Wisconsin Press, 1959, pp. 413-422.
4. The numerical treatment of differential equations. Berlin: Springer-Verlag OHG, 1960.

Courant, R., and D. Hilbert
1. Methods of Mathematical Physics. New York: Interscience Publishers, Inc., 1962, volume 2.

Duffin, R. J.
1. Lower bounds for eigenvalues. *Phys. Rev.*, (2) **71** (1947), pp. 827-828.
2. The maximum principle and biharmonic functions. *Journal of Math. Anal. and Applic.*, **3** (1961), pp. 399-405.

Dymkov, S. S.
1. The first boundary value problem for quasi-linear elliptic equations. *Dokl. Akad. Nauk. S.S.S.R.*, **115** (1957), pp. 220-222.

Earnshaw, S.
1. On the nature of the molecular forces which regulate the constitution of the luminiferous ether. *Cambridge Phil. Soc. Trans.*, **7** (1839), pp. 97-112.

Eidelman, S. D.
1. On the fundamental solution of parabolic systems. *Mat. Sbornik*, **95** (1961), pp. 73-136.

Feller, W.
1. Über die Lösungen der linearen partiellen Differentialgleichungen zweiter Ordnung vom elliptischem Typus. *Math. Ann.*, **102** (1930), pp. 633-649.

Fichera, G.
1. Su un principio di dualità per talune formole di maggiorazione. *Atti. Rend., Accad. Naz. Lincei*, (8) **19** (1955), pp. 411-418.
2. Sulle equazioni differenziali lineari ellittico-paraboliche del secondo ordine. *Atti. Accad. Naz. Lincei Memorie*, (8), **5** (1956), pp. 3-30.
3. On a unified theory of boundary value problems for elliptic-parabolic equations of second order. *Boundary Value Problems in Differential Equations*, Madison, Wis.: University of Wisconsin Press (1960), pp. 97-120.
4. El teorema del massimo modulo per l'equazione dell'elastostatica tridimensionale. *Arch. for Rat. Mech. and Anal.*, **7** (1961), pp. 373-387.

Fife, P. C.
1. Growth and decay properties of solutions of second order elliptic equations. *Annali della Scuola Norm. Sup. di Pisa*, Classe di Scienze **20** (1966), pp. 675-701.

Finn, R.
1. New estimates for equations of minimal surface type. *Arch. for Rat. Mech. and Anal.*, **14** (1963), pp. 337-375.

Finn, R., and D. Gilbarg
1. Three-dimensional subsonic flows and asymptotic estimates for elliptic partial differential equations. *Acta Math.*, **98** (1957), pp. 265-296.
2. Asymptotic behavior and uniqueness of plane subsonic flows. *Comm. on Pure and Appl. Math.*, **10** (1957), pp. 23-63.

Friedman, A.
1. On two theorems of Phragmèn-Lindelöf for linear elliptic and parabolic differential equations of the second order. *Pacific J. of Math.*, **7** (1957), pp. 1563-1575.
2. Remarks on the maximum principle for parabolic equations and its applications. *Pacific J. of Math.*, **8** (1958), pp. 201-211.
3. A strong maximum principle for weakly subparabolic functions. *Pacific J. of Math.*, **11** (1961), pp. 175-184.
4. Partial Differential Equations of Parabolic Type. Englewood Cliffs, N. J.: Prentice-Hall, Inc., 1964.

Garabedian, P.
1. Partial Differential Equations. New York: Interscience Publishers, Inc., 1964.

Gauss, C. F.
1. Allgemeine Theorie des Erdmagnetismus. *Beobachtungen des magnetischen Vereins im Jahre* 1838. Leipzig (1839). Also in Werke (Collected Works), **5**, p. 129.

Germain, P., and R. Bader
1. Sur le problème de Tricomi. *Rend. del Circ. Mat. di Palermo*, (2) **2** (1953), pp. 53-70.

Gilbarg, D.
1. The Phragmèn-Lindelöf theorem for elliptic partial differential equations. *J. of Rat. Mech. and Anal.*, **1** (1952), pp. 411-417.
2. Uniqueness of axially symmetric flow with free boundaries. *J. of Rat. Mech. and Anal.*, **1** (1952), pp. 309-320.
3. Comparison methods in the theory of subsonic flows. *J. of Rat. Mech. and Anal.*, **2** (1953), pp. 233-251.
4. Some hydrodynamical applications of function theoretic properties of elliptic equations. *Math. Zeitschr.*, **72** (1959), pp. 165-174.

Gilbarg, D., and J. B. Serrin
1. On isolated singularities of solutions of second order elliptic differential equations. *J. d'Analyse Math.*, **4** (1954-56), pp. 309-336.

Gilbarg, D., and M. Shiffman
1. On bodies achieving extreme values for the critical Mach number. *J. of Rat. Mech. and Anal.*, **3** (1954), pp. 209-230.

Giraud, G.
1. Géneralizations des problèmes sur les opérations du type elliptiques. *Bull. des Sciences Math.*, **56** (1932), pp. 248-272, 281-312, 316-352, 384.
2. Problèmes de valeurs á la frontière relatifs á certaines données discontinues. *Bull. de la Soc. Math. de France*, **61** (1933), pp. 1-54.

Glagelova, R. Ja.
1. The three cylinder theorem and its applications. *Dokl. Akad. Nauk S.S.S.R.*, **163** (1965), pp. 801-804. Translated in *Soviet Math.*, **6** (1965), pp. 1004-1008.

Gloistehn, H.
1. Monotoniesätze bei Differentialgleichungen zweiter Ordnung. *Arch. for Rat. Mech. and Anal.*, **6** (1960), pp. 399-408.
2. Monotoniesätze und Fehlerabschätzungen für Anfangswertaufgaben mit Hyperbolischer Differentialgleichung. *Arch. for Rat. Mech. and Anal.*, **14** (1963), pp. 384-404.

Haar, A.
1. Sur l'unicité des solutions des équations aux dérivées partielles. *Comptes Rendus de l'Academie des Sciences, Paris*, **187** (1928), p. 23.

Habetler, G. J., and M. A. Martino
1. Existence theorems and spectral theory for the multigroup diffusion model. *Proc., Symp. in Appl. Math. XI. Nuclear Reactor Theory*. American Math. Soc. (1961), pp. 127-139.

Hadamard, J.
1. Extension à l'équation de la chaleur d'un théorème de A. Harnack. *Rend. del Circ. Mat. di Palermo*, (2) **3** (1954), pp. 337-346.

Hartman, P.
1. Ordinary Differential Equations. New York: John Wiley & Sons, Inc., 1964.

Hartman, P., and R. Sacksteder
1. On maximum principles for non-hyperbolic partial differential operators. *Rend. del Circ. Mat. di Palermo*, (2) **6** (1957), pp. 218-232.

Hartman, P., and A. Wintner
1. On a comparison theorem for self-adjoint partial differential equations of elliptic type. *Proc. Amer. Math. Soc.*, **6** (1955), pp. 62-65.

Heins, M. H.
1. On the Phragmèn-Lindelöf principle. *Trans. of the Amer. Math. Soc.*, **60** (1946), pp. 238-244.
2. Selected Topics in the Classical Theory of Functions of a Complex Variable. New York: Holt, Rinehart and Winston, Inc., 1962.

Heinz, E.
1. Über die Lösungen der Minimalflächengleichung. *Nachr. der Akad. der Wissenschaften, Göttingen, Math.-Phys. Kl.* (1952), pp. 51-56.

Hellwig, G.
1. Partial Differential Equations. Waltham, Mass.: Blaisdell Publishing Company, 1964.

Hersch, J.
1. Sur la fréquence fondamentale d'une membrane vibrante: évaluations par défaut et principe de maximum. *Zeitschr. für Angew. Math. und Phys.*, **11** (1960), pp. 387-413.
2. Physical interpretation and strengthening of M. H. Protter's method for vibrating nonhomogeneous membranes; its analogue for Schroedinger's equation. *Pacific J. of Math.*, **11** (1961), pp. 971-980.

Hooker, W.
1. Lower bounds for the first eigenvalue of elliptic equations of order two and four. *Technical Report 10*, Math. Dept. Univ. of Calif., Berkeley (1960).

Hopf, E.
1. Elementare Bemerkungen über die Lösungen partieller Differentialgleichungen zweiter Ordnung vom elliptischen Typus. *Sitzungsber. d. Preuss. Akad. d. Wiss.*, **19** (1927), pp. 147-152.
2. A remark on linear elliptic differential equations of the second order. *Proc. of the Amer. Math. Soc.*, **3** (1952), pp. 791-793.
3. Remarks on the preceding paper of D. Gilbarg. *J. of Rat. Mech. and Anal.*, **1** (1952), pp. 419-424.

Huber, A.
1. On the uniqueness of generalized axially symmetric potential. *Ann. of Math.*, **60** (1954), pp. 351-358.

Il'in, A. M., A. S. Kalashnikov, and O. A. Oleĭnik
1. Linear second order parabolic equations. *Uspekhi Math. Nauk S. S. S. R.*, **17** (1962), pp. 3-146. Translated in *Russian Math. Surveys*, **17** (1962), No. 3, pp. 1-143.

Juberg, R. K.
1. Several observations concerning a maximum principle for parabolic systems. *Technical Report*, Univ. of Calif., Irvine (1967).

Kadlec, J., and R. Výborný
1. Strong maximum principle for weakly nonlinear parabolic equations. *Comment. Math. Univ. Carolinae*, 6 (1965), pp. 19-20.

Kaplan, S.
1. On the growth of solutions of quasi-linear parabolic equations. *Comm. on Pure and Appl. Math.*, 16 (1963), pp. 305-330.

Kellogg, O. D.
1. Foundations of Potential Theory. New York: Frederick Ungar Publishing Co., 1929.

Kolodner, I., and R. Pederson
1. Pointwise bounds for solutions of semilinear parabolic equations. *Journal of Differential Equations*, 2 (1966), pp. 353-364.

Korn, A.
1. Lehrbuch der Potentialtheorie. 2. Berlin: Ferd. Dümmlers Verlagsbuchhandlung, 1901, volume 2.

Kreith, K.
1. A new proof of a comparison theorem for elliptic equations. *Proc. of the Amer. Math. Soc.*, 14 (1963), pp. 33-35.

Kružkov, S. N.
1. Certain properties of solutions of elliptic equations. *Dokl. Akad. Nauk S.S.S.R.*, 150 (1963), pp. 470-473. Translated in *Soviet Math.*, 4 (1963), pp. 686-690.

Krzyżański, M.
1. Sur les solutions de l'équation linéaire du type parabolique déterminées par les conditions initiales. *Ann. Soc. Polon. Math.*, 18 (1945), pp. 145-156.
2. Sur les solutions de l'équation linéaire du type elliptique, discontinues sur la frontière du domaine de leur existence. *Studia Math.*, 11 (1950), pp. 95-125.
3. Evaluations des solutions de l'équation aux dérivées partielles du type parabolique, déterminées dans un domaine non borné. *Ann. Polon. Math.*, 4 (1957), pp. 93-97.
4. Certaines inégalités relatives aux solutions de l' équation parabolique linéaire normale. *Bull. Acad. Polon. Sci. Math. Astr. Phys.*, 7 (1959), pp. 131-135.

Ladyzhenskaya, O. A.
1. The solution of the first boundary value problem in the large for quasi-linear parabolic equations. *Trudy Moskov. Mat. Obšč.*, 7 (1958), pp. 147-177.

Ladyzhenskaya, O. A., and N. N. Ural'tzeva
1. Quasi-linear elliptic equations and variational problems with many independent variables. *Uspekhi Mat. Nauk.*, 16 (1961), pp. 19-90. Translated in *Russian Math. Surveys*, 16 (1961), No. 1, pp. 17-91.

2. Boundary value problems for linear and quasilinear parabolic equations. *Izv. Akad. Nauk S.S.S.R.*, **26** (1962), pp. 5-52 and 753-780.

Landis, E. M.

1. Some questions in the qualitative theory of elliptic and parabolic equations. *Uspekhi Mat. Nauk*, (85) **14** (1959), pp. 21-85. Translated as *Amer. Math. Soc. Translations*, ser. 2, No. 20.
2. A three-sphere theorem. *Dokl. Akad. Nauk S.S.S.R.*, **148** (1963), pp. 277-279. Translated in *Soviet Math.*, **4** (1963), pp. 76-78.
3. Some problems of the qualitative theory of second order elliptic equations. (Case of several variables.) *Uspekhi Mat. Nauk*, **18** (1963), pp. 3-62. Translated in Russian Math. Surveys, **18** (1963), pp. 1-62.

Lavrentiev, M.

1. On certain properties of univalent functions and their applications to wake theory. *Mat. Sbornik*, **46** (1938), pp. 391-458.

Lees, M.

1. A boundary value problem for nonlinear ordinary differential equations. *J. of Math. and Mech.*, **10** (1961), pp. 423-430.

Leighton, W.

1. Comparison theorems for linear differential equations of second order. *Proc. of the Amer. Math. Soc.*, **13** (1962), pp. 603-610.

Leja, F.

1. Remarques sur le travail de M. Picone. *Ann. de la Soc. Polon. de Math.*, **21** (1948), pp. 170-172.

Levi, E. E.

1. Sull' equazione del calore. *Reale Accad. dei Lincei, Roma. Rendiconti*, (5) **162** (1907), pp. 450-456.
2. Sull' equazione del calore. *Annali di Mat. Pura ed Appl.*, **14** (1908), pp. 187-264.

Lichtenstein, L.

1. Beiträge zur Theorie der linearen partiellen Differentialgleichungen zweiter Ordnung vom elliptischen Typus. *Rend. del Circ. Mat. di Palermo*, **33** (1912), pp. 201-211.
2. Neue Beiträge zur Theorie der linearen partiellen Differentialgleichungen zweiter Ordnung vom elliptischen Typus. *Math. Zeitschr.*, **20** (1924), pp. 194-212.

Littman, W.

1. A strong maximum principle for weakly L-subharmonic functions. *J. of Math. and Mech.*, **8** (1959), pp. 761-770.
2. Generalized subharmonic functions: monotonic approximations and an improved maximum principle. *Annali della Scuola Norm. Sup. di Pisa.*, (3) **17** (1963), pp. 207-222,

Łojckzyk-Królikiewicz, I.
1. Sur l'unicité et les limitations des solutions des problèmes de Fourier relatifs aux équations paraboliques à coefficients non bornés. *Ann. Polon. Math.*, **15** (1964), pp. 33-41.

McNabb, A.
1. Comparison and existence theorems for multicomponent diffusion systems. *J. of Math. Anal. and Appl.*, **3** (1961), pp. 133-144.
2. Strong comparison theorems for elliptic equations of second order. *J. of Math. and Mech.*, **10** (1961), pp. 431-440.

Meyer, A. G.
1. Schranken für die Lösungen von Randwertaufgaben mit elliptischer Differentialgleichung. *Arch. for Rat. Mech. and Anal.*, **6** (1960), pp. 277-298.

Meyers, N., and J. B. Serrin
1. The exterior problem for second order elliptic partial differential equations. *J. of Math. and Mech.*, **9** (1960), pp. 513-538.

Miller, K.
1. Three circle theorems in partial differential equations and applications to improperly posed problems. *Arch. for Rat. Mech. and Anal.*, **16** (1964), pp. 126-154.
2. An eigenfunction expansion method for problems with overspecified data. *Annali Scuola Norm. Sup. di Pisa*, (3) **19** (1965), pp. 397-405.

Miranda, C.
1. Formule di maggiorazione e teorema di esistenza per le funzioni biarmoniche di due variabili. *Giorn. Mat. Battaglini*, **78** (1948-49), pp. 97-118.
2. Sulla proprietà di minimo e di massimo delle soluzioni delle equazioni a derivate parziali del secondo ordine di tipo ellittico. *Atti. Rend. Accad. Naz. dei Lincei*, (8) **10** (1951), pp. 117-120.
3. Equazioni alle derivate parziali di tipo ellitico. Berlin: Springer-Verlag OHG, 1955.
4. Teorema del massimo modulo e teorema di esistenza e di unicità per il problema di Dirichlet relative alle equazioni ellitiche in due variabili. *Annali di Mat. Pura ed Appl.*, **46** (1958), pp. 265-311.

Mlak, W.
1. Differential inequalities of parabolic type. *Ann. Polon. Math.*, **3** (1957), pp. 349-354.

Morawetz, C.
1. Note on a maximum principle and a uniqueness theorem for an elliptic-hyperbolic equation. *Proc. of the Roy. Soc. of London*, Ser. A, **236** (1956), pp. 141-144.

Moser, J.
1. On Harnack's theorem for elliptic differential equations. *Comm. on Pure and Appl. Math.*, **14** (1961), pp. 577-591.
2. A Harnack inequality for parabolic differential equations. *Comm. on Pure and Appl. Math.*, **17** (1964), pp. 101-134 and **20** (1967), pp. 231-236.

Moutard, T.
1. Notes sur les équations aux dérivées partielles. *J. de l'École Polytechnique*, **64** (1894), pp. 55-69.

Müller, M.
1. Über die Eindeutigkeit der Integrale eines Systems gewöhnlicher Differentialgleichungen und die Konvergenz einer Gattung von Verfahren zur approximation dieser Integrale. *Sitzungsber., Heidelberger Akad. d. Wiss., Math.-natural. Kl.* (1927), **9**. abh.

Nagumo, M., and S. Simoda
1. Sur la solution bornée de l'équation aux dérivées partielles du type elliptique. *Proc. Japan. Acad.*, **27** (1951), pp. 334-339.
2. Note sur l'inégalité différentielle concernant les équations du type parabolique. *Proc. Japan. Acad.*, **27** (1951), pp. 536-539.

Neumann, C.
1. Über die Methode des Arithmetischen Mittels. *Abhand. der Königl. Sächsischen Ges. der Wissenshaften, Leipzig*, **10** (1888), pp. 662-702.

Nickel, K.
1. Einige Eigenschaften von Lösungen der Prandtlschen Grenzschicht-Differentialgleichungen. *Arch. for Rat. Mech. and Anal.*, **2** (1958), pp. 1-31.
2. Fehlerabschätzung bei parabolischen Differentialgleichungen. *Math. Zeit.*, **71** (1959), pp. 268-282.

Nirenberg, L.
1. A strong maximum principle for parabolic equations. *Comm. on Pure and Appl. Math.*, **6** (1953), pp. 167-177.
2. Estimates and existence of solutions of elliptic equations. *Comm. on Pure and Appl. Math.*, **9** (1956), pp. 509-531.

Nitsche, J. C. C.
1. Elementary proof of Bernstein's theorem on minimal surfaces. *Annals of Math.*, **66** (1957), pp. 543-544.
2. On new results in the theory of minimal surfaces. *Bull. of the Amer. Math. Soc.*, **71** (1965), pp. 195-270.

Oleĭnik, O. A.
1. On properties of some boundary problems for equations of elliptic type. *Math. Sbornik*, N. S. **30** (72) (1952), pp. 695-702.
2. On the equations of a boundary layer. *Seminari 1962-1963 di Analisi, Algebra, Geometria, e Topologia. Istituto Nazionale di Alta Matematica*, Edizioni Cremonese, Rome (1965), **1**, pp. 372-387.
3. On some degenerate quasilinear parabolic equations. *Seminari 1962-1963 di Analisi, Algebra, Geometria, e Topologia. Istituto Nazionale di Alta Matematica*, Edizioni Cremonese, Rome (1965), **1**, pp. 355-371.
4. A problem of Fichera. *Dokl. Akad. Nauk S. S. S. R.*, **157** (1964), pp. 1297-1301. Translated in *Soviet Math.*, **5** (1964), pp. 1129-1133.

5. Alcuni resultati sulle equazioni lineari e quasi lineari ellitico-paraboliche a derivate parziali del secondo ordine. *Atti Rend., Accad. Naz. dei Lincei*, (8) **40** (1966), pp. 775-784.
6. On the existence, uniqueness, stability and approximation of solutions of Prandtl's system for the nonstationary boundary layer. *Atti Rend., Accad. Naz. dei Lincei*, (8) **41** (1966), pp. 32-40.

Oleĭnik, O. A., and S. N. Kružkov
1. Quasi-linear second order parabolic equations with many independent variables. *Uspekhi Math. Nauk S. S. S. R.*, **16** (1961), pp. 115-156.

Oleĭnik, O. A., and T. D. Wentzel
1. The first boundary value problem and Cauchy's problem for quasi-linear equations of parabolic type. *Dokl. Akad. Nauk S. S. S. R.*, **97** (1954), pp. 605-608.
2. The first boundary value problem and Cauchy's problem for quasi-linear equations of parabolic type. *Math. Sbornik*, N. S. **41** (83) (1957), pp. 105-128.

Paraf, A.
1. Sur le problème de Dirichlet et son extension au cas de l'équation linéaire générale du second ordre. *Annales de la Faculté des Sciences de Toulouse*, ser. I, **6** (1892), H, pp. 1-75.

Payne, L. E.
1. Some explicit inequalities for uniformly elliptic operators. *Duke Math J.*, **31** (1964), pp. 485-489.

Payne, L. E., and D. Sather
1. On a singular hyperbolic operator. *Duke Math. J.*, **34** (1967), pp. 147-162.

Petrovsky, I. G.
1. *Lectures on Partial Differential Equations*. New York: Interscience Publishers, Inc., 1954.

Pfluger, A.
1. Quelques théorèmes sur une classe de fonctions pseudo-analytiques. *Comptes Rendus de l'Acad. des Sci., Paris*, **231** (1950), pp. 1022-1023.

Picard, E.
1. *Traité d'Analyse*. Paris: Gauthier-Villars, 1905, volume 2.

Picone, M.
1. Maggiorazione degli integrali di equazioni lineari ellitico-paraboliche alle derivate parziali del secondo ordine. *Atti Rend. Accad. Naz. dei Lincei*, (6) **5** (1927), pp. 138-143.
2. Maggiorazione degli integrali delle equazioni totalmente paraboliche alle derivate parziali de secondo ordine. *Annali di Mat. Pura ed Appl.*, (4) **7** (1929), pp. 145-192.
3. Sul problema della propagazione del calore in un mezzo privo di frontiera, conduttore, isotropo, é omogeneo. *Math. Ann.*, **101** (1929), pp. 701-712.

4. Nuove formule di maggiorazione per gli integrali delle equazioni lineari a derivate parziali del secondo ordine ellitico-paraboliche. *Atti. Rend. Accad. Naz. dei Lincei*, (6) **28** (1938), pp. 331-338.
5. Intorno alla teoria di una classica equazione a derivate parziali. *Annales de la Soc. Polon. de Math.*, **21** (1948), pp. 161-169.

Pini, B.
1. Sui punti singolari delle soluzioni delle equazioni paraboliche lineari. *Ann., Univ. di Ferrara, Sezione* VII, *Scienze Mat.*, **2** (1953), pp. 2-12.
2. Sulla soluzione generalizzata di Wiener per il primo problema di valori al contorno nel caso parabolico. *Rend., Sem. Mat., Univ. Padova*, **23** (1954), pp. 422-434.

Polya, G., and G. Szegö
1. Isoperimetric Inequalities in Mathematical Physics. Princeton, N. J.: Princeton University Press, 1951.

Protter, M. H.
1. A maximum principle for hyperbolic equations in a neighborhood of an initial line. *Trans. of the Amer. Math. Soc.*, **87** (1958), pp. 119-129.
2. Lower bounds for the first eigenvalue of elliptic equations. *Annals of Math.*, **71** (1960), pp. 423-444.
3. Properties of solutions of parabolic equations and inequalities. *Canad. J. of Math*, **13** (1961), pp. 331-345.

Protter, M. H., and H. F. Weinberger
1. On the capacity of composite conductors. *Journ. of Math and Physics*, **44** (1965), pp. 375-383.
2. On the spectrum of general second order operators. *Bull. of the Amer. Math. Soc.*, **72** (1966), pp. 251-255.

Pucci, C.
1. Sulla maggiorazione dell' integrale di un equazione differenziale lineare ordinaria del secondo ordine. *Atti. Rend. Accad. Naz. dei Lincei*, (8) **10** (1951), pp. 300-306.
2. Alcune limitazioni per gli integrali delle equazioni differenziali a derivate parziali, lineari, del secondo ordine, di tipo ellitico-parabolico. *Atti. Rend. Accad. Naz. dei Lincei*, (8) **11** (1951), pp. 334-339.
3. Maggiorazione dell' soluzione di un problema al contorno, di tipo misto, relativo a una equazione a derivate parziali, lineare, del secondo ordine. *Atti. Rend. Accad. Naz. dei Lincei*, (8) **13** (1952), pp. 360-366.
4. Bounds for solutions of Laplace's equation satisfying mixed conditions. *J. of Rat. Mech. and Anal.*, **2** (1953), pp. 299-302.
5. Proprietà di massimo e minimo delle soluzioni di equazioni a derivate parziali del secondo ordine di tipo ellitico e parabolico. *Atti. Rend. Accad. Naz. dei Lincei*, (8) **23** (1957), pp. 370-375 and (8) **24** (1958), pp. 3-6.
6. Maximum and minimum first eigenvalues for a class of elliptic operators. *Proc. of the Amer. Math. Soc.*, **17** (1966), pp. 788-795.
7. Operatori ellitici estremanti. *Annali di Mat. Pura ed Appl.*, (4) **72** (1966), pp. 141-170.

8. Limitazioni per soluzioni di equazioni ellitiche. *Annali di Mat. Pura ed Appl.* (4) **74** (1966), pp. 15-30.

Redheffer, R. M.
1. Maximum principles and duality. *Monatshefte der Math.*, **62** (1958), pp. 56-75.
2. On the inequality $\Delta u \geq f(u, |\text{grad } u|)$. *J. of Math. Anal. and Appl.*, **1** (1960), pp. 277-299.
3. An extension of certain maximum principles. *Monatshefte für Math.*, **66** (1962), pp. 32-42.
4. Bemerkungen über Monotonie und Fehlerabschätzung bei nichtlinearen partiellen Differentialgleichungen. *Arch. for Rat. Mech. and Anal.*, **10** (1962), pp. 427-457.

Rosenbloom, P. C., and D. V. Widder
1. A temperature function which vanishes initially. *Amer. Math. Monthly*, **65** (1958), pp. 607-609.

Sather, D.
1. Maximum properties of Cauchy's problem in three-dimensional space-time. *Arch. for Rat. Mech. and Anal.*, **18** (1965), pp. 14-26.
2. A maximum property of Cauchy's problem in n-dimensional space-time. *Arch. for Rat. Mech. and Anal.*, **18** (1965), pp. 27-38.
3. A maximum property of Cauchy's problem for the wave operator. *Arch. for Rat. Mech. and Anal.*, **21** (1965-66), pp. 303-309.
4. Maximum and monotonicity properties of initial-boundary value problems for hyperbolic equations. *Pacific J. of Math.*, **19** (1966), pp. 141-157.

Schechter, M.
1. On the Dirichlet problem for second order equations with coefficients singular at the boundary. *Comm. on Pure and Appl. Math.*, **13** (1960), pp. 321-328.

Schröder, J.
1. Invers-monotone Operatoren. *Arch. for Rat. Mech. and Anal.*, **10** (1962), pp. 276-295.
2. Monotonie-Eigenschaften bei Differentialungleichungen. *Arch. for Rat. Mech. and Anal.*, **14** (1963), pp. 38-60.

Serrin, J. B.
1. Uniqueness theorems for two free boundary problems. *Amer. J. of Math.*, **74** (1952), pp. 492-506.
2. Two hydrodynamical comparison theorems. *J. of Rat. Mech. and Anal.*, **1** (1952), pp. 563-572.
3. On plane and axially symmetric free boundary problems. *J. of Rat. Mech. and Anal.*, **2** (1953), pp. 563-575.
4. Comparison theorems for subsonic flows. *J. of Math. and Phys.*, **33** (1954), pp. 27-45.
5. On the Phragmèn-Lindelöf principle for elliptic differential equations. *J. of Rat. Mech. and Anal.*, **3** (1954), pp. 395-413.

6. On the Harnack inequality for linear elliptic equations. *J. d'Analyse Math.*, 4 (1954-56), pp. 292-308.
7. The Harnack inequality for elliptic partial differential equations in more than two independent variables. *Notices of the Amer. Math. Soc.* 5 (1958), pp. 52-53.
8. Local behavior of solutions of quasi-linear equations. *Acta Math.* 111 (1964), pp. 247-302.
9. A priori estimates for solutions of the minimal surface equation. *Arch. for Rat. Mech. and Anal.*, 14 (1963), pp. 337-375.

Stampacchia, G.

1. Le problème de Dirichlet pour les équations elliptiques du second ordre à coefficients discontinus. *Annales de l'Institut Fourier, Grenoble*, 15 (1965), pp. 189-258.

Stys, T.

1. On the unique solvability of the first Fourier problem for a parabolic system of linear differential equations of second order. *Prace Mat.*, 9 (1965), pp. 283-289.

Szarski, J.

1. Sur les systèmes majorants d'équations différentielles ordinaires. *Annales de la Soc. Polon. de Math.*, 23 (1950), pp. 206-223.
2. Sur les systèmes d'inégalités différentielles ordinaires remplies en dehors de certains ensembles. *Annales de la Soc. Polon. de Math.*, 24 (1951), pp. 1-8.
3. Sur la limitation et l'unicité des solutions d'un système non-linéaire d'équations paraboliques aux dérivées partielles du second ordre. *Ann. Polon. Math.*, 2 (1955), pp. 237-249.
4. Sur un système non-linéaire d'inégalités différentielles paraboliques. *Ann. Polon. Math.*, 15 (1964), pp. 15-22.
5. Differential inequalities. Warsaw: Polish Scientific Publishers, 1965.

Szeptycki, P.

1. Existence theorem for the first boundary problem for a quasilinear elliptic system. *Bull., Acad. Polon. des Sciences*, 7 (1959), pp. 419-424.

Täcklind, S.

1. Sur les classes quasianalytiques des solutions des équations aux dérivées partielles du type parabolique. *Nova Acta Societatis Scientiarum Upsalensis*, (4) 10 (1936), pp. 1-57.

Tikhonov, A. N.

1. Théorèmes d'unicité pour l'équation de la chaleur. *Mat. Sbornik*, 42 (1935), pp. 199-216.

Titchmarsh, E. C.

1. Theory of Functions. London: Oxford University Press, 1932.

Tricomi, F. G.

1. Equazioni a Derivate Parziali. Rome: Edizioni Cremonese, 1957.

Trudinger, N. S.
1. On Harnack type inequalities and their application to quasilinear elliptic equations. *Comm. on Pure and Appl. Math.*, **20** (1967).
2. Pointwise estimates and quasilinear parabolic equations. *Comm. on Pure and Appl. Math.*, **21** (1968).

Uhlmann, W.
1. Fehlerabschätzungen bei Anfangswertaufgaben gewöhnlicher Differentialgleichungssysteme 1. Ordnung. *Zeitschr. für Angew. Math. und Mech.*, **37** (1957), pp. 88-111.

Velte, W.
1. Eine Anwendung des Nirenbergschen Maximumprinzips für parabolische Differentialgleichungen in der Grenzschichttheorie. *Arch. for Rat. Mech. and Anal.*, **5** (1960), pp. 420-431.

Vyborný, R.
1. On properties of solutions of some boundary value problems for equations of parabolic type. *Dokl. Akad. Nauk. S.S.S.R.*, **117** (1957), pp. 563-565.

Walter, W.
1. Fehlerabschätzung bei hyperbolischen Differentialgleichungen. *Arch. for Rat. Mech. and Anal.*, **7** (1961), pp. 249-272.
2. Fehlerabschätzungen und Eindeutigkeitssätze für gewöhnliche und partielle Differentialgleichungen. *Zeit. für Angew. Math. und Mech.*, **42** (1962), pp. T49-T62.
3. Differential- und Integral-Ungleichungen. Berlin: Springer-Verlag OHG, 1964.

Ważewski, T.
1. Sur la limitation des intégrales des systèmes d'équations differentielles linéaires ordinaires. *Studia Math.*, **10** (1948), pp. 48-59.

Weinberger, H. F.
1. Upper and lower bounds for torsional rigidity. *J. of Math. and Phys.*, **32** (1953), pp. 54-62.
2. A maximum property of Cauchy's problem. *Ann. of Math.*, **64** (1956), pp. 505-513.
3. A maximum property of Cauchy's problem in three-dimensional space-time. *Proc., Symp. in Pure Math.* IV. *Partial Diff. Eq.*, American Math. Soc., (1961), pp. 91-99.
4. A first course in partial differential equations. Waltham, Mass.: Blaisdell Publishing Company, 1965.

Weinstein, A.
1. Discontinuous integrals and generalized potential theory. *Trans. of the Amer. Math. Soc.*, **63** (1948), pp. 342-354.
2. On a Cauchy problem with subharmonic initial values. *Annali di Mat. Pura ed Appl.*, **43** (1957), pp. 325-340.

3. Hyperbolic and parabolic equations with subharmonic data. *Symposium on the Numerical Treatment of Partial Differential Equations with Real Characteristics*, Provisional International Computation Centre, Rome (1959), pp. 74-86.

Wentzel (Ventcel'), T. D.
1. The first boundary value problem and Cauchy's problem for quasi-linear parabolic equations with many space variables. *Math. Sbornik.*, 41 (83) (1957), pp. 499-520.
2. Quasilinear parabolic systems with increasing coefficients. *Vestnik Moskov. Univ., Ser. I Math. Mech.*, (1963) No. 6, pp. 34-44.

Westphal, H.
1. Zur Abschätzung der Lösungen nichtlinearer parabolischer Differentialgleichungen. *Math. Zeitschr.*, 51 (1949), pp. 690-695.

Widder, D. V.
1. Positive temperatures on an infinite rod. *Trans. of the Amer. Math. Soc.*, 55 (1944), 85-95.

Zaremba, S.
1. Sur l'unicité de la solution du probléme de Dirichlet. *Bull. de l'Acad. des Sciences de Cracovie*, (1909), pp. 561-563.

INDEX

Adjoint operator, 201
Admissible domain, 202, 219
Agmon, S., 156, 157, 230, 232, 239
Aleksandrov, A., 156
Approximation in boundary value problems, 14 ff., 76 ff., 156
Approximation in initial value problems, 24 ff, 194
Area mean-value theorem, 137
Aronson, D., 186, 193, 194

Bader, R., 239
Barta, J., 92, 157
Bernoulli relation, 153
Bernstein, S., 152, 157
Bers, L., 140, 152, 157, 158, 194
Besala, P., 193, 194
Bessel function, 26, 213
Bitsadze, A., 239
Blohina, G., 157
Bodanko, W., 193
Bohn, E., 157
Boundary conditions, 12
   mixed, 71
Boundary estimates for derivatives, 141 ff.
Boundary layer, 194
Boundary value problems, 12 ff.
   approximation in, 76 ff.
   first, 68
   second, 70
   third, 70
Bounds for derivatives, 24 ff., 138, 139, 141 ff., 144 ff., 156
Bounds for eigenvalues, 37 ff., 91, 157

Calabri, E., 156
Capacity, 122 ff.

Capacity (*cont.*):
   of a conductor, 125
   of a sphere, 125
Carroll, R., 238, 239
Cauchy problem, 208
Causality, principle of, 160
Chain rule, 58, 225
Characteristic, 200, 218
Characteristic coordinates, 225
   curve, 67, 200, 218
   triangle, 195, 201, 219
Clark, C., 157
Closed interval, 1
Coddington, E., 49
Collatz, L., 50, 157, 194
Comparison functions, 34
Comparison theorems, 42 ff., 155, 158, 232 ff.
Compressible fluid, 153
Conjugate point, 9
Conormal derivative, 67, 202
Consecutive zeros, 42, 46
Continuous dependence on data, 16, 36, 80
Convex function, 129, 180
Courant, R., 153, 156, 234
Curve, characteristic, 67

Derivative:
   conormal, 67, 202
   directional, 65
   normal, 67
Derivatives of harmonic functions, 137 ff.
Diagonalization, 59
Diagonal matrix, 59
Diameter of a domain, 144

Difference quotient, 24
Directional derivative, 65
  outward, 65
Dirichlet problem, 68
Divergence, 52
  theorem, 52, 81
Domain, vii
Duffin, R., 92, 156, 157
Dymkov, S., 156

Earnshaw, S., 156
Eidelman, S., 194
Eigenfunction, 37, 89
Eigenvalue, 37, 89
  first, 38, 40, 41
  problem, 37 ff., 89 ff.
  smallest, 38, 40, 41, 91
Eigenvalues, 89 ff.
  lower bounds for, 37, 38, 91, 157
  of a matrix, 59
Eigenvector of a linear transformation, 59
Elastic range, 148
Electrostatic capacity, 122
Electrostatic potential, 145
Elliptic operator, 56, 60, 150, 186
Elliptic-parabolic equation, 194
Elliptic system, 156, 192
Elliptic, uniformly, 56
Equilibrium temperature, 71
Euler-Poisson-Darboux equation, 239

Feller, W., 111, 157
Fichera, G., 156, 194
Fife, P., 157
Finn, R., 152, 158
First boundary value problem, 68
First eigenvalue, 38
Flow:
  fluid, 147, 153 ff.
  subsonic, 153
Fluid, compressible, 153
Fluid flow, 147, 153 ff.
Friedman, A., 157, 158, 175, 193, 194

Garabedian, P., 156, 194, 234
Gauss, C., 156

Generalized maximum principle, 8 ff., 72 ff.
Germain, P., 239
Gilbarg, D., 104, 105, 106, 118, 121, 155, 156, 157, 158
Giraud, G., 156
Glagelova, A., 193
Gloistehn, H., 50, 239
Goursat problem, 231
Gradient, 52
Green's first identity, 81
Green's function, 81 ff.
  for the Laplace operator, 85
Green's identities, 81 ff.
Green's second identity, 81, 92
Green's third identity, 84

Habetler, G., 192
Hadamard, J., 194
Hadamard three-circles theorem, 98, 128 ff., 157, 178
Harmonic, 51
Harmonic function, 51
  polynomial, 80
Harnack inequalities, 106 ff., 157, 194
Hartman, P., 49, 157, 194
Heat equation, 159 ff.
Heine-Borel theorem, 110
Heins, M., 157
Hellwig, G., 194
Hermite polynomial, 216
Hersch, J., 92, 157
Hilbert, D., 153, 156, 234
Hooker, W., 92, 157
Hopf, E., 61 ff., 105, 156, 157, 164
Horizontal point of inflection, 5
Huber, A., 157
Hyperbolic operator in several dimensions, 239
Hyperbolic operator in two dimensions, 200 ff.

Il'in, A., 194
Infimum, 37
Inflection, point of, 5
Initial-boundary value problems, 213 ff.
Initial conditions, 11
Initial value problem, 10 ff., 208 ff.

Jackson, L., 157
John, F., 140, 158, 194
Juberg, R., 194

Kadlec, J., 194
Kaplan, S., 194
Kellogg, O., 157
Kolashnikov, A., 194
Kolodner, I., 194
Korn, A., 156
Kreith, K., 157
Kružkov, S., 194
Krzyżański, M., 106, 193

Ladyzhenskaya, O., 156, 157, 158
Landis, E., 136, 157, 158, 193, 194
Laplace operator, 51
Laplacian, 51
Lavrentiev, M., 158
Lees, M., 50
Leighton, W., 50
Leja, F., 106
Levi, E., 193
Levinson, N., 49
Lichtenstein, L., 111, 156, 157
Limit inferior, 67, 94
Limit superior, 54, 94
Linear operator, vii
Liouville's theorem, 111, 120, 130, 132, 140, 157, 194
Littman, W., 156
Lojckzyk-Królikiewicz, I., 193
Lower comparison functions, 34

Martino, M., 192
Matrix:
  diagonal, 59
  eigenvalues of, 59
  orthogonal, 57
McNabb, A., 156, 194
Mean value, 53
Mean value theorem, 47, 53
Meyer, A., 157
Meyers, N., 105, 157
Miller, K., 157
Minimal surface equation, 152
Minimum principle, 3, 53, 64, 161
Miranda, C., 140, 156, 158

Mixed boundary conditions, 71
Mlak, W., 193, 194
Monge-Ampere equation, 152, 153
Morawetz, C., 239
Moser, J., 121, 157, 194
Moutard, T., 156
Müller, M., 50

Nagumo, M., 156, 194
Neumann, C., 86, 156
Neumann function, 86
Neumann problem, 70
Nickel, K., 194
Nirenberg, L., 156, 157, 164, 193, 230, 232, 239
Nitsche, J., 152
Nonlinear operators, 47 ff., 50, 149 ff., 157, 186 ff., 194, 239
  ellipticity of, 150, 187
  parabolic, 187
Normal derivative, 67

Obstacle, 147
Oleĭnik, O., 156, 194
One-dimensional maximum principle, 2
One-dimensional parabolic operator, 163 ff.
Open interval, 1
Operator, adjoint, 200
Operator, linear, vii
Operators, nonlinear, 47 ff., 50, 149 ff., 157, 186 ff., 194, 239
Orthogonal matrix, 57
  transformation, 57
Oscillation theorems, 42 ff.
  Sturmian, 43
Outward directional derivative, 65

Parabolic operator:
  general, 173 ff., 187
  one-dimensional, 163 ff.
  uniformly, 163, 173
Parabolic systems, 188 ff., 194
Paraf, A., 156
Payne, L., 239
Pederson, R., 194
Petrovsky, I., 194
Pfluger, A., 157

Phragmèn-Lindelöf principle, 93 ff., 157, 181 ff., 194
Picard, E., 156
Picone, M., 106, 156, 193, 194
Piecewise smooth, 81
Pini, B., 194
Plastic region, 148
Poisson equation, 68
  formula, 107
Polya, G., 157
Potential function, 147
Principal part of an operator, 60
Principle of causality, 160
Protter, M., 91, 157, 230, 232, 239
Pucci, C., 50, 92, 156, 157, 193, 194

Quasilinear, 154, 155

Radiation kernel, 211
Redheffer, R., 156, 194
Removable singularity, 102, 186, 194
Riemann's function, 210 ff.
  method, 196
Robin function, 86
Rosenbloom, P., 181
Rotation, 57

Sacksteder, R., 194
Sather, D., 215, 238, 239
Schechter, M., 140, 156, 158, 194
Schröder, J., 50, 194, 239
Second boundary value problem, 70
Serrin, J., 104, 105, 106, 111, 117, 118, 121, 152, 155, 157, 158, 194
Shear stresses, 148
Shiffman, M., 158
Simoda, S., 156, 194
Singularity, removable, 102, 186, 194
Smallest eigenvalue, 38, 40, 41, 91
Sonic speed, 154
Stampacchia, G., 156, 157
Stokes' theorem, 196
Stream function, 153
Streamlines, 153
Stress, magnitude of, 148
Stress function, 148
Stretching, 59
Strict inequality, 1

Sturmian oscillation theorem, 43
Stys, T., 192, 194
Subharmonic, 54
Subharmonic function, 54
Subsonic flow, 153, 154
Superharmonic, 54
Supremum, 37
Swanson, C., 157
Szarski, J., 50, 194
Szegö, G., 158
Szeptycki, P., 156, 192, 194

Täcklind, S., 181, 193
Temperature, 71, 159 ff.
Temperature, boundary, 71
Temperature, equilibrium, 71
Temperature distribution, 71
Third boundary value problem, 70
Three-circles theorem, 98, 128 ff., 157, 178
Three-curves theorem, 178 ff.
Three-cylinder theorem, 193
Three-parabolas theorem, 182
Three-spheres theorem, 131
Tikhonov, A., 180, 181, 193
Titchmarsh, E., 157
Total charge, 122
Tricomi, F., 239
Tricomi operator, 208, 239
Trudinger, N., 157, 194
Two-characteristic problem, 218 ff.
Two-dimensional hyperbolic operator, 200 ff.

Uhlmann, W., 50
Uniformly elliptic, 56, 58
Uniformly hyperbolic, 200
Uniformly parabolic, 163, 173
Upper comparison functions, 34
Ural'tzeva, N., 158

Velte, W., 194
Výborný, R., 193, 194

Walter, W., 50, 157, 194, 239
Wave equation, 195 ff.
Wave equation in higher dimensions, 234 ff.

Wave operator, 197 ff.
Ważewski, T., 50
Weakly coupled elliptic system, 192
Weakly coupled parabolic system, 188 ff., 194
Weinberger, H., 91, 156, 157, 194, 218, 236, 238, 239
Weinstein, A., 236, 239

Wentzel (Ventcel'), T., 194
Westphal, H., 194
Widder, D., 181, 193
Wintner, A., 157

Zaremba, S., 106
Zeros, consecutive, 42, 46